Conference Proceedings of the Society for Experimental Mechanics Series

Series Editor

Kristin B. Zimmerman
Society for Experimental Mechanics, Inc.,
Bethel, USA

The Conference Proceedings of the Society for Experimental Mechanics Series presents early findings and case studies from a wide range of fundamental and applied work across the broad range of fields that comprise Experimental Mechanics. Series volumes follow the principle tracks or focus topics featured in each of the Society's two annual conferences: IMAC, A Conference and Exposition on Structural Dynamics, and the Society's Annual Conference & Exposition and will address critical areas of interest to researchers and design engineers working in all areas of Structural Dynamics, Solid Mechanics and Materials Research.

Janko Slavic · Dan Rohe
Editors

Computer Vision & Laser Vibrometry, Vol. 6

Proceedings of the 43rd IMAC, A Conference and Exposition on Structural Dynamics 2025

River Publishers

Published, sold and distributed by:

River Publishers
Broagervej 10
9260 Gistrup
Denmark

www.riverpublishers.com

ISBN 97887-438-0151-1 (Hardback)

ISBN 97887-438-0163-4 (eBook)

https://doi.org/10.13052/97887-438-0151-1

Conference Proceedings of the Society for Experimental Mechanics

An imprint of River Publishers

Preface

Computer Vision & Laser Vibrometry represent one of twelve volumes of technical papers presented at the 43rd IMAC, A Conference and Exposition on Structural Dynamics, organized by the Society for Experimental Mechanics, and held February 10-13, 2025. The full proceedings also include volumes on Nonlinear Structures & Systems; Dynamics of Civil Structures; Model Validation and Uncertainty Quantification; Dynamic Substructuring & Transfer Path Analysis; Special Topics in Structural Dynamics & Experimental Techniques; Dynamic Environments Testing; Sensors & Instrumentation and Aircraft/Aerospace Testing Techniques; Topics in Modal Analysis & Parameter Identification I&II; Data Science in Engineering; and Structural Health Monitoring & Machine Learning.

This volume of proceedings shares advances in the area of computer vision, laser vibrometry, digital image correlation, photogrammetry, and optical technique and applications of these techniques for dynamic measurements, structural dynamics, and structural health monitoring.

The organizers would like to thank the authors, presenters, session organizers, and session chairs for their participation in this track.

Editor:
Janko Slavič - *University of Ljubljana, Slovenia*; Dan Rohe–*Sandia National Laboratories*, NM, USA.

Contents

Chapter 1
Spatiotemporally-resolved Kinematic and Stress Measurements of Interfacial Cavitation in Soft Matter via DIC

Jin Yang, Alexander McGhee, Zixiang Tong, Lehu Bu, Sicong Wang, Griffin Radtke, Mauro Rodriguez, and Christian Franck

Abstract Inertial cavitation is a common phenomenon found in nature and many engineering systems. When harnessed carefully, laser or ultrasound-focused, energy-driven cavitation can be a very beneficial tool in a wide range of medical and materials applications, including laser surgery, lithotripsy, drug delivery, and more recently, soft material characterization. Recently we have developed a novel material property characterization method, called Inertial Microcavitation Rheometry (IMR), to investigate laser-induced inertial cavitation (LIC) in soft matter, where the surrounding material is subjected to ballistic and ultra-high strain rates ($10^3 \sim 10^8$ s^{-1}). Through IMR, we can precisely quantify the nonlinear viscoelastic, finite deformation constitutive behavior of soft materials at ultra-high strain rates. Following this, we will present our recent findings on the dynamics of laser-induced inertial cavitation (LIC) near the gel-water interface. Historically, studies of cavitation dynamics at liquid-solid interfaces have been limited to observations of surface deformations and cavitation bubble morphology due to challenges in measuring subsurface behavior. However, understanding the intricate dynamics of cavitation at these interfaces has significant implications for engineering and medical applications. By utilizing Digital Image Correlation (DIC), we provide high-fidelity and high-throughput full-field measurements on the spatiotemporal deformation behavior and wave propagation within soft materials near interfaces due to laser-induced inertial cavitation at extremely high rates. Our results provide critical insight into how soft biological tissues respond to the immense forces generated by the violent collapse of a cavitation event. These measurements will be particularly useful for minimizing collateral damage to non-target tissues in cavitation-based medical therapies.

Keywords cavitation · soft materials · high speed imaging · interface · DIC

Jin Yang · Zixiang Tong · Lehu Bu · Sicong Wang
Department of Aerospace Engineering and Engineering Mechanics, The University of Texas at Austin, Austin, TX 78712
e-mail: jin.yang@austin.utexas.edu; zachtong@utexas.edu; lhbu@utexas.edu; sicongw@utexas.edu

Alexander McGhee
Department of Biomedical Engineering, The University of Arizona, Tucson, AZ 85721
e-mail: mcgheealex@arizona.edu

Griffin Radtke
John A. Paulson School of Engineering and Applied Sciences, Harvard University, Cambridge, MA 02138
e-mail: gradtke@fas.harvard.edu

Mauro Rodriguez
School of Engineering, Brown University, Providence, RI 02912
e-mail: mauro_rodriguez@brown.edu

Christian Franck
Department of Mechanical Engineering, University of Wisconsin-Madison, Madison, WI, 53706
e-mail: cfranck@wisc.edu

Jin Yang
To whom correspondence should be addressed
e-mail: jin.yang@austin.utexas.edu

Janko Slavic et al. (eds.), *Computer Vision & Laser Vibrometry, Vol. 6*, Conference Proceedings of the Society for Experimental Mechanics Series, https://doi.org/10.13052/9788743801511-1

Introduction

Cavitation is a fascinating phenomenon observed across a wide range of biological and engineering systems, from the prey-stunning capability of the snapping shrimp to cavitation erosion on metallic pumps and propellers, and cavitation-induced cell and tissue damage [1–6]. When harnessed properly, energy-driven, focused cavitation can serve as a powerful tool in a diverse set of applications, including tissue phantom modeling, laser surgery, lithotripsy, DNA injection, and more recently, high- to ultra-high strain-rate soft material characterization [7–15]. Inertial cavitation can be triggered by various external stimuli, such as ultrasound, laser, shock waves, hydrodynamic variations, electrical discharges, and microwave radiation. Particularly, laser-induced inertial cavitation (LIC) utilizes a high-energy laser focused on a small region of liquid or soft solid materials, rapidly vaporizing the liquid and forming a bubble. As this bubble expands and collapses, it generates powerful shock waves and high temperatures, resulting in diverse effects and applications. LIC exploits the complex interactions between light, heat, and mechanical forces. The laser parameters can be precisely controlled to tailor the cavitation process for specific outcomes, making LIC a versatile and effective tool for scientific research and industrial applications.

Recently, by leveraging the laser-induced inertial cavitation in soft biological materials, we developed a novel experimental technique called Inertial Microcavitation Rheometry (IMR), which enables the extraction of bulk soft biological materials' viscoelastic properties at extremely high strain rates ($> 10^3\ \mathrm{s}^{-1}$) [15–19], allowing us to resolve and predict inertial cavitation in soft materials away from surfaces or impedance-changing interfaces.

In many biologically and clinically relevant applications, cavitation often occurs or is deliberately produced along material interfaces, particularly at soft matter-liquid interfaces. Examples include cancer cell removal, targeted drug delivery, cataract removal, histotripsy, noninvasive ocular surgeries, and even blast-related traumatic brain injuries. Whether cavitation is generated intentionally or occurs as an unintended side effect, it is critical to understand its impact on surrounding tissues. Therefore, quantitatively describing, modeling, and predicting the complex dynamics of cavitation at these interfaces is vital to achieving successful outcomes in these applications.

Unlike cavitation in bulk materials, inertial cavitation near interfaces and surfaces presents more complex physics due to the asymmetry of the material domain [20–23]. The dynamics of inertial cavitation near interfaces often involve complex, non-spherical bubble shapes and instability patterns, which can lead to high-speed re-entrant or emerging jets at the interface. These instabilities, along with highly dynamic pressure and stress fields, depend on factors such as the distance between the nucleation site and the interface, the maximum bubble radius, and the impedance differences across the interface. Soft materials exhibit large, nonlinear deformations in response to these pressure and stress fields, sometimes resulting in material failure. This creates a strong, nonlinear coupling between the solid and fluid, which remains poorly understood.

To identify the key physical phenomena involved in inertial cavitation at impedance-mismatched boundaries, we developed a novel experimental procedure combining three state-of-the-art techniques. We integrated (i) laser-induced cavitation (LIC) [17] with (ii) the internal, embedded speckle pattern technique [24, 25] to generate optically-focused, single cavitation bubbles at various stand-off distances from a speckled gel-water interface, and analyzed the results using (iii) SpatioTemporal Adaptive Quadtree mesh Digital Image Correlation (STAQ-DIC) [26]. These experiments provided the first high-resolution, quantitative spatiotemporal kinematic and stress fields of inertial cavitation along a soft hydrogel-water interface. Specifically, our data offered rich insight into the complex subsurface deformations and wave propagation within the surrounding soft solid material during interfacial cavitation. We measured for the first time the LIC-induced material damage and residual stress near the gel-water interface caused by laser pulse ablation. These highly controlled experimental investigations enable for a detailed analysis of single inertial cavitation events near a water-gel interface with unprecedented resolution. Beyond single cavitation events, repeated laser pulses can accumulate cavitation-induced damage as the number of pulses increases. We also aim to quantify the residual strain and stress caused by these cumulative effects.

Materials and Methods

Material preparation

In this work, we synthesized 10 wt% gelatin hydrogels using porcine gelatin (gel strength 300, Type A, Sigma-Aldrich). A speckle pattern of microscale inkjet toner particles was embedded in the middle plane for each hydrogel sample following the protocol developed by McGhee et al. [17]. After solidification, the gel samples were stored at 4°C for more than 24 hours. Before the laser-induced cavitation experiment, the samples were acclimated to room temperature for one hour to

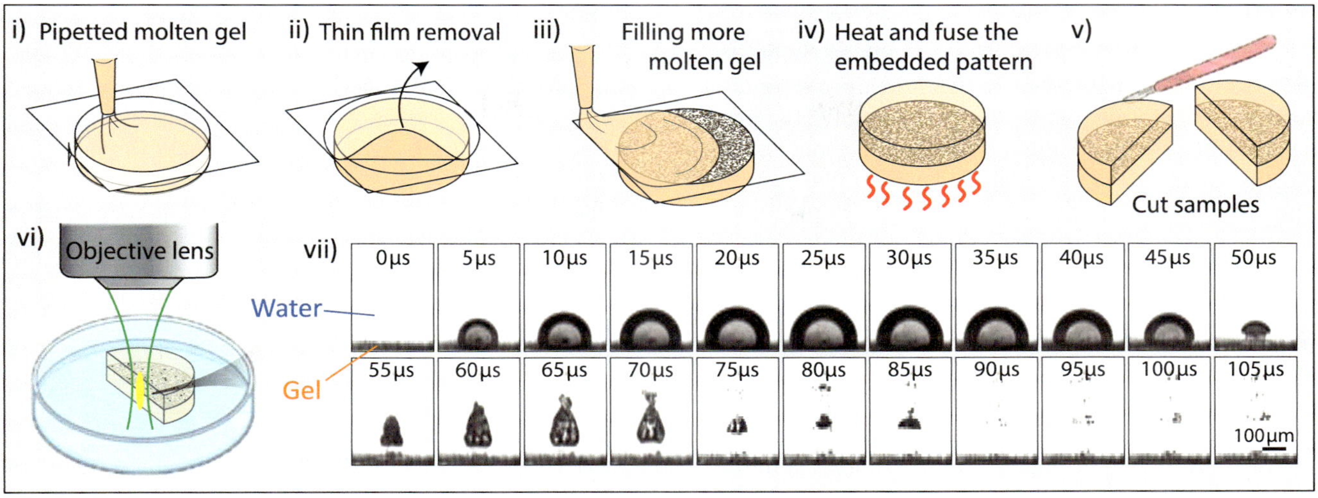

Fig. 1 Experimental setup for the laser-induced inertial cavitation (LIC) experiments at the gel-water interface. (i-v) Experimental procedures for preparing the embedded speckle pattern in a hydrogel sample; (ii) A single nanosecond pulsed-laser focusing within gel samples; (vii) Selected experimental images of laser-induced inertial cavitation events (adapted from Ref [27]).

reach thermal equilibrium. Then a gel-water interface was created by a scalpel cutting through the gel perpendicular to the speckle pattern (see Fig. 1(i-v)).

The prepared 10 wt% gel's quasistatic, ground-state shear modulus was determined using an ARES-G2 rheometer with a 25 mm stainless steel smooth plate force transducer (TA Instruments, DE). The gel presents rate-dependent material properties. To characterize the high strain-rate, viscoelastic properties of our gelatin hydrogels, as most relevant to the laser-induced inertial cavitation experiments investigated here, we also employed our previously established Inertial Microcavitation Rheometry (IMR) technique [15–18]. The measured quasistatic and dynamic shear moduli are $G_{\infty} = 3.08 \pm 0.01$ kPa and $G = 29.85 \pm 6.20$ kPa, respectively. The measured viscosity is $\mu = 0.055 \pm 0.061$ Pa·s [28].

Laser-induced inertial microcavitation

We utilized the same laser-induced inertial microcavitation (LIC) setup as previously described in Refs [15–18]. Briefly, a single ~6 nanosecond-long laser pulse is focused via a 10× microscope objective directly onto the embedded speckle pattern plane within each gel sample, and positioned at various distances from the interface, as illustrated in Fig. 1(vi). While the z-position of the cavitation bubble is controlled by the objectives focal point, the x, y position can be precisely controlled via the motorized microscope stage. In our experiments, various combinations of maximum bubble radius and distance from the interface were tested. To normalize these parameters, we utilize the same dimensionless variable ($\gamma = d/R_{\max}$) as first introduced by Brujan et al. [20, 21], where the bubble standoff distance from the interface is denoted by d, and the maximum bubble radius by $R_{\max}$. Here, positive and negative values of γ correspond to initial cavitation nucleation sites in either the water (positive γs) or gel (negative γs), respectively. In this paper, we focus on studying when the cavitation events are induced very close to the gel-water interface, i.e., $\gamma \sim 0$.

Each cavitation event was recorded using an HPV-X2 high-speed camera (Shimadzu Corporation, Kyoto, Japan) at 1 million frames per second. Pulsed illumination for each frame was achieved through a Cavilux Smart UHS laser illumination system with a 20 ns pulse duration (Shimadzu Corporation, Kyoto, Japan) and synchronized with the high-speed camera via a trigger. Selected frames depicting a typical cavitation event are shown in Fig. 1(vii).

Beyond single cavitation events, we also repeatedly shot laser pulses at the same spot, with an estimated time interval of 1 minute between any two consecutive laser pulses. This time interval was chosen based on the assumption that, after 1 minute, the transient effects of the previous LIC event would have reached a near-equilibrium state, where the changes in surrounding material strain and stress become significantly slower than the cavitation dynamics timescale. Although much slower diffusion processes may still occur, they take on the order of 10 minutes or more to fully equilibrate—much longer than the transient timescale of cavitation dynamics.

SpatioTemporally Adaptive mesh Digital Image Correlation (STAQ-DIC) analysis

We employed our previously developed SpatioTemporally Adaptive Quadtree mesh Digital Image Correlation (STAQ-DIC) method [26] to analyze material deformations resulting from cavitation event. This method compares the deformed frames with the undeformed frame to directly resolve the incremental change in the deformation of the surrounding material for each frame [29]. In addition to displacement and strain fields, we also calculated velocity fields by dividing the incremental displacement by the time elapsed between two successive image frames. We used 20 pixels $\times$ 20 pixels as the DIC post-processing window size, and 16 pixels $\times$ 16 pixels as the DIC window spacing. We also apply adaptive quadtree meshes where the finest DIC window spacing is further refined to 4 pixels $\times$ 4 pixels near the bubble wall [30].

Results and Analysis

Captured bubble dynamics using ultra-high-speed imaging

In this study, we investigate cavitating bubbles at the gel-water interface where stand-off distances are effectively zero, i.e., $\gamma \sim 0$ (the effect of stand-off distances is our planned future work). Repeated laser pulses were focused on the same spot, and the resulting bubble dynamics, along with the surrounding soft material deformations, were captured using ultra-high-speed imaging at 1 million frames per second. This allowed us to quantify the accumulated material damage from twenty successive cavitation events. For our case study, we used 10 wt% gelatin, and the applied laser pulse energy was 337.7 $\pm$ 5.9 μJ.

Figure 2 presents a high-speed imaging sequence showing the evolution of the 10 th laser pulse-induced cavitation bubble near a 10 wt% gelatin hydrogel-water interface over a time span of 2 μs to 207 μs. The images reveal the dy-

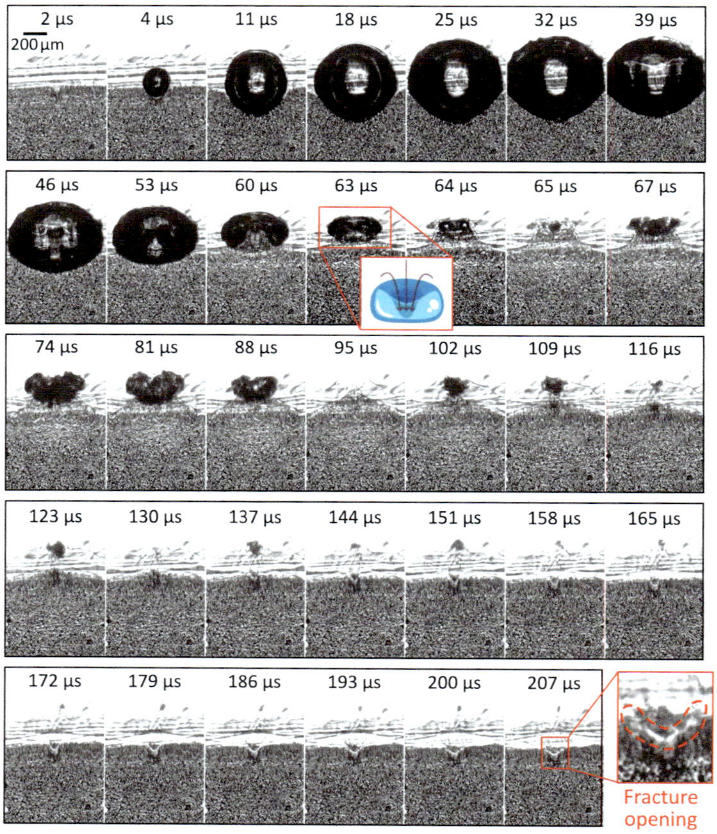

Fig. 2 High-speed imaging sequence showing the evolution of the 10 th laser pulse-induced cavitation bubble near a gelatin hydrogel-water interface. The inset at 63 μs provides a detailed view of the formed cavitation jet interacting with the interface. The inset at 207 μs shows the cavitation induced surface fracture opening.

namics of bubble expansion, collapse, and subsequent jet formation, all significantly influenced by the proximity of the gelatin interface. During the initial stages, from 2 μs to 39 μs, the cavitation bubble undergoes spherically symmetrical growth. As the bubble expands and approaches the interface, its interaction with the boundary becomes more pronounced, resulting in the formation of a cavitation jet around 63 μs. The inset at this time point provides a detailed view of the jet directed toward the interface, highlighting the strong hydrodynamic interaction between the collapsing bubble and the gelatin substrate.

As the bubble collapses further, the imaging sequence between 74 μs and 207 μs shows the aftermath of the jet's impact on the surface. The inset at 207 μs highlights the surface damage caused by the cavitation event, revealing the localized deformation and potential erosion at the gelatin-water boundary. This progression indicates the destructive potential of cavitation-induced jets in soft materials, which is relevant for applications involving cavitation near biological tissues, soft interfaces, or hydrogel-based materials.

Experimentally measured kymographs of hyperelastic von Mises strain and stress generated by each laser-induced inertial cavitation event

For each DIC image sequence, we extracted a 10-pixel-wide vertical slice along the bubble's axis of symmetry from the full-field results. These slices were then concatenated to create kymographs, which visualize the spatiotemporal evolution of the von Mises strain and stress fields within the surrounding gelatin hydrogels, as shown in Fig. 3 and Fig. 4.

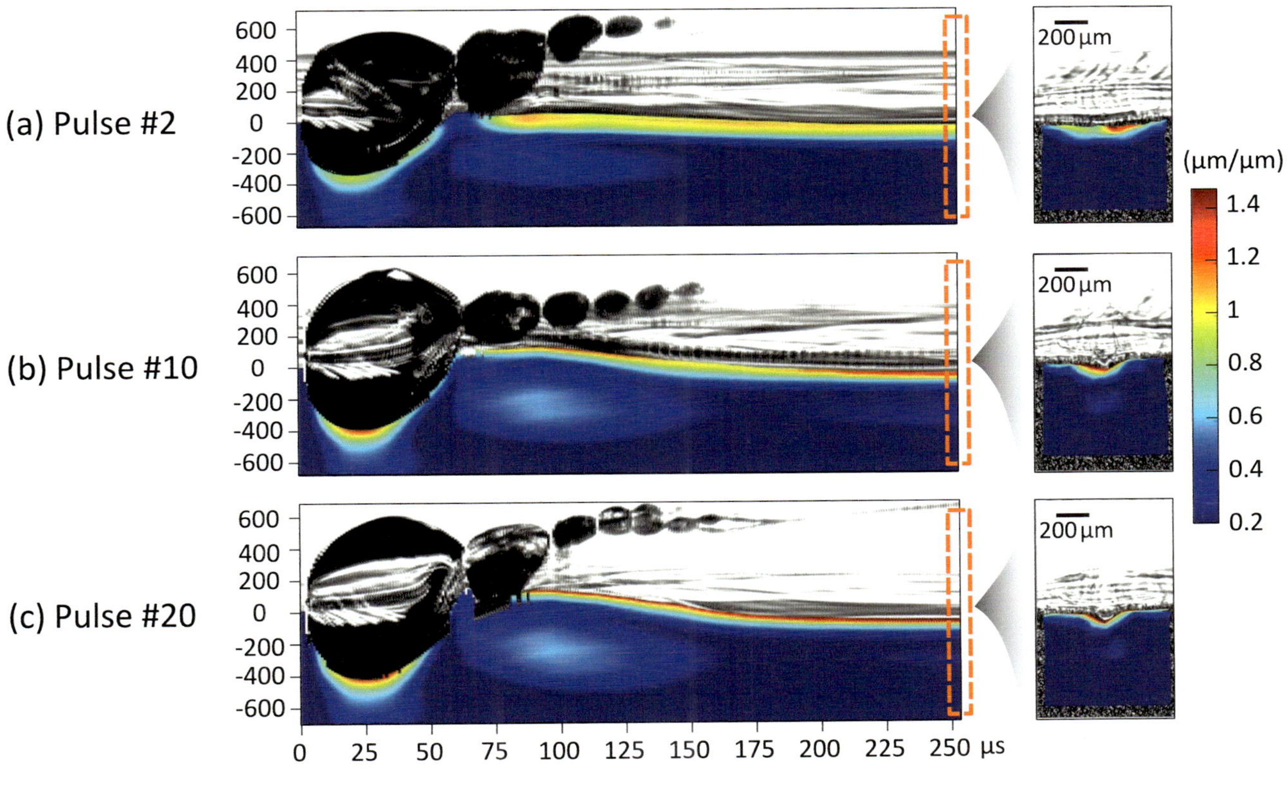

Fig. 3 Comparison of the von Mises strain kymographs from DIC results for laser-induced cavitation at pulse #2 (a), pulse #10 (b), and pulse #20 (c). The strain maps illustrate the distribution and evolution of localized strain, with inset zooms highlighting strain distributions at the last frame when the surface damage reaches an equilibrium state. Each DIC result compares the net strain change of each laser pulse-induced cavitation event, indicating cumulative damage from repeated cavitation events.

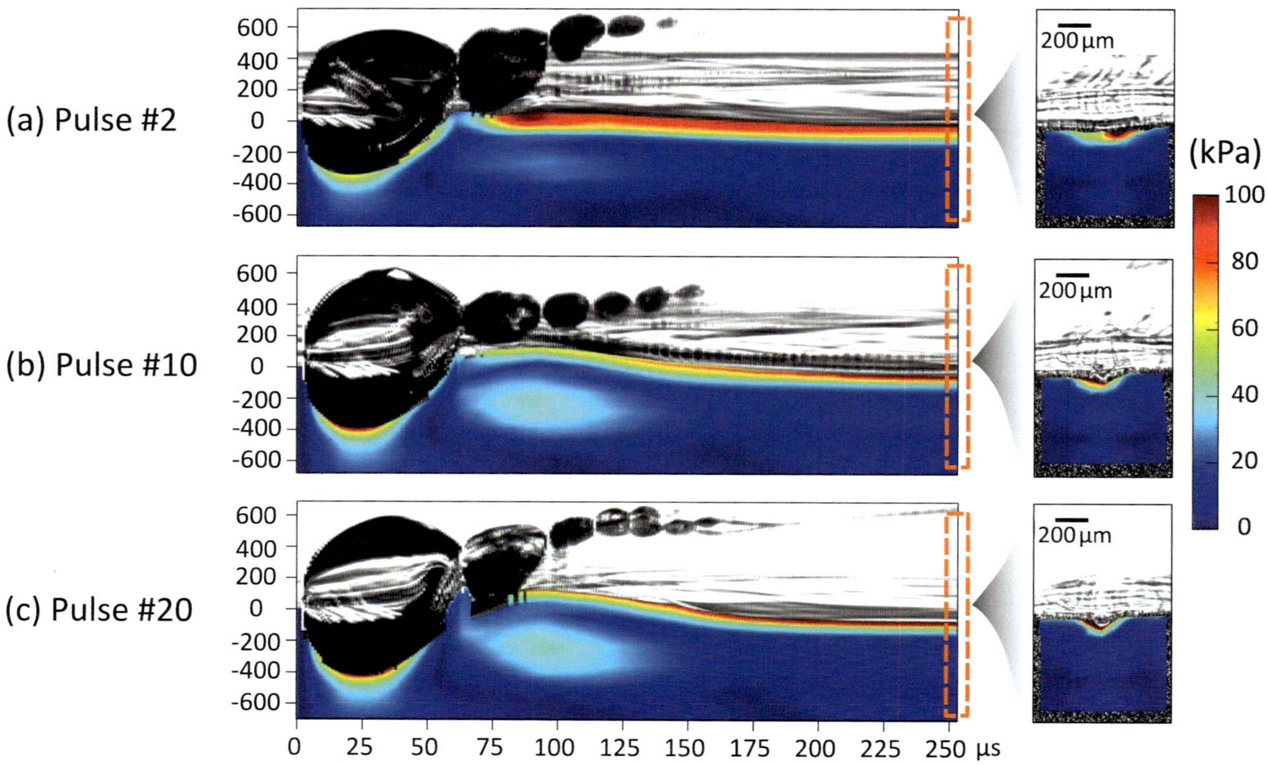

Fig. 4 Comparison of the von Mises stress kymographs from DIC results for laser-induced cavitation at pulse #2 (a), pulse #10 (b), and pulse #20 (c). The stress maps illustrate the distribution and evolution of localized stress, with inset zooms highlighting stress distributions at the last frame when the surface damage reaches an equilibrium state. Each DIC result compares the net stress change of each laser pulse-induced cavitation event, indicating cumulative damage from repeated cavitation events.

Here we assume the bubble dynamics is axisymmetric whose displacement field is denoted as $\mathbf{u} = [u_r, u_\theta, u_z]^\top$. The infinitesimal strain is defined as

$$\boldsymbol{\varepsilon} = \begin{bmatrix} \frac{\partial u_r}{\partial r} & 0 & \frac{1}{2}\left(\frac{\partial u_r}{\partial z} + \frac{\partial u_z}{\partial r}\right) \\ 0 & \frac{u_r}{r} & 0 \\ \frac{1}{2}\left(\frac{\partial u_r}{\partial z} + \frac{\partial u_z}{\partial r}\right) & 0 & \frac{\partial u_z}{\partial z} \end{bmatrix} \tag{1}$$

The von Mises stain is defined as

$$\varepsilon_{VM} = \sqrt{\frac{1}{2}\left[\left(\varepsilon_1 - \varepsilon_2\right)^2 + \left(\varepsilon_2 - \varepsilon_3\right)^2 + \left(\varepsilon_3 - \varepsilon_1\right)^2\right]}, \tag{2}$$

where ε_1, ε_2, and ε_3 are three principal values of infinitesimal strain $\boldsymbol{\varepsilon}$.

We model the surrounding hydrogel as an incompressible neo-Hookean material with shear modulus G. The resulted hyperelastic Cauchy stress follows

$$\boldsymbol{\sigma} = G\,\mathrm{dev}(\mathbf{B}), \tag{3}$$

where dev$(\cdot)$ takes the deviatoric part, and $\mathbf{B}$ is the left Cauchy-Green deformation tensor. The von Mises stress is further defined as

$$\sigma_{VM} = \sqrt{\frac{1}{2}\left[\left(\sigma_1 - \sigma_2\right)^2 + \left(\sigma_2 - \sigma_3\right)^2 + \left(\sigma_3 - \sigma_1\right)^2\right]}, \tag{4}$$

where σ_1, σ_2, and σ_3 are three principal values of Cauchy stress tensor $\boldsymbol{\sigma}$.

Figures 3 and 4 compare the von Mises strain and stress fields generated by laser-induced cavitation across multiple cavitation events at pulse #2, pulse #10, and pulse #20. The first frame of each image sequence was set as the reference image, with subsequent frames treated as deformed images. The high-speed imaging, combined with the corresponding kymographs derived from DIC measurements, reveals the dynamic interactions between the cavitation bubbles and the gelatin hydrogel-water interface. For each pulse, the strain maps display the "net gain" in strain distribution and the evolution of the localized strain along the interface, with the highest concentrations occurring near the bubble collapse regions. The insets in the right of each subplot highlight the strain distribution in the final stage of each pulse, where the surface damage appears to stabilize and reaches a near equilibrium state.

In this context, near equilibrium refers to the point at which the temporal changes in the surrounding material strain and stress states become significantly slower than the cavitation dynamics timescale. While much slower diffusion processes can still occur, they take approximately $O(10)$ minutes to reach equilibrium, which is considerably longer than the transient timescale of the cavitation dynamics.

The results also emphasize the cumulative nature of surface damage caused by repeated cavitation events. As seen in the progression from pulse #2 to pulse #20, the net strain change increases with successive pulses, indicating that strain does not fully relax between events. This suggests that cavitation-induced surface damage accumulates over time, particularly in the regions closest to the hydrogel-water interface.

Cumulative Material Damage from Repeated Laser Pulses

In this subsection, we examine how repetitive laser pulses ablate tissue and accumulate surrounding material damage. Figure 5 illustrates the progression of surface deformation at the gelatin hydrogel-water interface due to repeated laser-induced cavitation events over 20 consecutive pulses. Specifically, Figure 5(a) provides a schematic of the experimental setup, showing the laser focus point positioned near the interface. Figure 5(b) presents high-speed imaging of the hydrogel surface, with zoomed-in insets tracking the development of localized surface deformations after each laser pulse.

As the number of pulses increases, the deformation becomes more pronounced, particularly after pulse #7, suggesting that the repeated cavitation events induce cumulative surface damage (see Fig. 5). The hydrogel's surface exhibits progressive material degradation and surface depression, which is highlighted by the magnified regions. This result highlights the critical role that repeated cavitation plays in surface erosion and the importance of understanding how soft materials respond to such cyclic loading conditions. The observations can be applied to a variety of fields, including biomedical applications where cavitation occurs near biological tissues or at soft material interfaces.

Figure 6 illustrates the displacement fields generated by repeated laser-induced cavitation events at the gelatin hydrogel-water interface. Figure 6(a) shows the z-displacement maps (vertical displacement) for pulses #2 to #20, while Fig. 6(b) displays the x-displacement maps (horizontal displacement) for the same sequence. We also note that the absolute value of the x-coordinate in the xz-plane shown in Fig. 6 corresponds to the radial coordinate r in the 3D cylindrical coordinate system. The color maps depict how cavitation bubbles influence surface deformation. Early in the sequence, the displacements are localized near the laser focus point, but as the number of pulses increases, the deformation spreads across a larger area, indicating an accumulation of surface damage.

The z-displacement maps in Fig. 6(a) show significant vertical movement at the interface, particularly as the cavitation bubble expands and collapses, generating surface deformations that grow more pronounced with each pulse. Similarly, the x-displacement maps in Fig. 6(b) reveal the horizontal shifting of the interface, which becomes more evident in later pulses. This progressive deformation, seen in both the vertical and horizontal directions, highlights the cumulative effects of repeated cavitation, suggesting that the surface damage is not confined to the immediate vicinity of the cavitation event, but spreads over time, impacting the overall integrity of the hydrogel surface. These results underscore the importance of understanding cavitation-induced surface degradation, particularly in applications where soft material interfaces are exposed to cyclic cavitation loading.

Figure 7 illustrates the strain fields generated by repeated laser-induced cavitation events at the gelatin hydrogel-water interface. Initially, the strain fields (both e_{zz} and e_{rr}) remain minimal, as seen from pulse #2 to pulse #7, with low strain values represented by the blue regions. As cavitation events progress, there is a noticeable increase in strain, particularly after pulse #10, where a more significant deformation occurs near the bubble collapse region including compression strain in e_{rr} component and tension strain in e_{zz} component. These results demonstrate and quantify the cumulative nature of material damage and surface deformation as laser-induced cavitation is applied repeatedly.

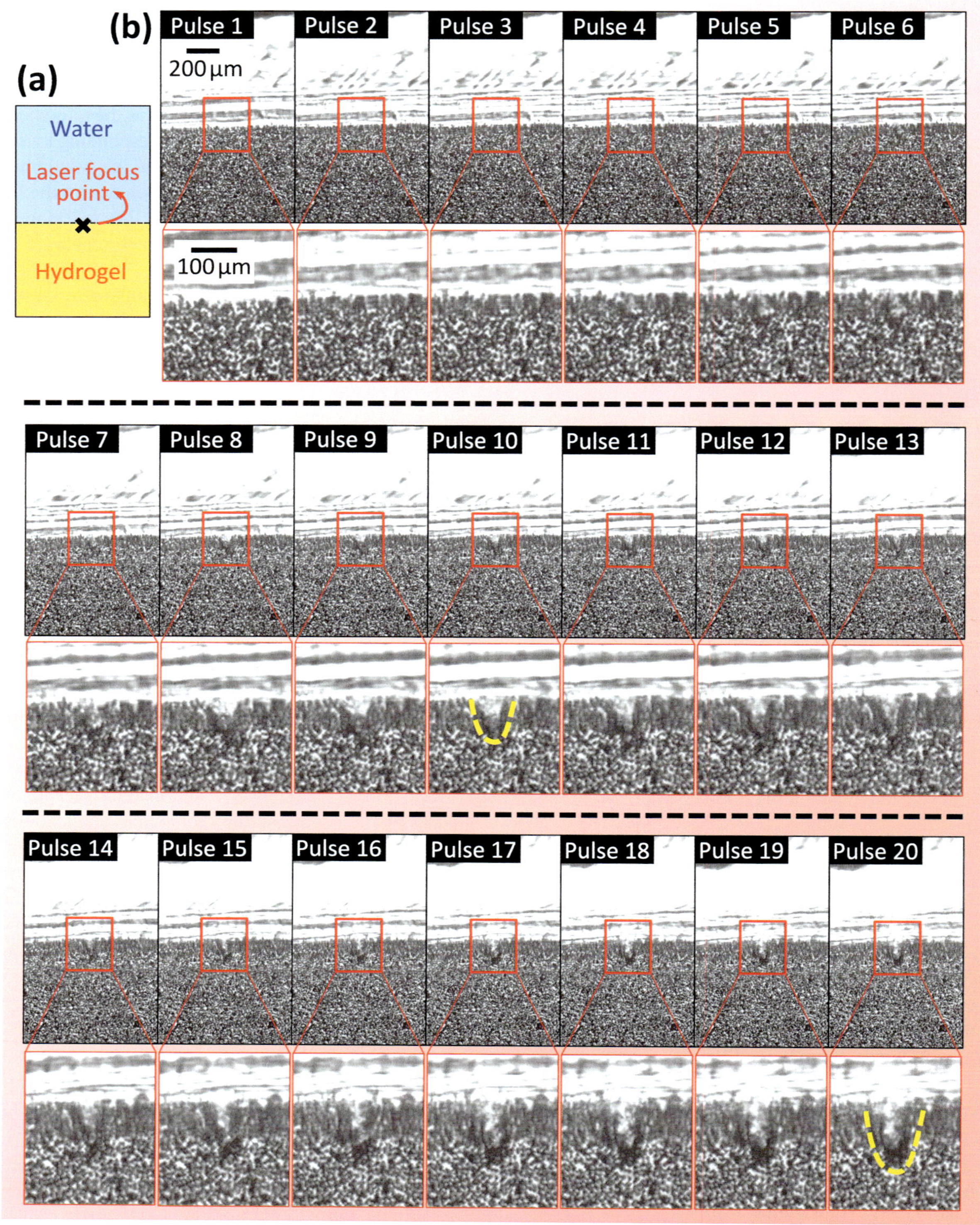

Fig. 5 (a) Schematic of the experimental setup showing the laser focus point at the gelatin hydrogel-water interface. (b) High-speed imaging sequence displaying the cumulative effects of laser-induced cavitation from pulse #1 to pulse #20. Each frame captures the localized surface response at the hydrogel interface, with magnified insets highlighting regions of interest. Yellow dashed-line markers indicate the cavitation-induced surface fracture openings after pulse #10 and pulse #20, visualizing the accumulation of damage with repeated cavitation events.

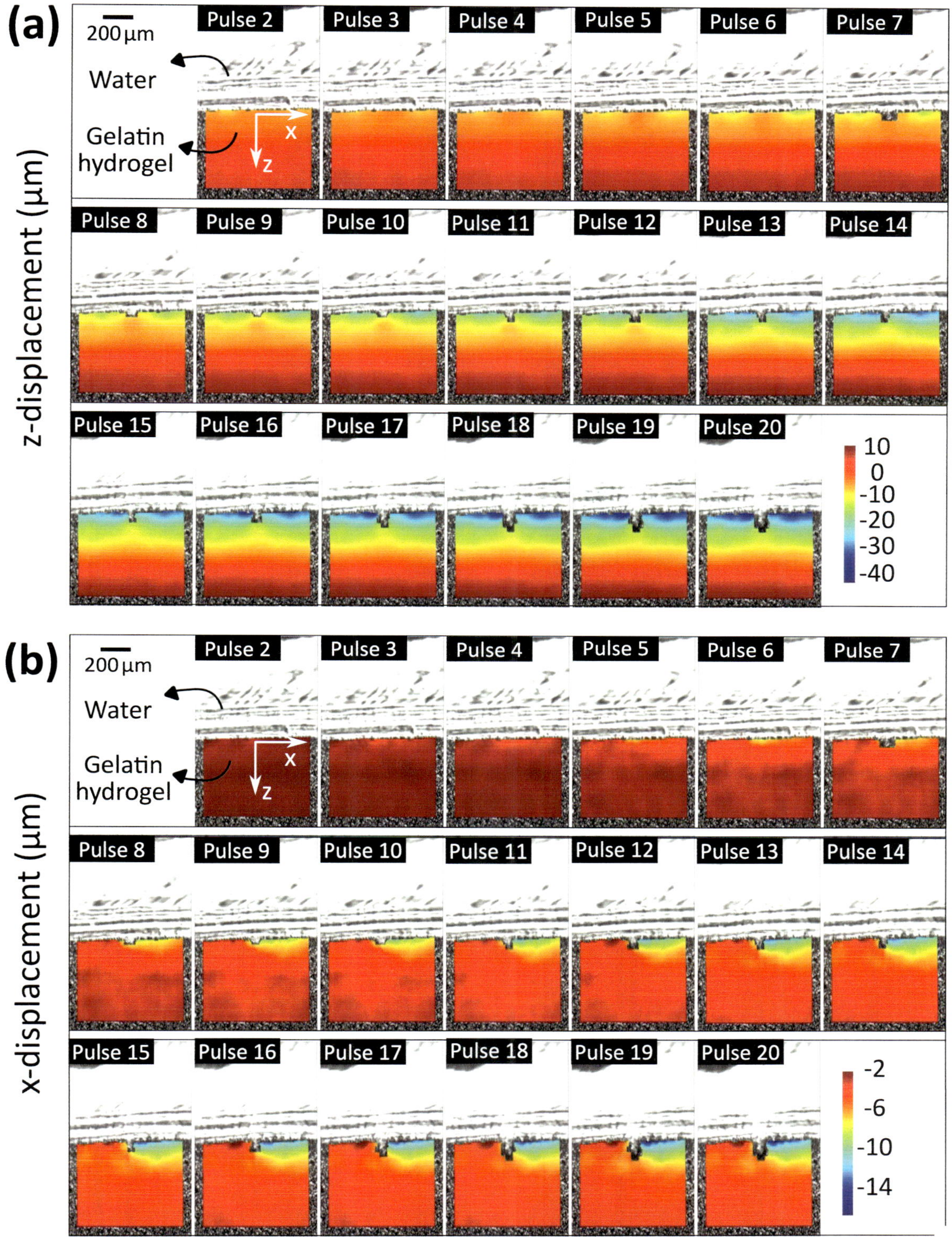

Fig. 6 (a) z- and (b) x-displacement maps showing the evolution of surface deformation in the gelatin hydrogel-water interface due to laser-induced cavitation across pulses #2 to #20. Each frame captures the strain fields generated by successive cavitation events, with positive and negative values represented by the color scale.

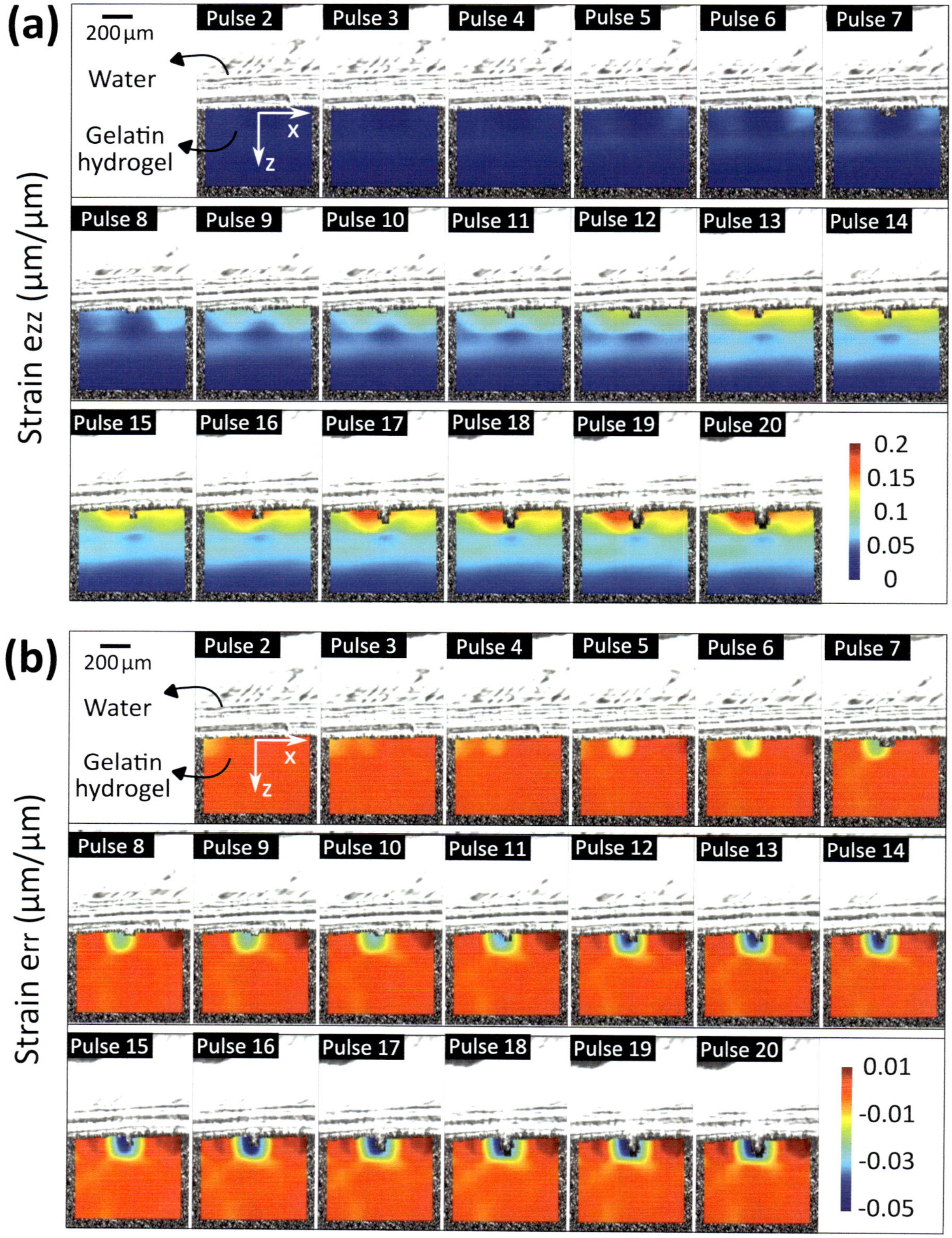

Fig. 7 (a) Strain e_{zz} and (b) e_{rr} distribution maps showing the evolution of surface deformation in the gelatin hydrogel-water interface due to laser-induced cavitation across pulses #2 to #20. Each frame captures the deformation fields generated by successive cavitation events, with positive and negative displacement values represented by the color scale.

Conclusion

In conclusion, the experimental investigation of laser-induced cavitation near a gelatin hydrogel-water interface reveals critical insights into the dynamic interaction between cavitation bubbles and soft material surfaces. The displacement and strain maps obtained through Digital Image Correlation (DIC) offer detailed views of the full-field deformation behavior, with progressive damage accumulation observed across successive cavitation events. As the number of pulses increases, the hydrogel surface experiences significant strain and deformation, leading to permanent surface alterations.

Here we highlight our study on the cumulative nature of cavitation-induced damage, particularly in soft materials exposed to repeated cavitation events. The detailed strain and displacement maps show how cavitation bubbles induce localized surface deformations, which intensify with each subsequent pulse. These findings are essential for understanding the long-term impact of cavitation on soft material interfaces, with implications for a range of applications, including biomedical devices, material testing, and tissue interaction during therapeutic laser procedures. The results underscore the importance of developing strategies to mitigate cavitation-induced damage in soft materials, particularly when used in environments subject to repeated cavitation exposure.

Acknowledgments We gratefully acknowledge support from the US Office of Naval Research under PANTHER award number N000142112044 through Dr. Timothy Bentley. J.Y. and M.R.J. acknowledge the support provided by the U.S. National Science Foundation (NSF) under Grant No. 2232427 and No. 2232428. J.Y. acknowledges the start-up fund from the University of Texas at Austin and the Haythornthwaite Foundation for the Research Initiation Grant awarded by the Applied Mechanics Division of the American Society of Mechanical Engineers (ASME).

References

1. Versluis, M., Schmitz, B., von der Heydt, A., and Lohse, D. "How snapping shrimp snap: Through cavitating bubbles". *Science*, 289:2114–2117 (2000)
2. Philipp, A. and Lauterborn, W. "Cavitation erosion by single laser-produced bubbles". *Journal of Fluid Mechanics*, 361:75–116 (1998)
3. Movahed, P., Kreider, W., Maxwell, A., Hutchens, S., and Freund, J. "Cavitation-induced damage of soft materials by focused ultrasound bursts: A fracture-based bubble dynamics model". *The Journal of the Acoustical Society of America*, 140:1374–1386 (2016)
4. Quinto-Su, P., Dijkink, R., Prabowo, F., Gunalan, K., Preiser, P., and Ohl, C. "Interaction of red blood cells with arrays of laser-induced cavitation bubbles". In *Proceedings of the 7th International Symposium on Cavitation* (2009)
5. Vlaisavljevich, E., Maxwell, A., Mancia, L., Johnsen, E., Cain, C., and Xu, Z. "Visualizing the histotripsy process: Bubble cloud-cancer cell interactions in a tissue-mimicking environment". *Ultrasound in Medicine and Biology*, 42:2466–2477 (2016)
6. Mancia, L., Vlaisavljevich, E., Yousefi, N., Rodriguez, M., Ziemlewicz, T., Lee, F., Henann, D., Franck, C., Xu, Z., and Johnsen, E. "Modeling tissue-selective cavitation damage". *Physics in Medicine & Biology*, 64 (2019)
7. Brujan, E. and Vogel, A. "Stress wave emission and cavitation bubble dynamics by nanosecond optical breakdown in a tissue phantom". *Journal of Fluid Mechanics*, 558:281–308 (2006)
8. Brennen, C. "Cavitation in medicine". *Interface Focus*, 5:20150022 (2015)
9. Wan, M., Feng, Y., and ter Haar, G. *Cavitation in Biomedicine*. Springer (2015)
10. Bailey, M., Khokhlova, V., Sapozhnikov, O., Kargl, S., and Crum, L. "Physical mechanisms of the therapeutic effect of ultrasound (a review)". *Acoustical Physics*, 49:369–388 (2003)
11. Hutson, M. and Ma, X. "Plasma and cavitation dynamics during pulsed laser microsurgery in vivo". *Physical Review Letters*, 99:158104 (2007)
12. Cherian, A. and Rau, K. "Pulsed-laser-induced damage in rat corneas: time-resolved imaging of physical effects and acute biological response". *Journal of Biomedical Optics*, 13:024009 (2008)
13. Blomley, M., Cooke, J., Unger, E., Monaghan, M., and Cosgrove, D. "Microbubble contrast agents: a new era in ultrasound". *BMJ*, 322:1222–1225 (2001)
14. Ibsen, S., Schutt, C., and Esener, S. "Microbubble-mediated ultrasound therapy: a review of its potential in cancer treatment". *Drug Design, Development and Therapy*, 7:375–385 (2013)
15. Estrada, J., Barajas, C., Henann, D., Johnsen, E., and Franck, C. "High strain-rate soft material characterization via inertial cavitation". *Journal of the Mechanics and Physics of Solids*, 112:291–317 (2018)
16. Yang, J., Cramer, H., and Franck, C. "Extracting non-linear viscoelastic material properties from violently-collapsing cavitation bubbles". *Extreme Mechanics Letters*, 39:100839 (2020)
17. McGhee, A., Yang, J., Bremer, E., Xu, Z., Cramer, H., Estrada, J., Henann, D., and Franck, C. "High-speed, full-field deformation measurements near inertial microcavitation bubbles inside viscoelastic hydrogels". *Experimental Mechanics*, 63(1):63–78 (2023)
18. Yang, J., Cramer, H., Bremer, E., Buyukozturk, S., Yin, Y., and Franck, C. "Mechanical characterization of agarose hydrogels and their inherent dynamic instabilities at ballistic to ultra-high strain-rates via inertial microcavitation". *Extreme Mechanics Letters*, 51:101572 (2022)
19. Yang, J. and Franck, C. "Strain stiffening effects of soft viscoelastic materials in inertial microcavitation". In *Dynamic Behavior of Materials, Volume 1: Proceedings of the 2019 Annual Conference on Experimental and Applied Mechanics*, pages 175–179. Springer (2020)
20. Brujan, E.A., Nahen, K., Schmidt, P., and Vogel, A. "Dynamics of laser-induced cavitation bubbles near an elastic boundary". *Journal of Fluid Mechanics*, 433:251–281 (2001)

21. Brujan, E.A., Nahen, K., Schmidt, P., and Vogel, A. “Dynamics of laser-induced cavitation bubbles near elastic boundaries: influence of the elastic modulus”. *Journal of Fluid Mechanics*, 433:283–314 (2001)
22. Henzel, T., Nijjer, J., Chockalingam, S., Wahdat, H., Crosby, A., Yan, J., and Cohen, T. “Interfacial cavitation”. *PNAS Nexus*, 1(4):pgac217 (2022)
23. Yang, J., Yin, Y., Cramer, H., and Franck, C. “The penetration dynamics of a violent cavitation bubble through a hydrogel-water interface”. In Amirkhizi, A., Notbohm, J., Karanjgaokar, N., and DelRio, F., editors, *Challenges in Mechanics of Time Dependent Materials. Mechanics of Biological Systems and Materials & Micro-and Nanomechanics*, volume 2, pages 65–71. Springer International Publishing, Cham (2022)
24. McGhee, A., Bennett, A., Ifju, P., Sawyer, G., and Angelini, T. “Full-field deformation measurements in liquid-like-solid granular microgel using digital image correlation”. *Experimental Mechanics*, 58:137–149 (2018)
25. McGhee, A., McGhee, E., Famiglietti, J., and Schulze, K. “Dynamic subsurface deformation and strain of soft hydrogel interfaces using an embedded speckle pattern with 2d digital image correlation”. *Experimental Mechanics*, 61(6):1017–1027 (2021)
26. Yang, J., Rubino, V., Ma, Z., Tao, J., Yin, Y., McGhee, A., Pan, W., and Franck, C. “Spatiotemporally adaptive quadtree mesh (STAQ) digital image correlation for resolving large deformations around complex geometries and discontinuities”. *Experimental Mechanics*, 62:1191–1215 (2022)
27. Yang, J., McGhee, A., Tong, Z., Radtke, G., Rodriguez Jr., M., and Franck, C. “Inertial interface cavitation creates complex, flow-like structures within a soft solid”. bioRxiv, 2025-02 (2025).
28. Bremer-Sai, E., Yang, J., McGhee, A., and Franck, C. “Ballistic and blast-relevant, high-rate material properties of physically and chemically crosslinked hydrogels”. *Experimental Mechanics*, 64:587–592 (2024)
29. Yang, J. and Bhattacharya, K. “Augmented lagrangian digital image correlation”. *Experimental Mechanics*, 59:187–205 (2019)
30. Yang, J. and Bhattacharya, K. “Fast adaptive mesh augmented lagrangian digital image correlation”. *Experimental Mechanics*, 61(4):719–735 (2021)

Chapter 2
Continuous Scanning LDV: Comparison among different Vibrometer configuration and technologies

Milena Martarelli, Alessia Caputo, Joerg Sauer, and Paolo Castellini

Abstract Laser Doppler Vibrometry (LDV) is a mature measurement technique with numerous applications in civil, industrial and research fields. Despite this, developments in the technique are continuously progressing, both in terms of the measurement system and hardware, and in terms of measurement strategies and data processing. Compared to traditional point-by-point scanning, the technique of continuously moving the laser beam across a surface has the inherent potential for high spatial resolution combined with high scanning speed, opening up new applications beyond structural dynamics evaluation. In this paper, the Continuous Scanning strategy was applied using vibrometers with different wavelengths and technologies in order to evaluate the effects on the measurement results. In particular, results were compared with interferometers utilizing red and infrared wavelengths, along with traditional demodulation methods and multi-path interferometry (QTec). The findings highlight that the activation of QTec significantly improves the signal-to-noise ratio and reduces phase jumps, thus enhancing measurement stability and accuracy. The results obtained were carefully compared, focusing on the measurement uncertainty of the operational vibrational modes.

Keywords Laser Doppler Vibrometry · Continuous Scanning · QTec · Signal Quality · Operational Modal Analysis

Introduction

Measuring vibration without contact using a laser based on the Doppler effect is a technique already proposed 1964 [1]. Since the Doppler shift f_D expressed for reception in direct reflection

$$f_D = \frac{2nv}{\lambda} \tag{1}$$

where n is the index of refraction (1 in air) and v is the instantaneous velocity of the target and λ the laser wavelength, is in real world experiments eight orders of magnitude smaller than the laser light frequency. An interferometer has proven to be the best method to precisely determine the Doppler shift. Building upon this principle, Laser Doppler Vibrometer (LDV) equipped with a couple of moving mirrors automate the positioning of the laser beam on the surface of the object under test, allowing the reconstruction of the vibration spatial distribution of the object by scanning over a dense grid of points in a fast way. This vibration measurement strategy is named Scanning LDV (SLDV) and is widely used for modal analysis.

A strategy that overcomes this limitation is the so called Continuous SLDV (CSLDV) that consists of using the mirrors moving continuously in order to make the laser beam sweeping across the surface of the object under test. By means of this technique the vibration spatial distribution is therefore measured in one shot and, consequently, in a limited time [2]. Unfortunately the scanning of the laser beam over a rough surface exacerbated the well-known drawback of optical interferometry which is the speckle noise [3]. In the case of beam like structures, the CSLDV is typically exploited in the line scan mode, e.g. the laser beam is made to scan back and forward along a line, see Figure 1 (a). In this circumstance

Milena Martarelli · Alessia Caputo · Paolo Castellini
Department of Industrial Engineering and Mathematical Sciences, Polytechnic University of Marche, Ancona, Italy
e-mail: m.martarelli@staff.univpm.it; · a.caputo@staff.univpm.it; · p.castellini@staff.univpm.it

Joerg Sauer
Polytec Gmbh, Polytec-Platz 1-7, 76337 Waldbronn, Germany
e-mail: j.sauer@polytec.de

Janko Slavic et al. (eds.), *Computer Vision & Laser Vibrometry, Vol. 6*, Conference Proceedings of the Society for Experimental Mechanics Series, https://doi.org/10.13052/9788743801511-2

the laser beam passes periodically on the same speckle and therefore the effect of the speckle noise is boosted. In the case of plate like structures, the adopted CSLDV strategy is the area scan, see Figure 1 (b), that consists of driving the mirrors moving the laser beam with two sinewaves at frequencies mutually prime. The laser beam will then move along a curvilinear line, i.e. a Lissajous pattern, and it will never pass through the same point [4]. In this circumstance, opposite of the line scan, the speckle noise will be reduced. Despite this, techniques aimed at mitigating the effect of speckle noise make it possible to improve the quality of data collected via CSLDV.

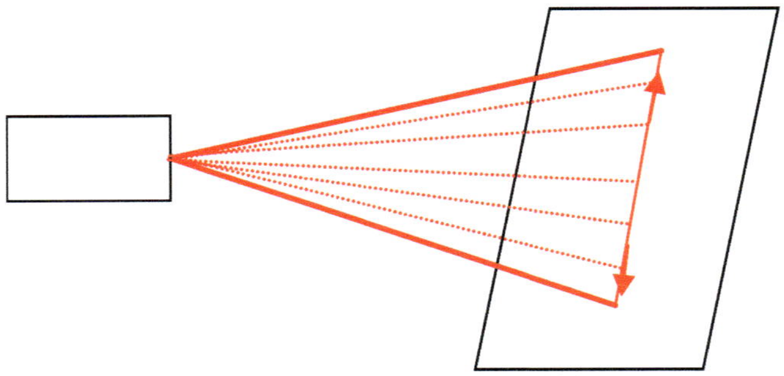
(a) Line scan

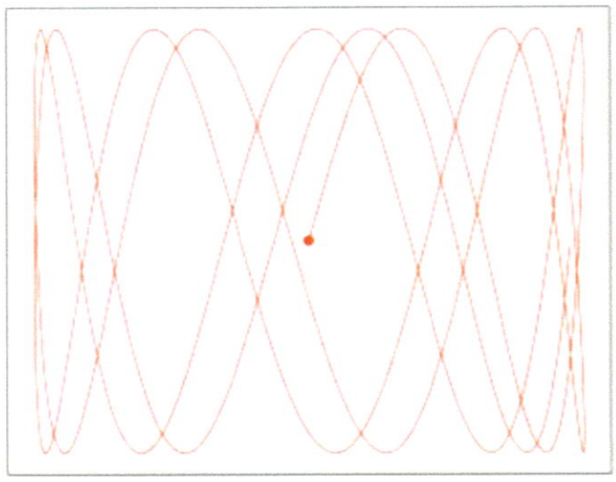
(b) Area scan

Fig. 1 Schematic representation of CSLDV strategies: **(a)** Line Scan – back-and-forth laser movement increases speckle noise; **(b)** Area Scan – Lissajous pattern reduces speckle effects by avoiding repeated points.

The CSLDV philosophy is based on the fact that when the laser beam scans continuously along a trajectory over a vibrating structure at a known excitation frequency with a motion low known, the LDV output is modulated in amplitude by the Operational Deflection Shape (ODS) of the structure itself at that excitation frequency. The easiest method for recovering the ODS from the LDV output is demodulation. The conventional and most diffuse method is instead the polynomial reconstruction. This latter is founded on the hypothesis that the ODS can be expressed by an analytic formulation based on a polynomial series. The polynomial coefficients are linked to the LDV output spectrum which has a shape typical of modulated signals. In practice it consists of a series of sidebands equi-spaced and centered to the excitation frequency. The polynomial coefficients can be calculated from the amplitude and phase of those sidebands by applying a Chebeyshev transformation matrix. There are several alternative methods for reconstructing ODSs from CSDLV data, like re-sampling the time history synchronously with the fundamental period (i.e. the scanning frequency) in order to recover a sort of "lifted" time histories at a finite number of points on the structure [5], [6]. Another method is the so called mode matching strategy which is based on the fact that mode shapes of the structure under test are known a priori (from a numerical model, an analytical formulation or previous measurements). It is then possible to settle a procedure that searches for similarities between those known mode shapes (the candidate mode shapes) and ODSs that actually modulate the signal measured and identify the actual mode shapes excited and the corresponding frequency [7]. Another strategy deserving to be mentioned is a blind identification of the polynomial coefficients based on the solution, in least square sense, of a set of linear equation defining the relations between the contributions of each cluster of frequencies (central frequency + sidebands) potentially defining the spectral pattern of a certain ODS to the complex spectrum measured [8]. In this paper the conventional method founded on the polynomial reconstruction of ODS by applying the Chebeyshev transformation matrix will be applied to demonstrate the efficacy of the Qtec technology in improving the ODS reconstruction from CSLDV data. The improvement of CLSDV technique open up new field of application of the method not only in modal analysis but also in structural damage detection where the measurement noise can compromise and sometimes invalidate the detection of defects as they typical have a minimum contribution to the overall vibration of the whole structure [9],[10].

Material and Methods

The experimental set up

The experimental setup consists of a rectangular aluminum plate of dimensions of 180x350 mm and thickness of 2 mm and clamped at one of the shortest sides. The plate was forced to vibration by means of an electrodynamic shaker (TIRA model TV 50009) connected to the structure via a stinger and a force cell (PCB model 208C01) for measuring the force transmitted to the plate. The vibration measurement was performed while the structure was forced at one single resonance; the resonances of the plate has then been measured in advance via a sine sweep test. The shaker has therefore been set to provide sinusoidal excitation at the resonance frequencies in sequence from mode 1 to mode 6, see Table 1. The table reports also the Modal Participation (MP) factor of each mode to the overall vibration. For the sake of synthesis, in the following only the mode shapes 2, 3, 5 and 6 will be considered whose spatial distribution is visualized in Figure 2 which represents the modes nodal lines. The vibration has been measured via an SLDV (Polytec model PSV-500 with selectable QTec option) in continuous scanning mode. Specifically the laser beam was made to sweep over the entire surface of the plate by piloting the driving mirrors with two sinewaves at 1.1 Hz, in the shortest direction, and 20 Hz, in the longest one. The experimental setup is shown in Figure 3; the shaker and the force cell are not visible because they are behind the plate but the driving point (the connection of the stinger attached to the shaker and the plate) is noticeable as the black dot on the surface of the plate.

Table 1 Natural frequency and mode participation of plate

	Frequency [Hz]	MP [%]
Mode 1	12.8	18.4
Mode 2	53.6	6.4
Mode 3	74.8	13.2
Mode 4	80.6	13.5
Mode 5	168.2	7.3
Mode 6	196.8	13.8

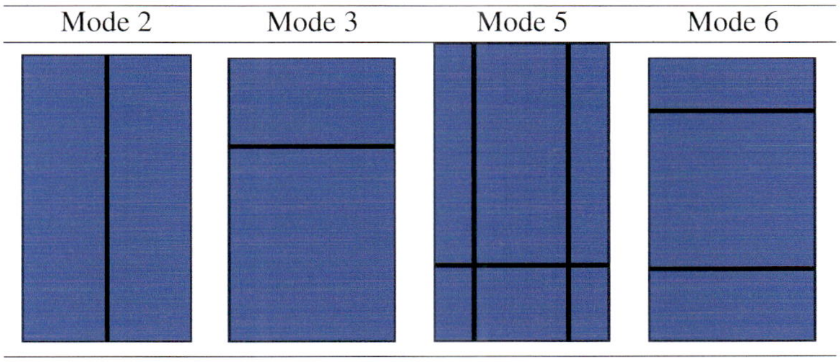

Fig. 2 Nodal lines of mode shapes considered in the analysis.

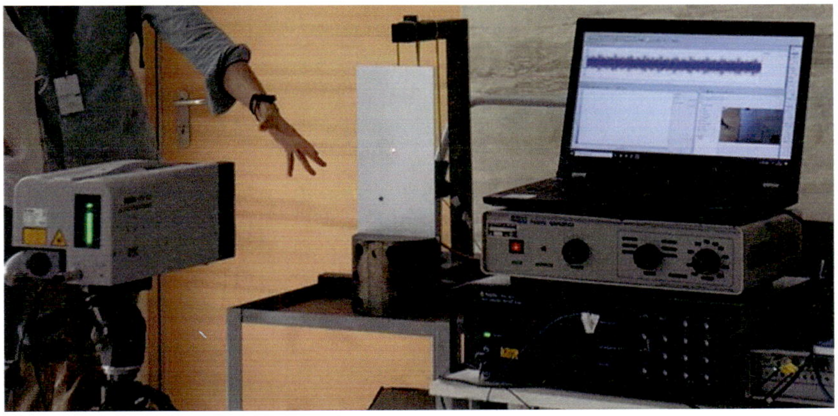

Fig. 3 Experimental set up for vibration measurement

ODS reconstruction from CLSDV data

Consider a line scan in the x-direction along a beam like structure. If the vibration is measured while the structure is forced at one resonance ω_j, and if the laser beam is moving along the structure, the LDV output is modulated in amplitude by the ODS of the structure itself at the excitation frequency ω_j. Consider that the ODS analytic formulation is a polynomial, as equation 2:

$$v(x) = \sum_{i=1}^{N} \mathrm{V}_i \mathrm{x}^{(i-1)} \tag{2}$$

being v the vibration velocity at the resonance frequency ω_j, V the polynomial coefficients, x the linear coordinate and N the polynomial order. The coefficients of such polynomial can be recovered from the amplitude and phase of the sidebands of the LDV output spectrum Si by applying the Chebeyshev transformation matrix $[T]$ as in equation 3:

$$\{\mathrm{V}\} = [\mathrm{T}]\{\mathrm{S}\} \tag{3}$$

QTec technology

For unambiguous detection of the sign of the velocity, the LDV is equipped with an optical-acoustic modulator (Bragg-cell) that adds a constant offset to the Doppler shift. This system allows to improved phase accuracy and signal-to-noise [11]. The most prominent noise source are "pseudo vibrations" caused by speckle transitions or changes of the speckle pattern caused by the coherent illumination of an optical rough surface (Figure 4) [12].

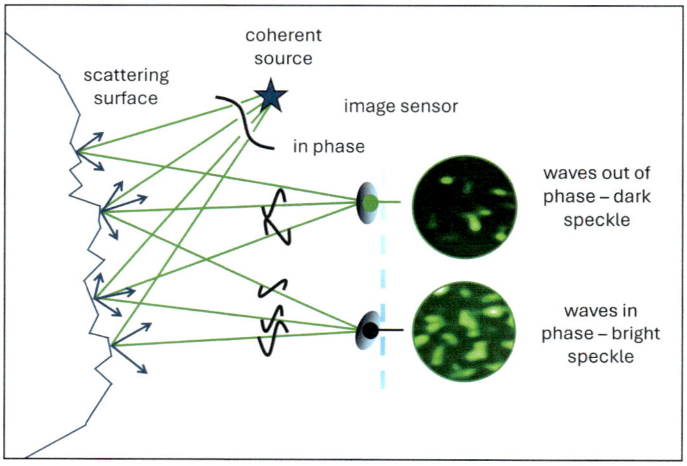

Fig. 4 Constructive and destructive interference depending on the observer position

Changes of speckle patterns cause signal drop-outs which express themselves in spikes in the velocity signal and phase jumps in the order of the wavelength λ in the displacement signal (Figure 5). At the same time the received signal strength (RSSI) drops. Due to a destructive interference little or no light arrives at the photo detector.

Speckle pattern changes are occurring with practically all real would measurements. Scanning Laser Doppler vibrometers in point-by-point scanning mode where the laser position is determined by galvo driven mirrors are exposed to speckle pattern changes while the scanning mirrors are moving or in decay. This makes wait times between each point mandatory if the SNR shall be optimized. This leads to the conclusion that the measurement time for scanning LDV could be drastically reduced if the speckle noise can be avoided. Inherent and not avoidable with the above-described interferometer configuration is speckle noise with rotating measurement samples like shafts and with continuous scanning laser Doppler vibrometer experiments. Since the speckle pattern is depending on the observer position relative to the illumination direction with coherent light adding observer position (multi-path) and using the received signals from all positions to recombine a speckle-drop-out free signal (Figure 6) [11].

The basic principle is used in WiFi receivers for a long time successfully and is called diversity reception. The implementation of a multi-path interferometer was proposed in the study of Martarelli et al. and is today implemented in the Polytec QTec® Laser Doppler Vibrometers used for generating the data for this paper [2]. The phase jumps and corresponding spikes of speckle changes are completely avoided with multi-path diversity combining if the RSSI signal is above the general FM threshold that allows signal detection and demodulation.

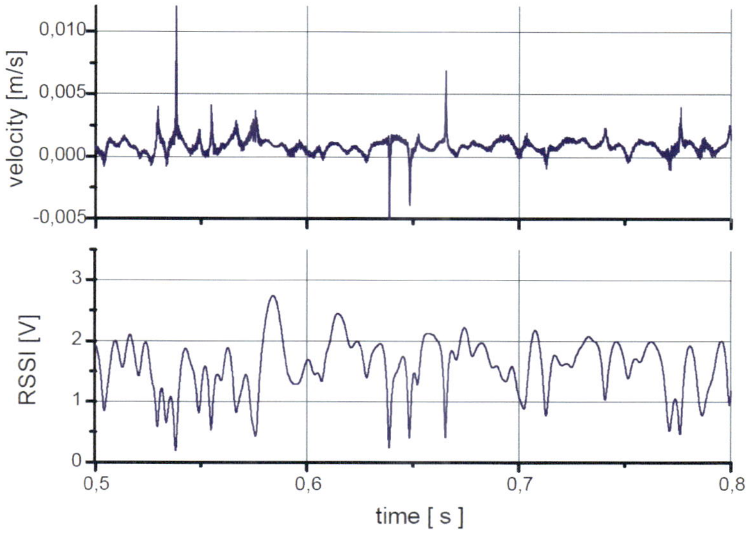

Fig. 5 Velocity signal spikes and corresponding RSSI signal drop a dark speckle

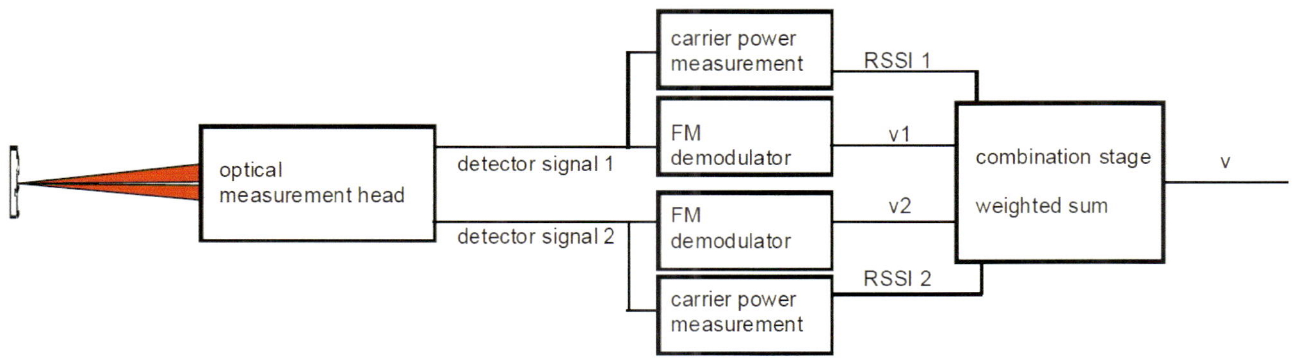

Fig. 6 General signal flow for a dual-path optical diversity system

Table 2 Increase in SNR and phase shift produced by the activation of the QTec option.

Mode	SNR [dB]		Phase shift θ [rad]		
	QTec off	QTec on	QTec off	QTec on	$\Delta\theta$
2	64.3	64.8	1.92	-0.04	1.96
3	26.4	34.5	1.48	2.73	1.24
5	13.4	14.2	2.07	2.94	0.87
6	20.31	20.92	-1.75	1.62	3.38

Results

In Figure 7 it can be observed the time histories and relative amplitude spectra measured by the CSLDV at the fourth resonance frequencies considered (the ones corresponding to the 2nd, 3rd, 5th and 6th ODS). Examining the signals related to the 2nd excitation frequency (53.6 Hz), it can be remarked that the time history is slightly more noisy when the QTec is off while the main difference is in the phase spectrum around the sideband frequencies. Observing the signals at the 3rd excitation frequency (74.8 Hz) the increase of noisiness in the time history acquired when the QTec is off is even more evident. The phase spectrum is instead less observable. The same can be said for the other two signals at the highest excitation frequencies. In order to express those differences in a synthetic way, two condensed metrics have been identified: the Signal

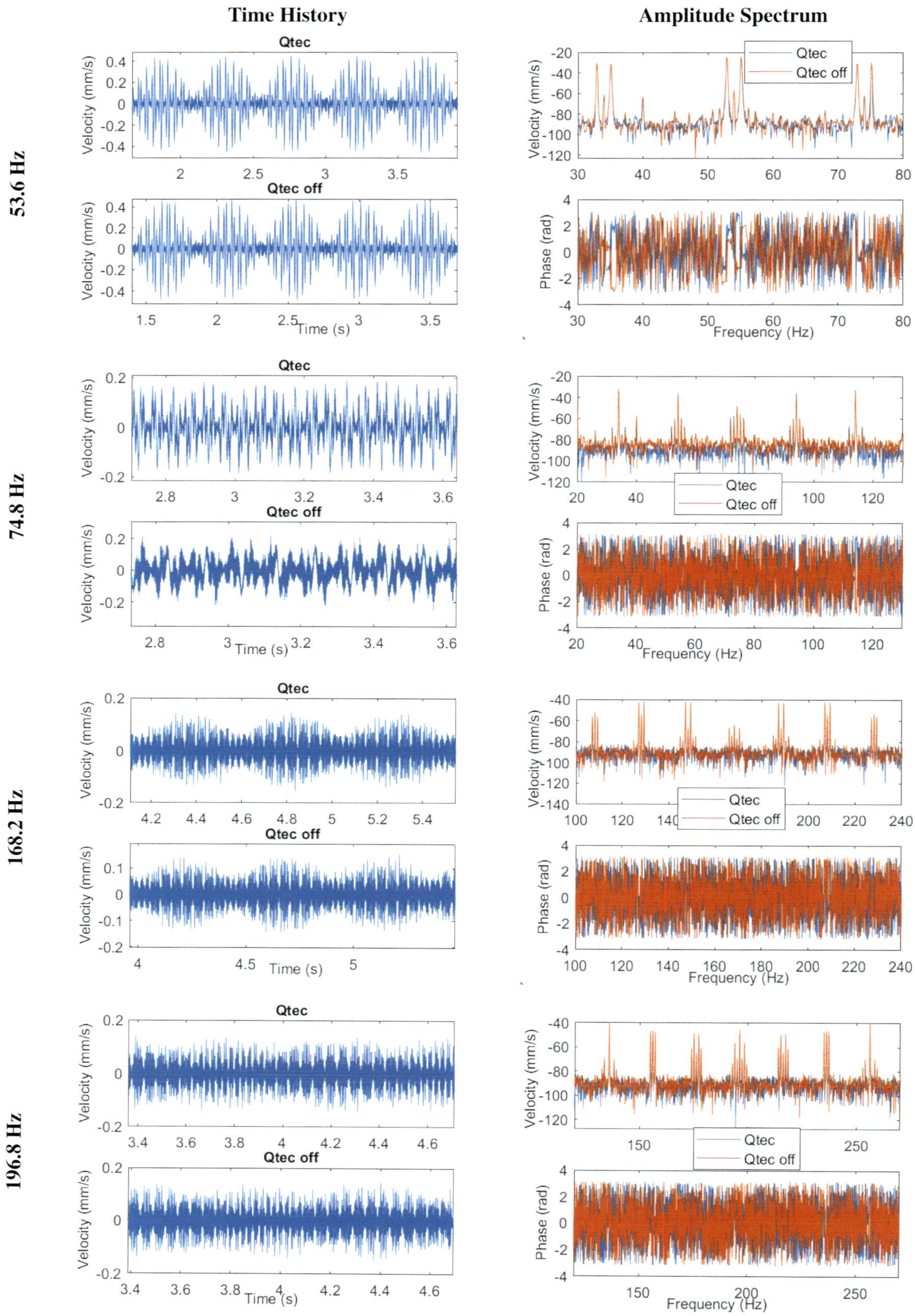

Fig. 7 CLSDV time histories and amplitude spectra at four excitation frequencies

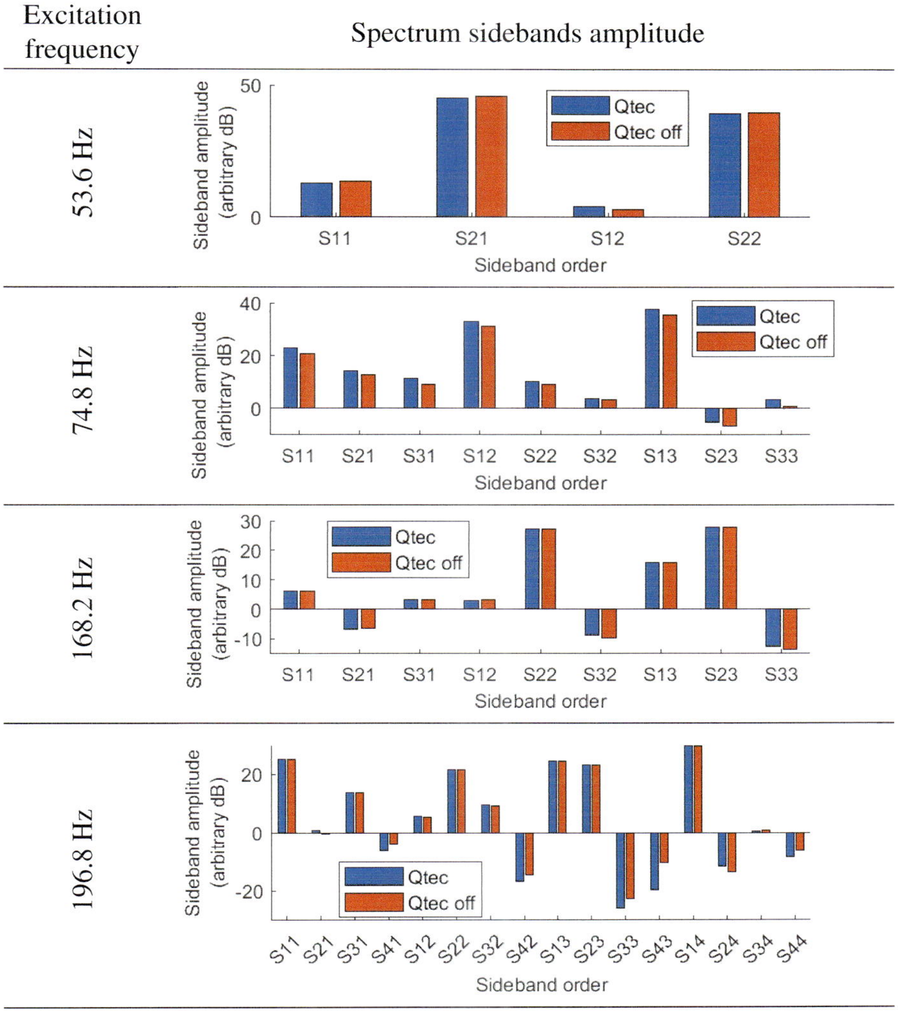

Fig. 8 Spectrum sidebands amplitude with Qtec activated and deactivated at the four excitation frequencies

to Noise Ratio (SNR) and the phase shift ($\Delta\theta$). The SNR of the signals acquired with the QTec activated and deactivated has been calculated as the ratio between the amplitude of the spectrum at the excitation frequency ωj and the mean value of the spectrum at all the frequencies except the excitation frequency and its sidebands. The phase shift has been determined as the difference between the phase of the spectrum acquired with the QTec option activated at the first sideband frequency and the same phase of the spectrum acquired with the QTec option deactivated. Those metrics are reported in Table 2, from which it can be deduced that the QTec does not reduce SNR which is always extremely high, except for the mode 3 there an increase of almost 8 dB is observed if the QTec is active. Concerning the sideband phase instead it is evident that the QTec modifies it of about 1 rad for the majority o the modes but for the 6th mode a phase shift higher than 3 rad is observed. It is expected that this phase variation will affect the mode shapes. Another way to highlight the effect of the QTec option is to observe the amplitude of the sideband spectra at the different excitation frequencies as reported in Figure 8. For modes 2 and 5, the torsional ones, the spectrum sidebands amplitude are almost not affected by the QTec mode. For mode 3 instead the sidebands calculated from the CSLDV data acquired with and without QTec differ the most and this confirm also the fact that the SNR in this case is 8 dB less when the QTec is deactivated. Considering mode 6, the sideband amplitudes are again different between the conditions of QTec on and off. This effect is very evident at the highest sidebands orders, which is justified by the fact that the high order sidebands are small and therefore more influenced by noise.

The final comparison can be done by observing the ODS reconstructed applying the Chebeyshev transformation to recover the 2D polynomials, which are reported in Figure 9. As evidenced by the previous analysis the ODSs suffering the most the QTec deactivation are the 3rd and the 6th ones. When the QTec mode is deactivated is visible in the ODS a not expected concavity along the x-direction that is due to high order sidebdnds distortion.

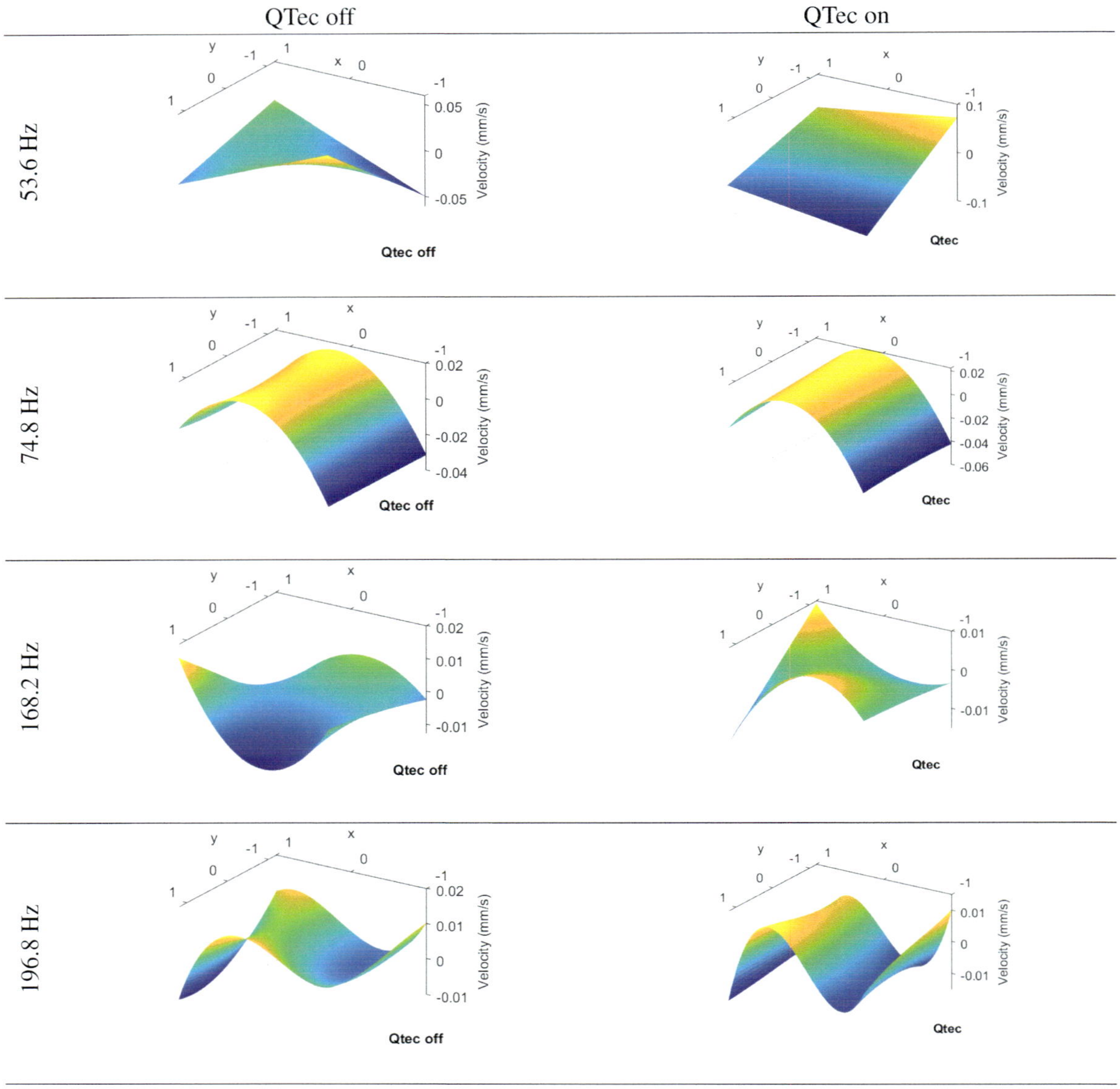

Fig. 9 Comparison of ODSs with QTec activated (right) and deactivated (left) at various frequencies

Conclusion

The experimental analysis demonstrated the advantages of QTec technology in vibration measurements using CSLDV. The observed improvement of the SNR with QTec engaged for the 3rd mode meets the expectation from experimental modal analysis experiments in point-by-point scanning mode. The SNR influence in the other modes was thus not as expected, indicating that QTec's effectiveness in suppressing phase jumps varies across different modes. The suppression of the actual phase jumps between speckles is effective also at the high speckle exchange rates of several 100 kHz as they occur at higher scan rates (Figure 10). While SNR is generally high without QTec, the technology is crucial for mitigating phase fluctuations, especially in the 3rd and 6th modes, which are sensitive to distortions. The analysis of ODS confirmed that QTec improves the accuracy of modal shape reconstruction, minimizing errors in structural analysis. Future work will address theoretical assumptions related to phase changes within the speckle that may limit SNR improvements. Overall, integrating QTec technology enhances measurement reliability and data integrity in complex vibration analysis.

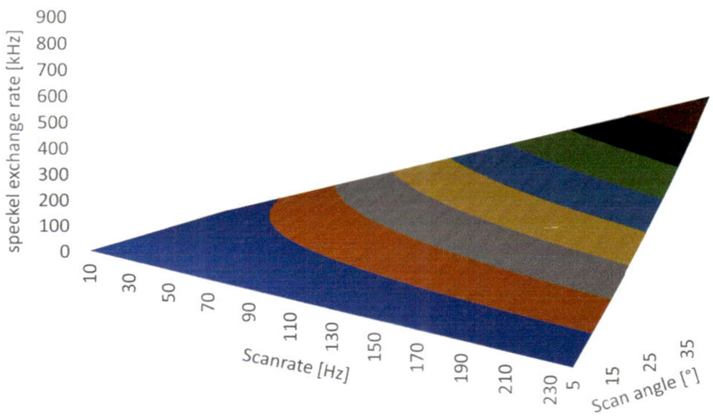

Fig. 10 Speckle exchange rates in CSLDV experiments depending on scan rate and angle for the PSV QTec with 1550 nm wavelength at 1m distance

References

1. Yeh, Y. and Cummins, H. "Localized fluid flow measurements with an he–ne laser spectrometer". *Applied Physics Letters*, 4(10):176–178 (1964)
2. Di Maio, D., Castellini, P., Martarelli, M., Rothberg, S., Allen, M., Zhu, W., and Ewins, D. "Continuous scanning laser vibrometry: A raison d'être and applications to vibration measurements". *Mechanical Systems and Signal Processing*, 156:107573 (2021)
3. Martarelli, M. and Ewins, D.J. "Continuous scanning laser doppler vibrometry and speckle noise occurrence". *Mechanical Systems and Signal Processing*, 20(8):2277–2289 (2006)
4. Stanbridge, A.B., Martarelli, M., and Ewins, D.J. "Measuring area vibration mode shapes with a continuous-scan ldv". *Measurement*, 35(2):181–189 (2004)
5. Allen, M.S. and Sracic, M.W. "A new method for processing impact excited continuous-scan laser doppler vibrometer measurements". *Mechanical Systems and Signal Processing*, 24(3):721–735 (2010)
6. Castellini, P., Sopranzetti, F., and Martarelli, M. "Re-sampling of continuous scanning ldv data for ods extraction". *Mechanical Systems and Signal Processing*, 72–73:667–677 (2016)
7. Chiariotti, P., Martarelli, M., and Castellini, P. "Mode matching of continuous scanning Laser Doppler vibration data in the frequency domain". *Optics and Lasers in Engineering*, 107:231–240 (2018)
8. Mignanelli, L., Chiariotti, P., Castellini, P., and Martarelli, M. *Blind Identification of Operational Deflection Shapes from Continuous Scanning Laser Doppler Vibrometry Data*, pages 105–111. Springer International Publishing (2016)
9. Martarelli, M., Chiariotti, P., Pezzola, M., and Castellini, P. "Delamination detection in composites by laser ultrasonics". In *AIP Conference Proceedings*, pages 405–412. AIP Publishing LLC (2014)
10. Chiariotti, P., Martarelli, M., and Revel, G. "Delamination detection by multi-level wavelet processing of continuous scanning laser doppler vibrometry data". *Optics and Lasers in Engineering*, 99:66–79 (2017)
11. Rothberg, S., Allen, M., Castellini, P., Di Maio, D., Dirckx, J., Ewins, D., Halkon, B., Muyshondt, P., Paone, N., Ryan, T., Steger, H., Tomasini, E., Vanlanduit, S., and Vignola, J. "An international review of laser doppler vibrometry: Making light work of vibration measurement". *Optics and Lasers in Engineering*, 99:11–22 (2017)
12. Rothberg, S. "Numerical simulation of speckle noise in laser vibrometry". *Applied Optics*, 45(19):4523 (2006)

Chapter 3
A Wirelessly Time-Synchronized Vision-Based System Network for Non-Contact Structural Health Monitoring

Miaomin Wang, Ki Young Koo, Fuyou Xu, and James Brownjohn

Abstract Structural health monitoring (SHM) projects often require numerous sensors across entire structures, leading to significant costs and complex installations. Vision-based systems offer a promising alternative by capturing data remotely across a wide field of view. However, these systems are limited by measurement inaccuracies over large areas and in restricted zones due to a shallow depth of field, which complicates their application in tasks requiring high-accuracy displacement measurements of large-scale structures and input-output measurements. This study proposes the use of multiple time-synchronized vision systems to address these challenges. We introduce the Flexible Vision Network (FVN), a wireless network of vision-based systems where each node's clock is synchronized with satellite atomic clocks via GPS modules. This synchronization allows for flexible deployment without limitations on spatial arrangement or node quantity. Each node in the FVN utilizes a template matching algorithm for structural displacement measurements. The FVN's effectiveness was demonstrated through 3D displacement measurement and modal testing on operational bridges.

Keywords Structural health monitoring · Computer vision · Wireless sensor · Time synchronization · Modal testing

Introduction

Vision-based systems provide a contactless alternative to traditional sensors, enabling the collection of input load and structural response data for structural health monitoring (SHM) [1]. They have therefore gained popularity in the SHM field, especially for measuring structural displacements in bridges, buildings, and dams. This is achieved through digital image processing techniques such as template matching [2–4], feature tracking [5], optical flow [6], and motion magnification [7–9].

Another SHM application involves detecting vehicles and pedestrians on bridges to estimate loads. For instance, Pan et al. [10] developed a vision-based framework to detect and track vehicles on the Jiangyin Bridge in China. However, accurately estimating load requires more than just position data. This has led to methods that combine vision-based detection with weigh-in-motion (WIM) systems and tire-road contact models [11, 12].

Single-camera systems face a trade-off between field of view (FOV) and measurement accuracy, which limits their effectiveness in large, multi-task scenarios. Multi-camera systems help address this issue but require time synchronization and reliable data transmission. Basic synchronization methods involve post-processing with audio cues [13] or overlapping FOVs [14], while more precise techniques use wireless triggers like radio frequency or stroboscopic lights. Unfortunately, temperature-sensitive crystal oscillators in cameras can introduce timing uncertainties [15]. Wired camera systems using GPIO and Ethernet cables allow synchronized data transmission but are limited by the processing capacity of the master computer and cable length [16].

Wireless networks like the SmartVision system [17] eliminate cable restrictions by using radio waves for synchronization. However, scaling these networks can be challenging due to bandwidth constraints and spatial limitations on node placement.

Miaomin Wang · Ki Young Koo · James Brownjohn
Faculty of Environment, Science and EconomyUniversity of Exeter, UK
e-mail: miaominwang@hotmail.com; k.y.koo@exeter.ac.uk; j.brownjohn@exeter.ac.uk

Fuyou Xu
School of Infrastructure EngineeringDalian University of Technology, China
e-mail: fuyouxu@hotmail.com

Janko Slavic et al. (eds.), *Computer Vision & Laser Vibrometry, Vol. 6*, Conference Proceedings of the Society for Experimental Mechanics Series, https://doi.org/10.13052/9788743801511-3

Synchronizing with other sensor types (e.g., accelerometers) is also difficult because of differences in time protocols across systems.

This paper introduces the Flexible Vision Network (FVN), a wireless vision sensor network that uses GNSS-based time synchronization, eliminating master node dependency. It offers flexibility in spatial deployment and supports multi-task SHM applications.

Flexible Vision Network (FVN)

Hardware of an FVN node

In FVN, all nodes share the same basic hardware. As shown in Figure 1, each FVN node comprises an image capture device, an image processing unit, a GNSS module, and other peripheral components. These nodes are designed to be both portable and economical.

Fig. 1 Hardware of a FVN node.

(1) Image Capture: The system uses the Raspberry Pi High Quality Camera, which provides a high resolution of 4056 × 3040 pixels at an affordable cost. It includes a variable focal length manual-zoom lens (8–50 mm), allowing customization for different applications. For capturing high-speed motion, the system uses the Raspberry Pi Global Shutter Camera, which avoids the motion blurs of rolling shutter cameras by capturing the entire frame at once.
(2) Processing Unit: The Raspberry Pi 4B is the main computing component, chosen for its balance of processing power and cost-efficiency. This model has a Broadcom BCM2711 Quad-core Cortex-A72 64-bit SoC and 8GB of RAM, which is sufficient for real-time image processing tasks.
(3) GNSS Module: For precise timekeeping, the setup includes a Uputronics Raspberry Pi GPS/RTC Expansion Board with a GNSS antenna. The board connects to the Raspberry Pi through UART and pulse-per-second (PPS) connections.
(4) Display and Power Supply: A compact 7-inch monitor is used for display. A 50,000 mAh power bank supplies power to the Raspberry Pi, GNSS module, and monitor. All components, except the camera, are enclosed in a specially printed case.
(5) Control Interface: Each node is operated using a wireless keyboard and mouse.

Internal clock updating of FVN nodes

Each Raspberry Pi independently aligns its internal clock with accurate atomic clocks in satellites. This setup enables all nodes to synchronize using a common external time reference, ensuring inter-node synchronization.

(1) A GNSS antenna, placed with a clear view of the sky, receives signals from multiple satellites. These signals include time data encoded in National Marine Electronics Association (NMEA) sentences and pulse-per-second (PPS) signals.
(2) These GNSS signals are then sent to the GNSS board via a cable, where the NMEA sentences and PPS signals are extracted.
(3) The NMEA sentences are transmitted to the Raspberry Pi through the UART pin, while the PPS signals are delivered via the PPS pin.
(4) Within the Raspberry Pi, a Network Time Protocol (NTP) daemon processes the NMEA sentences and PPS signals to obtain time information.

(5) The NTP daemon compares the time data with the Raspberry Pi's local clock and, if necessary, adjusts the local clock to match the GNSS time.

Through this configuration, each Raspberry Pi in the FVN nodes synchronizes with the satellite clocks, allowing the nodes to remain synchronized with each other. A consistent process is followed for image capture and timestamping: when a camera image arrives at the Raspberry Pi, it is timestamped and encoded into a video. This setup ensures that each node records videos with precise timestamps aligned to the satellite clocks. During data extraction from these videos, each piece of extracted data is time-stamped accurately. The data are then resampled to achieve alignment.

Image processing of FVN nodes

FVN nodes are primarily designed to measure structural displacement. The camera needs to be placed nearly perpendicular to the target on the structure. A scaling factor (SF) is applied to convert pixel displacement into actual physical displacement. This factor is calculated by dividing a known physical length by the pixel length in the image of the target.

To obtain structural displacement from the video, a zero-mean normalized cross-correlation coefficient (ZNCC) algorithm is used. The authors' previous paper describes this algorithm's workflow in detail [2].

Verification On Time Synchronisation Accuracy

In this test, two FVN nodes were used to measure the time difference between displacement signals from nodes with different setups. This was done to evaluate their synchronization performance. Two types of cameras, rolling shutter and global shutter, were used, with specifications listed in Table 1. The test setup is shown in Figure 2, and the node configurations are in Table 2.

Table 1 The specifications of the two cameras used in this study.

Camera	Sensor	Shutter type	Used resolution	Pixel size	Price (as of 2024)
Raspberry High Quality Camera	Sony IMX477	Rolling shutter	1280*720	1.55 μm	£50
Raspberry Global Shutter Camera	Sony IMX296	Global shutter	1280*720	3.45 μm	£50

Table 2 FVN node configurations in this validation test.

Case	Sensor	Frame rate	Processing mode
1	Rolling shutter	20 Hz	Real time
2	Rolling shutter	100 Hz	Post processing
3	Global shutter	50 Hz	Post processing

The two nodes measured the dynamic displacement of a shared target on a laboratory cantilever. Although indoors, the nodes received GNSS signals through the lab's partially glass roof. The camera frame rate was set to 20 Hz for real-time processing in Cases 1 and increased to 100 Hz for video recording in Cases 2. In Cases 3, the global shutter camera operated at 50 Hz.

The nodes began measuring at different times, and then the cantilever was manually excited to induce free decay vibration. The time difference between the displacement signals from the two nodes was calculated, and results are shown in Table 3. This test was repeated five times.

Table 3 Time difference between displacement signals from the two nodes in different cases.

Case	Time difference in various test trials (Unit: ms)					Mean
	Trial 1	Trial 2	Trial 3	Trial 4	Trial 5	
1	1.77	2.58	2.27	2.37	2.15	2.23
2	2.26	1.87	2.76	2.67	2.41	2.39
3	0.01	0.12	0.24	0.06	0.10	0.11

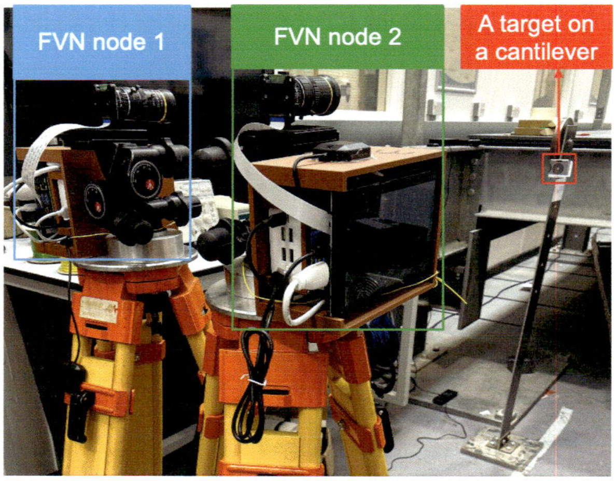

Fig. 2 The setup of the validation test.

Results showed that the camera type significantly affected the time difference between signals output by the two nodes. With the rolling shutter camera, the maximum time difference was 2.76 ms, and the average time differences were over 2 ms. With the global shutter camera, time differences were under 0.2 ms, and the maximum time difference was 0.24 ms.

Applications Of FVN

3D displacement measurement of a cross-sea bridge

Measuring 3D displacement of bridges typically requires a stereo vision system. However, as the measurement distance increases, accuracy in depth decreases. For long-span bridges, like measuring at midspan, a large baseline between the two cameras is necessary to maintain accuracy in depth. This setup poses significant challenges for camera communication with each other and with the computing unit.

Using FVN nodes, this task can be simplified. In this approach, two FVN nodes are used: one node measures horizontal and vertical displacements at midspan, and the other measures the longitudinal displacement of the bridge girder.

This method was applied to measure 3D displacement at the midspan of the Xinghaiwan Bridge, a cross-sea bridge in Dalian, China. It is a double-layer steel truss suspension bridge, with a main span of 460 m and side spans of 180 m, as shown in Figure 16a. The bridge tower has a height of 114.31 m.

During the measurement, the lower crossbeam of one bridge tower was used as a stationary area. Two FVN nodes were placed on this crossbeam, as shown in Figures 3(a) and 3(b). One node measured horizontal and vertical displacements, while the other measured longitudinal displacement at the end of the bridge girder. The longitudinal displacement along the girder was assumed to be consistent, as the rigid body movement was much larger than any extension in the longitudinal direction.

FVN Node 1, equipped with a 300 mm lens, was positioned 227 m from the midspan, with its optical axis aligned parallel to the longitudinal direction. The scaling factor was determined to be 3.718 mm/pixel. Since it was not feasible to install artificial markers under the main girder, the white drainage pipe at the midspan was selected as the target, as shown in Figure 3(b).

FVN Node 2, configured with a 120 mm lens, had its optical axis perpendicular to the bridge's longitudinal direction. It measured the longitudinal displacement of the main girder using bolts near the bridge girder's support as the target. This node was located 6.414 m from the target, with a calculated scaling factor of 0.263 mm/pixel.

The experiment took place on the morning. FVN Node 1 started at 09:29:36, while FVN Node 2 started at 09:17:33. Both systems were time-synchronized via their GNSS antennas. Data from the overlapping 100-min period were analyzed, with displacement signals shown in Figure 4. The test results confirmed the feasibility of using two FVN nodes for 3D displacement measurement on long-span bridges.

Fig. 3 The setup of the Xinghaiwan Bridge measurement.

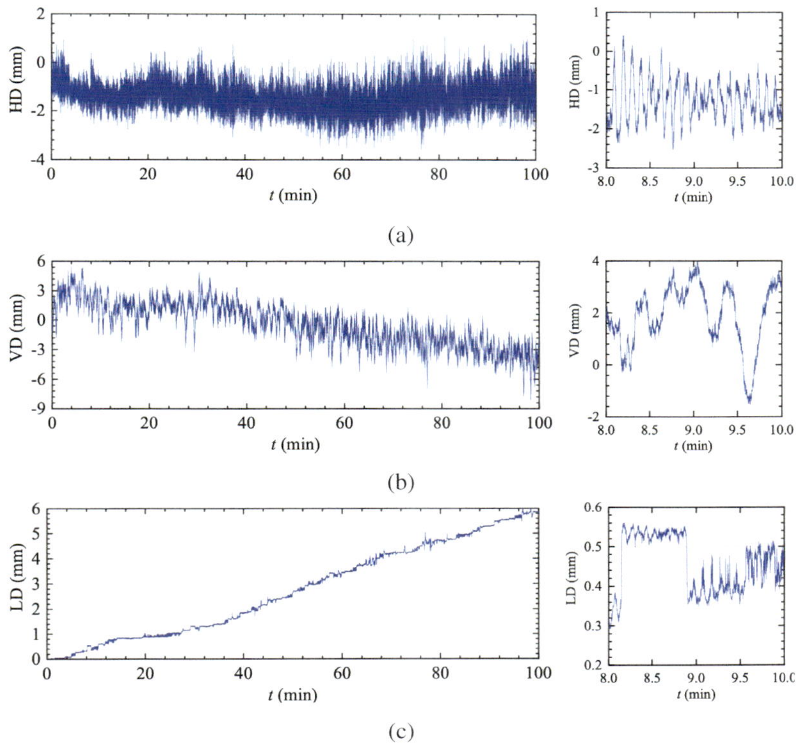

Fig. 4 Measurement result of the Xinghaiwan Bridge: (a) horizontal displacement (HD), (b) vertical displacement (VD), and (c) longitudinal displacement (LD).

3D modal test of a suspension bridge

A 3D modal test was conducted on the Beida Bridge, a vehicle suspension bridge in Dalian, China, using five FVN nodes. Measurement points were set on both sides of the main girder at seven 1/8-span sections, as shown in Figure 5(a). Taking advantage of the portability and easy installation of FVN nodes, these five nodes were placed on the bridge tower columns. Each camera was aligned to points of the same color, as illustrated in Figure 5(b). Artificial targets were used in this test, as shown in Figure 5(c). Wireless time synchronization was achieved through the GNSS modules in each system.

Fig. 5 The setup of the modal testing of the Beida Bridge (a) devices layout, (b) FVN nodes, and (c) targets.

A heavy truck crossed the bridge four times, exciting various vibration modes. The truck drove over one of the two lanes, inducing both vertical bending and torsional modes. Using the SSI-COV algorithm, the first five vertical bending modes and the first three torsional modes were identified, with frequencies and mode shapes displayed in Figure 6.

Another modal test on the Beida Bridge was conducted in 2017, providing mode frequencies and shapes for comparison. Table 4 compares the previous test results with the frequencies identified in this study, where "S", "A", and "V" denote

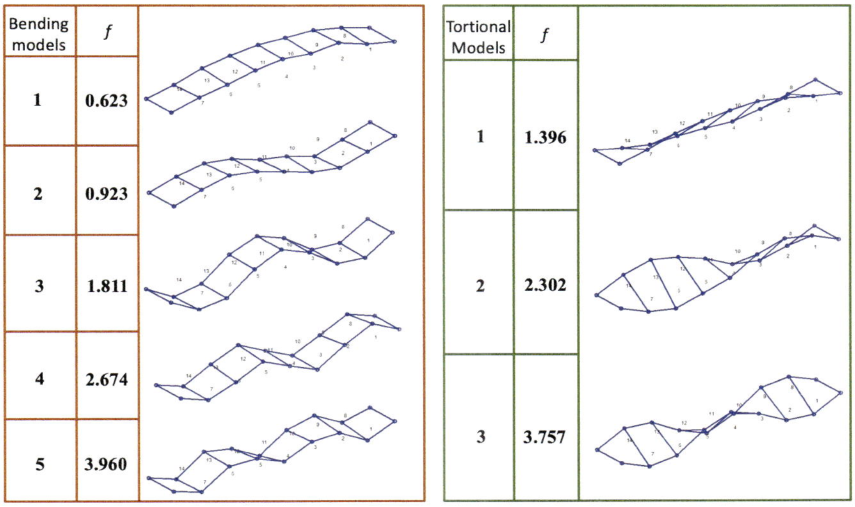

Fig. 6 Identified models of Beida Bridge.

Table 4 Identified mode frequencies based on test data at different time (Unit: Hz).

Models	Test in 2017	This study	Difference
SV1	0.63	0.62	1.59%
AV1	0.88	0.92	4.54%
SV2	–	1.81	–
AV2	–	2.67	–
SV3	–	3.96	–
ST1	–	1.40	–
AT1		2.30	–
ST2	–	3.76	–

symmetric, antisymmetric, and vertical bending of the girder, respectively. The mode frequencies from this test are close to the previous results. These findings demonstrate the effectiveness of FVN in 3D modal analysis of long-span bridges.

Conclusion

This study introduces a wireless vision sensor network, the Flexible Vision Network (FVN), for structural health monitoring (SHM). The key findings are:

(1) When two FVN nodes, each using a rolling shutter camera (Raspberry Pi High Quality camera), measured a common target, the time difference between the displacement signals was less than 2.76 ms. With global shutter cameras (Raspberry Pi Global Shutter cameras), the difference stayed below 0.24 ms. These results indicate significant time-stamping uncertainty in rolling shutter cameras.

(2) The use of two FVN nodes enables accurate 3D displacement measurement over long distances, making it practical for monitoring long-span bridges.

(3) It is feasible to use multiple FVN nodes to capture bending and tortional structural models of operational long-span bridges.

Acknowledgments The authors are grateful for the financial support of the Engineering and Physical Sciences Research Council (EPSRC), UK, via the Programme Grant EP/W005816/1.

References

1. M. Wang, W.K. Ao, J. Bownjohn, F.Y. Xu, A novel gradient-based matching via voting technique for vision-based structural displacement measurement, Mech Syst Signal Pr, 171 (2022).
2. M. Wang, K.Y. Koo, C. Liu, F. Xu, Development of a low-cost vision-based real-time displacement system using Raspberry Pi, Eng Struct, 278 (2023).
3. D. Feng, M.Q. Feng, Experimental validation of cost-effective vision-based structural health monitoring, Mech Syst Signal Pr, 88 (2017) 199–211.
4. Y. Xu, J. Zhang, J. Brownjohn, An accurate and distraction-free vision-based structural displacement measurement method integrating Siamese network based tracker and correlation-based template matching, Measurement, 179 (2021) 109506.
5. C. Dong, F.N. Catbas, A non-target structural displacement measurement method using advanced feature matching strategy, Advances in Structural Engineering, 22 (2019) 3461–3472.
6. D. Gorjup, J. Slavič, A. Babnik, M. Boltežar, Still-camera multiview Spectral Optical Flow Imaging for 3D operating-deflection-shape identification, Mech Syst Signal Pr, 152 (2021) 107456.
7. E. Cai, Y. Zhang, X. Lu, P. Li, T. Zhao, G. Lin, W. Guo, A target-free video structural motion estimation method based on multi-path optimization, Mech Syst Signal Pr, 198 (2023) 110452.
8. J.G. Chen, T.M. Adams, H. Sun, E.S. Bell, O. Büyüköztürk, Camera-based vibration measurement of the world war I memorial bridge in Portsmouth, New Hampshire, Journal of Structural Engineering, 144 (2018) 04018207.
9. E. Cai, Y. Zhang, S.T. Quek, Visualizing and quantifying small and nonstationary structural motions in video measurement, Computer-Aided Civil and Infrastructure Engineering, 38 (2023) 135–159.
10. Y. Pan, D. Wang, Y. Dong, B. Peng, A Novel Vision-Based Framework for Identifying Dynamic Vehicle Loads on Long-Span Bridges: A Case Study of Jiangyin Bridge, China, IEEE Transactions on Intelligent Transportation Systems, 23 (2022) 10441–10457.
11. H. Wu, X. Wu, W. He, F. Wang, L. Deng, A computer vision-assisted method for identifying wheel loads of moving vehicles from dynamic bridge responses, Mech Syst Signal Pr, 197 (2023) 110372.
12. X. Kong, J. Zhang, T. Wang, L. Deng, C.S. Cai, Non-contact vehicle weighing method based on tire-road contact model and computer vision techniques, Mech Syst Signal Pr, 174 (2022) 109093.
13. C. Dong, S. Bas, F.N. Catbas, A completely non-contact recognition system for bridge unit influence line using portable cameras and computer vision, Smart Struct. Syst, 24 (2019) 617–630.
14. M. Bolognini, G. Izzo, D. Marchisotti, L. Fagiano, M.P. Limongelli, E. Zappa, Vision-based modal analysis of built environment structures with multiple drones, Automation in Construction, 143 (2022) 104550.
15. J.A. Rice, K. Mechitov, S.-H. Sim, T. Nagayama, S. Jang, R. Kim, B.F. Spencer Jr, G. Agha, Y. Fujino, Flexible smart sensor framework for autonomous structural health monitoring, Smart structures and Systems, 6 (2010) 423–438.
16. M.M. Wang, W.K. Ao, J. Bownjohn, F.Y. Xu, Completely non-contact modal testing of full-scale bridge in challenging conditions using vision sensing systems, Eng Struct, 272 (2022).
17. S.A. V. Shajihan, T. Hoang, K. Mechitov, B.F. Spencer Jr., Wireless SmartVision system for synchronized displacement monitoring of railroad bridges, Comput-Aided Civ Inf, 37 (2022) 1070–1088.

Chapter 4
Safeguarding Masterpieces: Multi-point Vibrometry (MPV) Based Monitoring of Georgia O'Keeffe's Art

David Damiani, Dale Kronkright, and Vikrant Palan

Abstract The process of understanding the state of preservation of any work of art is complex. Like most individually fabricated objects, artworks are inevitably anisotropic structures, consisting of amended materials, often irregularly applied, each with slightly different aging characteristics, and each with individual histories of use, environmental exposure, travel, and repair. Much of the responsibility of preserving these works for future generations falls to museum conservators, who utilize a range of analytical and imaging techniques to understand the materials and techniques employed by the artist and to characterize the effects of time, environmental exposure, damage, and cumulative fatigue.

Laser Doppler vibrometry can be used by conservators to observe and document the responses of works of art to vibration excitations. In this case study, we introduce the concept of multi-point vibrometry (MPV) for making a full-field vibrational assessment of a Georgia O'Keeffe painting. Georgia O'Keeffe (Sun Prairie, Wisconsin,1887 – Santa Fe, New Mexico,1986), an American artist, created well over two thousand works of art, including oil and pastel paintings, charcoal and graphite drawings, watercolors, photographs and sculptures during her 60-year, professional career. A multi-point vibrometer is composed of multiple single-point vibrometers where the total number and system configuration is customized based on the needs of the test. A MPV system is preferred for applications where the excitations cannot be either continuous or recurring. A few examples of where this approach works well are one-of-a-kind specimens (paintings, space structures), non-stationary excitation or response (wind-tunnel testing, engine run ups, shock) events and cases with time-variant vibration characteristics.

MPV offers great promise for conservators because it allows for the measurement across entire surfaces with a single, low-energy impact. In this paper, we present the framework for a multi-point vibrometer including its construction, components, and a step-by-step process for acquiring data. The proposed approach is sensor agnostic and covers a frequency and amplitude range of 24 MHz and 2 meters respectively. The goal of the study is to demonstrate the use of vibrometry, especially MPV for preemptive maintenance efforts in works of art which in turn will extend its lifetime for generations to come.

Keywords MPV · Full field vibrometry · Impact testing

Background On Paintings Tested

This case study focuses on the vibrational behavior of three canvas paintings by the American modern artist Georgia O'Keeffe in the collection of the Museum of Fine Arts Boston. Like most canvas paintings, the three paintings have gross structural characteristics conceptually similar to a drumhead. A woven, linen canvas was stretched over a rectangular wooden framework, commonly known as a stretcher. The artist then brushed on pigmented oil paints to form the design surface. In some ways, the vibrational response of canvas paintings can be expected to have vibrational modes like those of a model rectangular diaphragm, held under tension at the edges. But the effect of individually applied oil paints contribute to the anisotropic response of each artwork to vibrational excitation. Because the chemical and mechanical properties of different

David Damiani · Vikrant Palan
Polytec, Inc., 16400 Bake Parkway, Irvine, CA 92618
e-mail: d.damiani@polytec.com; v.palan@polytec.com

Dale Kronkright
Georgia O'Keeffe Museum, 217 Johnson Street, Santa Fe, NM 87501
e-mail: dkronkright@okeeffemuseum.org

Janko Slavic et al. (eds.), *Computer Vision & Laser Vibrometry, Vol. 6*, Conference Proceedings of the Society for Experimental Mechanics Series, https://doi.org/10.13052/9788743801511-4

pigment particles, suspended in oxidizing oils such as linseed, walnut and poppy seed oils vary greatly, mechanical characteristics such as weight, MOE, tensile, compression, sheer and torsional strength of paint layers likewise vary greatly from color to color and brushstroke to brushstroke.

As a result of these variations in materials, vibrations experienced in handling and during road and air transit, at or near the resonant frequencies of a painting, result in idiosyncratic deterioration of cohesion within the paints and adhesion to the canvas below. These damages manifest as cracks through paint layers, cleaving and lifting paints from the underlying canvas, paint losses and other mechanical forms of deterioration. These fractures through previously continuous paint layers result in altered and measurable vibrational responses.

Despite these types of adhesion failures within paint layers, evidence of damage may not have completely manifested visibly on the surface of the painting. Aged, brittle, oxidized oil paints are the restraining layer in 100-year-old paintings and their isolated loss of cohesion and adhesion can be indicated by changes in displacement amplitudes and resonant frequencies detected using multipoint laser vibrometry. Likewise, measures taken by art conservators to repair fractured and cleaving paints inevitably stiffen areas where displacement amplitudes at damaged paints were formerly less retrained by the damage. These can likewise be measured to determine the effect of repair measures within the context of the surrounding undamaged original materials and structures.

The three paintings examined using MPV in this study were:

Fishhook from Hawaii, No. 2, 1939, an oil painting on canvas measuring 36 inches high by 24 inches wide. Collection: Museum of Fine Arts, Boston, Alfred Stieglitz Collection, Bequest of Georgia O'Keeffe, accession number 1987.540

Deer Skull with Pedernal, 1936, an oil painting on canvas measuring 36 inches high by 30 inches wide. Collection: Museum of Fine Arts, Boston. Gift of the William H. Lane Foundation, accession number 1990.432

In the Patio No. IV, 1948, an oil painting on canvas measuring 14 inches high and 30 inches wide. Collection of the Museum of Fine Arts, Boston. Gift of the William H. Lane Foundation, accession number 1990.434

Fig. 1 Above left, Fishhook from Hawaii, No. 2, 1939, collection: Museum of Fine Arts, Boston, 1987.540. Above center, Deer Skull with Pedernal, 1936, collection: Museum of Fine Arts, Boston, 1990.432. Above, right, In the Patio No. IV, 1948, collection: Museum of Fine Arts, Boston, 1990.434. © 2024 Museum of Fine Arts, Boston. Reproduced with permission.

History and structural characteristics of the paintings

Fishhook from Hawaii, No. 2, 1939 retains the original materials and structures used by the artist at the time of fabrication. The wood stretcher, commercially primed, fine weave, linen canvas, tacking edges and paints are essentially unaltered since leaving the artist's studio. A thin protective synthetic acrylic varnish was spray applied to the painted surface under O'Keeffe's direction after a light surface cleaning in 1948.Outside of occasional dusting and changes in framing materials, the painting remains as O'Keeffe originally presented it.

In this painting, O'Keeffe brush-applied a series of blue paints, light to dark in value, across the lower background and then spread and smoothed the paste-like paints with a palette knife in a manner like cake frosting. The result is a cured paint surface that is relatively thick but compliant to the vibrational behavior of the overall rectangular diaphragm. In a raking light image, it is possible to see a series of parallel, lifted, vertical cracks in very light-blue paints, associated with the region of the first drum-mode response near the center of the painting. There are a series of similarly shaped vertical cracks in the viewer upper right quadrant. The somewhat concentric shape of some of the cracks in both areas suggest that both areas historically received a soft vertical press or impact on the front of the painting.

Deer Skull with Pedernal, 1936 is similar in size and orientation to the Fishhook painting, just 6 inches wider and painted just three years earlier. In contrast to the Fishhook painting, however, Deer Skull with Pedernal, 1936 is painted using single applications of color, with the twill-weave canvas texture generally visible under the paints. There is a carefully brushed vertical form representing a dead juniper trunk that runs generally through the vertical middle of the painting and horizontal forms near the bottom of the composition representing the New Mexico landscape adjacent to O'Keeffe's Ghost Ranch home.

Sometime between 1970 and 1976, in preparation for traveling exhibitions, the back of the painting was adhered to a secondary canvas using a heat activated adhesive mixture of natural resins and waxes, commonly referred to as a "wax resin lining." The goal of this lining was to flatten unsightly creases and secure cracked paints that had developed over the previous 40 years of travel and exhibition. Thus, this painting has two canvas supports, stretched over a new, heavy weight wood stretcher. It is thus stiffer, with more mass than the similar Fishhook painting. Oddly, the viewer right half of the painting has a noticeably thinner coating of the wax-resin adhesive on the verso or back of the canvas, than the viewer left side, which is thicker and likely heavier and stiffer.

In the Patio No. IV, 1948 is a much smaller, narrower rectangular painting. It is painted on a heavy linen canvas, with various mixtures of the iron oxide pigments raw and burnt umber, with white pigments to create values of light to dark and to add various degrees of opacity. In an interesting departure from her practice at the near mid-point in her professional career, O'Keeffe appears to have added a natural resin varnish to her initial brown-toned oil paints, possibly to achieve a highly matte, non-reflective surface. Using an imaging technique known as ultraviolet induced visible luminescence, we can see the characteristic greenish yellow glow of natural resin amendments at the beginning edges of many brush strokes. Additionally, we can see that O'Keeffe added a second layer of lighter overpaints across the viewer left side of the composition, resulting in a heavier, stiffer paint film in these areas, when compared to the viewer right side. The bright blue sky at the top of the painting is heavily amended with zinc oxide white, making that region more rigid when compared to the areas of brown paints.

In about 1958, this work was also adhered to a secondary canvas using a wax-resin mixture to repair a dent and gouge in the viewer lower left corner and two impact dents in the viewer right side of the composition. This treatment would add stiffness and mass to the canvas which, along with the smaller dimension, results in higher natural frequency responses, when compared to an unlined or larger format paintings.

Multi-Point Vibrometer (MPV) Components And Process Flow

The MPV system is a flexible and modular design consisting of a series of single point laser doppler vibrometers (LDV). The system can be configured based on the number of sensors required, frequency, amplitude, and modularity needed. As seen in Fig. 2, the system is configured based on application requirements. For this test, the system used consisted of 16 Polytec VibroGo LDVs each capable of measuring up to 100 kHz and 2 m/s. All systems were connected to a central data acquisition and communication box.

The controller has 3 input ports per sensor: ethernet, data acquisition, and power (Fig. 3). The ethernet port controls digital communication to change ranges, set bandwidth, autofocus the lasers, and provide any other communication controls to the laser. The analog BNC provides the selected data stream from the sensor. For this test, velocity was the chosen output but displacement and acceleration could have also been output. The power port allows all sensors to be connected to a common power source and one outlet.

For this test, the sensors were rows of 4 in a stair shape as seen in Figure 3. The first three rows had all 4 systems on a row mounted on one beam for easy movement and setup. The last row was on individual tripods to allow for flexibility on harder to reach points. All sensors were configured to be in complimentary IP Address networks and powered on. Each sensor was then aligned to the painting under test.The first eight sensors were connected to the lead controller. The second eight sensors were connected to the secondary controller. From each controller a USB cable and ethernet cable is run to a central laptop through a USB/Ethernet hub. The information flow thus described is demonstrated in Figure 4.

From the MDE control software, the sensors were autofocused, ranges optimized, and acquisition parameters were set. In a separate software (MDE), a geometry file was created that represented the orientation of the sensors as aligned on the painting. This geometry was then imported into the MDE software so that it could be applied to the data once collected. Once all the parameters were set, the data was collected and monitored in the MPV software. Once data collection was complete, the data was combined with the geometry and available for analysis in an existing MPV analysis program. This program

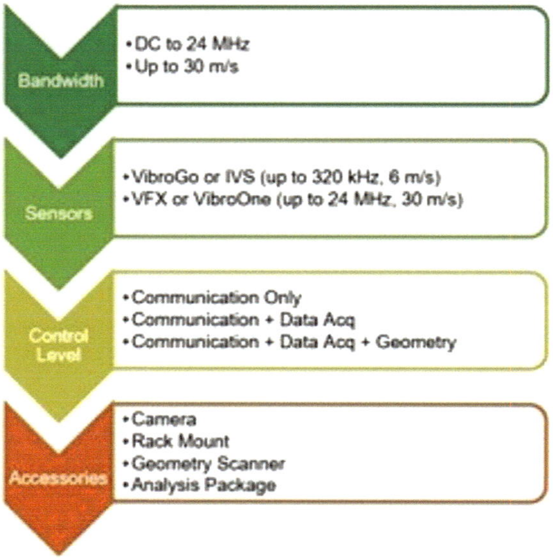

Fig. 2 System configuration options.

Fig. 3 MDE controller (left) and example test setup (right). © 2024 Museum of Fine Arts, Boston. Reproduced with permission.

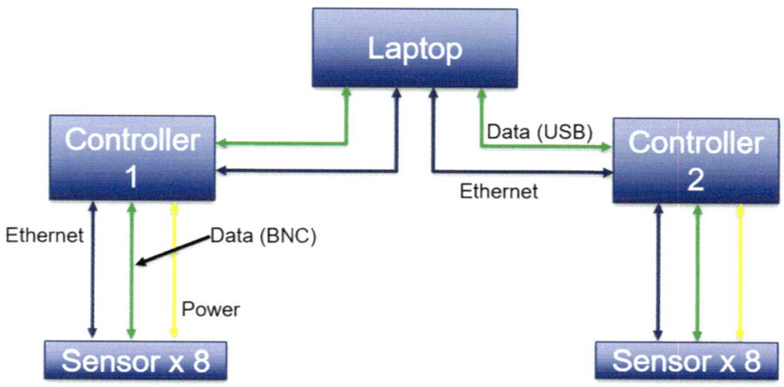

Fig. 4 Information flow and wiring diagram for multi-point vibrometer system.

allows for any signal processing necessary including but not limited to filtering, FFT transformations, and data extraction. The process flow is pictorially described in Figure 5.

Figure 6 show the software interface for the MPV software. This is the data acquisition portion of the software. Once the data is acquired, it is then exported to MDE software which handles the ODS analysis and further post-processing.

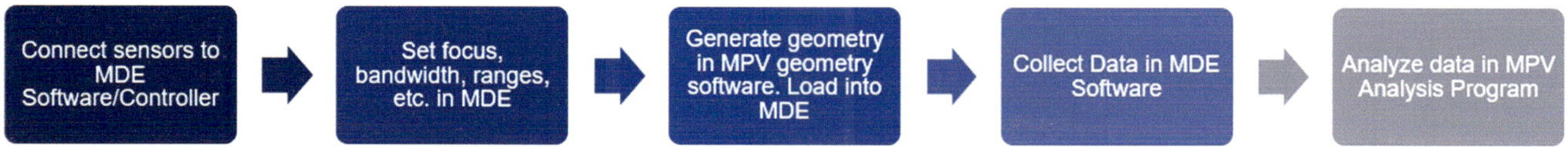

Fig. 5 MDE Process Flow.

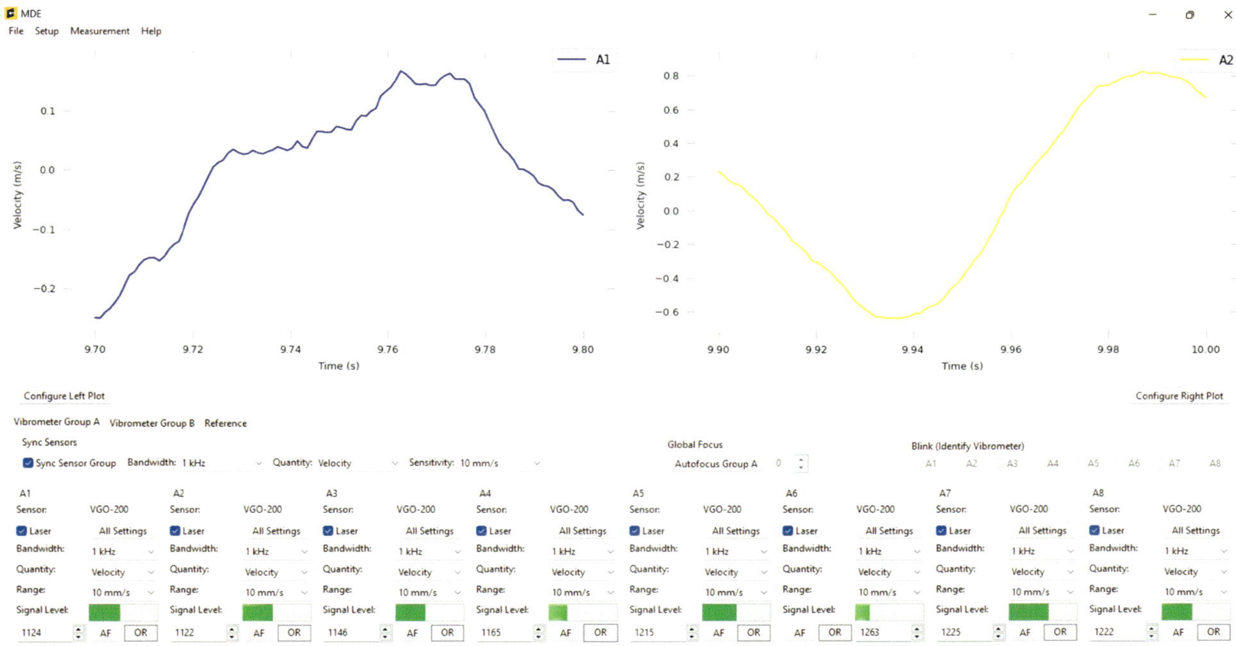

Fig. 6 MDE software interface. Top left and right: example measurements from two different sensors. This portion of the interface is configurable. Bottom: Vibrometer settings for one set of 8 sensors.

Measurement Results And Analysis

The MPV data user interface provides a dynamic presentation of a painting's response to vibration excitations across a range of frequencies selected in the FFT frequency domain. This makes it possible for conservators to evaluate displacement and velocity amplitudes across a work and identify idiosyncratic responses where there may be damage or repairs.

Fig. 7 Painting with measurement locations – 16 locations in total. *In the Patio IV*, 1948.

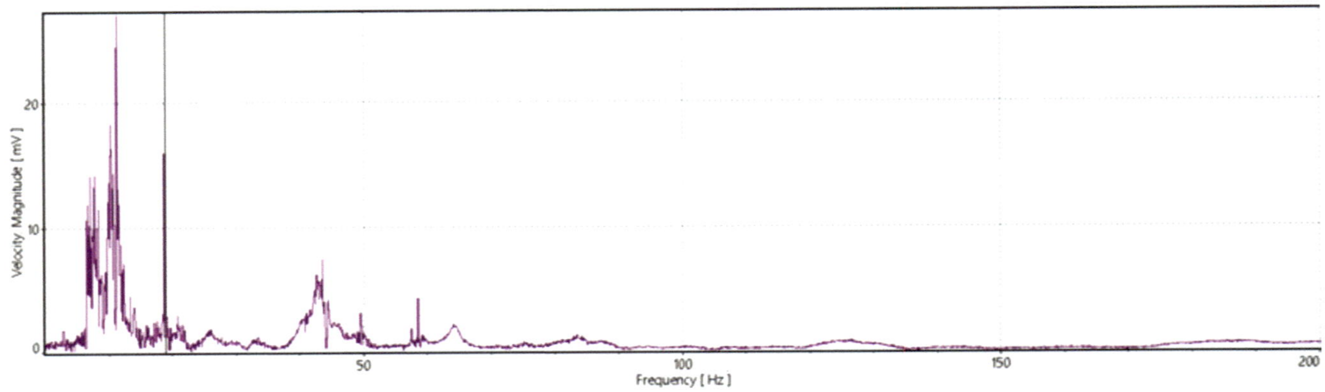

Fig. 8 Average frequency spectrum of *In the Patio IV* over all points measured. The sensitivity for the spectrum is 5 mm/s/V.

MPV analyses provided important insights for all three of the tested paintings. For example, the smallest painting, *In the Patio IV*, 1948, has the highest natural frequency of the three paintings at 11.4 Hz, (Fig. 7). The full spectrum response shows that most of the response is below 100 Hz and primarily below 50 Hz (Fig. 8). A previously repaired vertical crack, near the center of the painting and to the right of the central, rectangular door form, showed a stiffer response when compared to the areas immediately adjacent to the left, which exhibited displacements approximately four times as great. (Fig. 9). This indicated that this crack repair remains secure and has not re-opened.

Deer Skull with Pedernal, 1936, (Fig. 10) has a vibration response that is rather typical of a large canvas painting (Fig. 11), with a natural frequency at 8.6 Hz (Fig. 12). At this frequency, the center of the painting is highly compliant and exhibits displacement and velocity magnitudes 8 times greater than the surrounding edges, which are restrained by being secured to the surrounding wooden stretcher. Above the center of the painting, the painted skull form, outlined by top-most, central MPV reference points (Fig. 13), is painted predominantly in lead carbonate white, a paint that results in a very stiff, cured, oil paint layer. As might be expected, the velocity and displacement of the area covered by the stiff skull paints are like those seen at the retrained edges of the painting. Additionally, at 13.5 Hz, 23.67 Hz and of 26.1 Hz, a vertical displacement can be seen along the richly textured, brown and black paints of the dead juniper trunk form that runs vertically down the

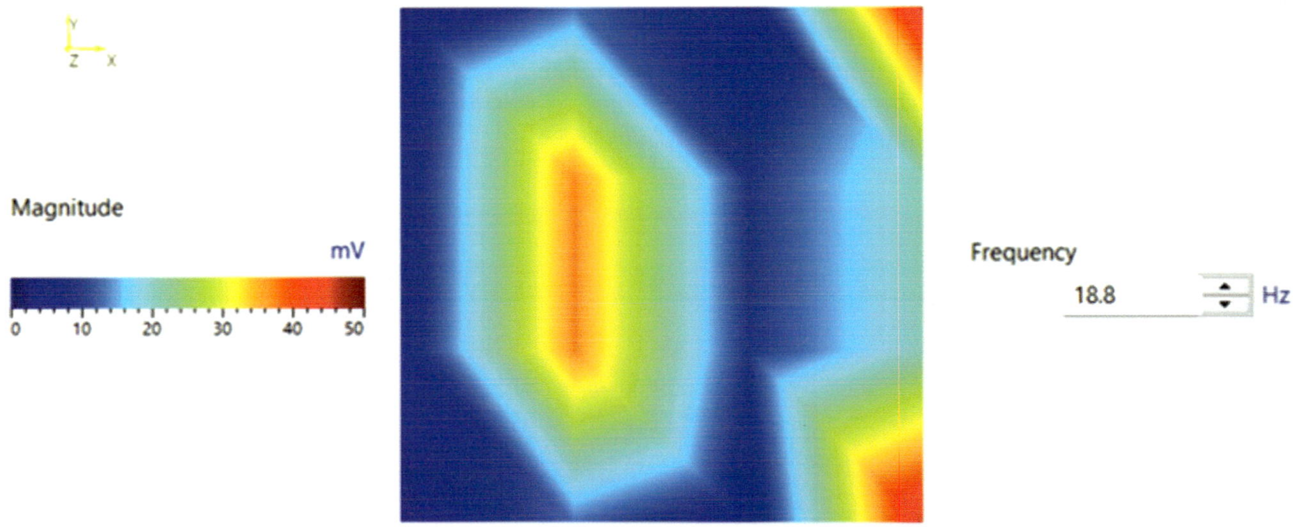

Fig. 9 Operation deflection shape indicating the effect of a repaired crack. This is manifested by a 4x increase in displacement. The sensitivity for this deflection shape is 5 mm/s/V.

Fig. 10 Painting with measurement locations – 16 locations in total. *Deer Skull with Pedernal*, 1936.

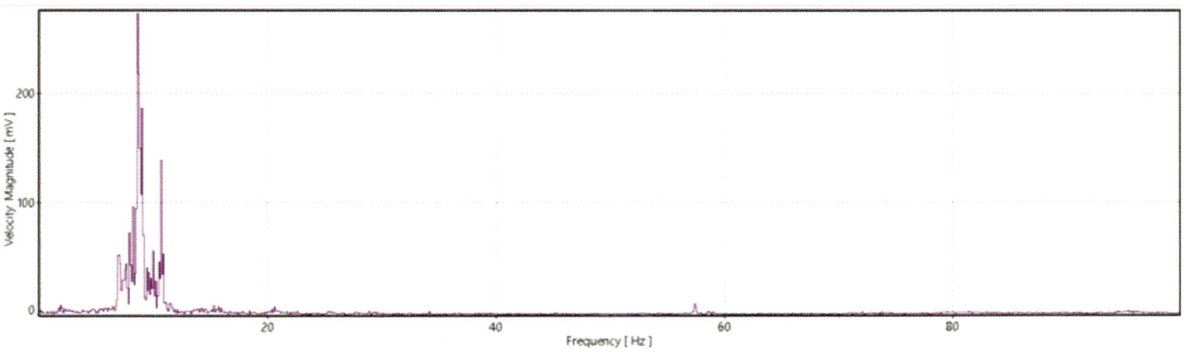

Fig. 11 Average frequency spectrum of *Skull with Pedernal* over all points measured. The sensitivity for conversion is 2.5 mm/s/V.

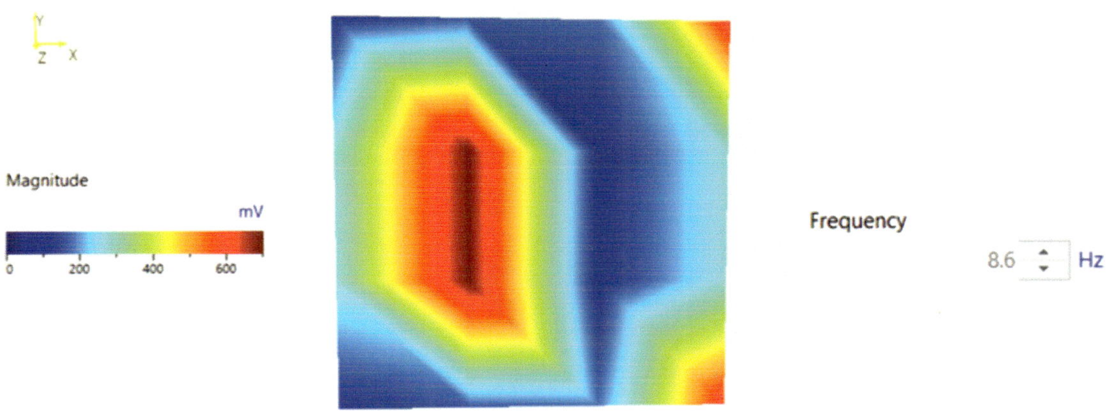

Fig. 12 First natural frequency. *Deer Skull with Pedernal*, 1936. The sensitivity for this deflection shape is 5 mm/s/V.

Fig. 13 Top portion of the painting done using lead carbonate white, explaining the increased stiffness in the response. *Deer Skull with Pedernal*, 1936.

center of the painting, (Fig. 14). Also interesting is the fact that vertical cracks though the paints, repaired by the 1970's era lining project, have preferential vibrational displacements at around 16 Hz, (Fig. 15) with greater displacement and velocity amplitudes than the areas immediately below them. These displacements may indicate fatigue failures in the 1970's repairs at these two locations.

In contrast, *Fishhook from Hawaii*, No. 2, 1939, (Fig. 16) has a very idiosyncratic vibrational response, despite its similar size. For this work, the artist selected a very fine weave, linen canvas, exhibiting a natural frequency of 10.3 Hz, (Fig. 18). Because of the greater density of canvas thread crimps, and because it has a slightly stiffer white priming layer under the paints, the painting exhibits a higher natural frequency than that of Deer Skull with Pedernal, 1936. What is most interesting is that *Fishhook from Hawaii*, No. 2, 1939 has a series of fatigue creases and cracks through the priming layers and paints at the top-center and lower left- and right-side edges. Even though the edges of a canvas painting are generally the most restrained and therefore the least compliant, being immediately adjacent to the edge where the canvas is tacked to the side of the stretcher frame, the canvas and paints at these cracked regions exhibit the greatest displacement and velocity, six-times greater than the grid points on the edges immediately above and below the actively cracking areas. In addition, the shape

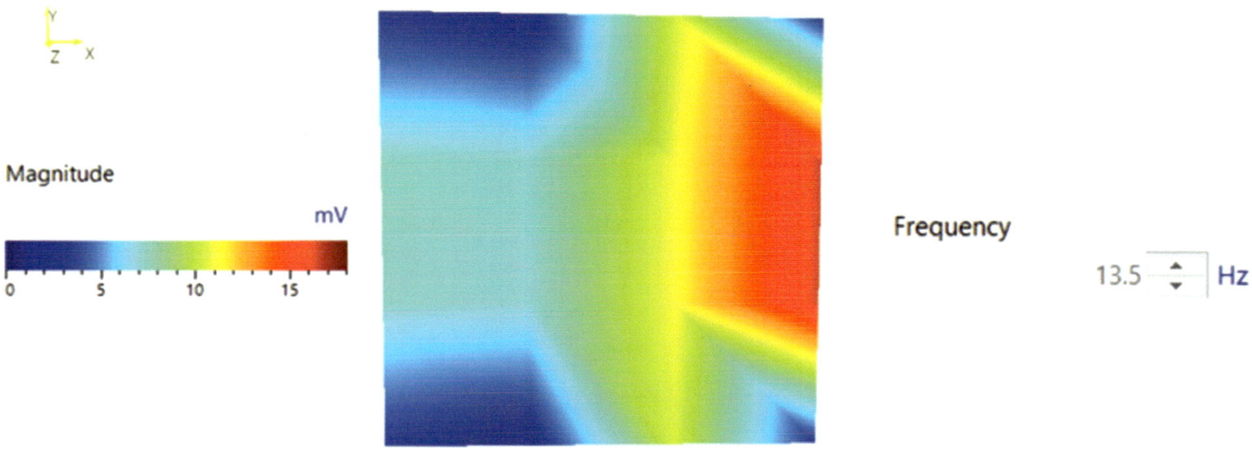

Fig. 14 Vertical "band" of displacement manifested because of the paint use. *Deer Skull with Pedernal*, 1936. The sensitivity for this deflection shape is 5 mm/s/V.

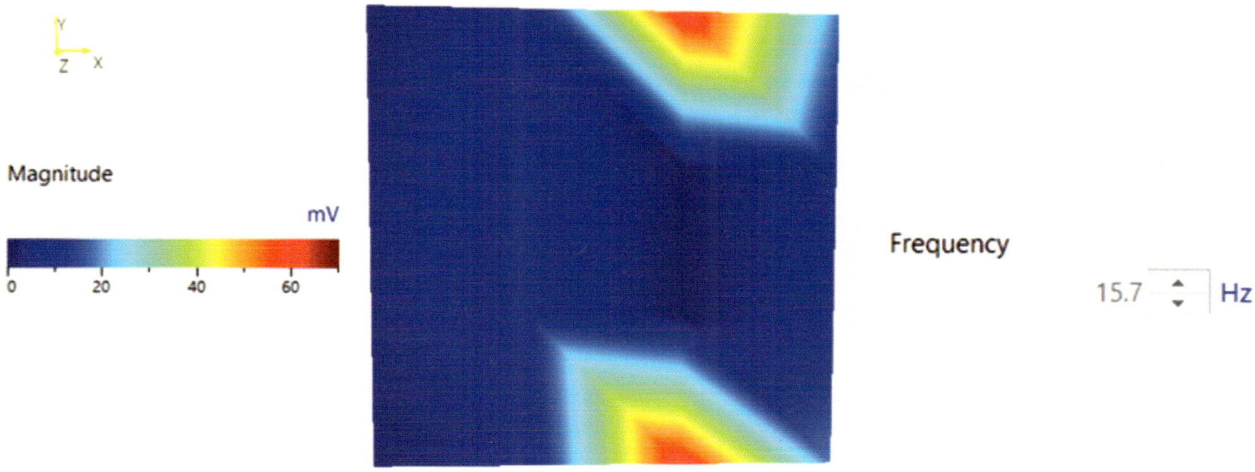

Fig. 15 Efficacy of the crack repair project. Asymmetric displacement reflects the differences in the quality of repair. *Deer Skull with Pedernal*, 1936. The sensitivity for this deflection shape is 5 mm/s/V.

Fig. 16 Painting with measurement locations – 16 locations in total. *Fishhook from Hawaii*, No. 2, 1939.

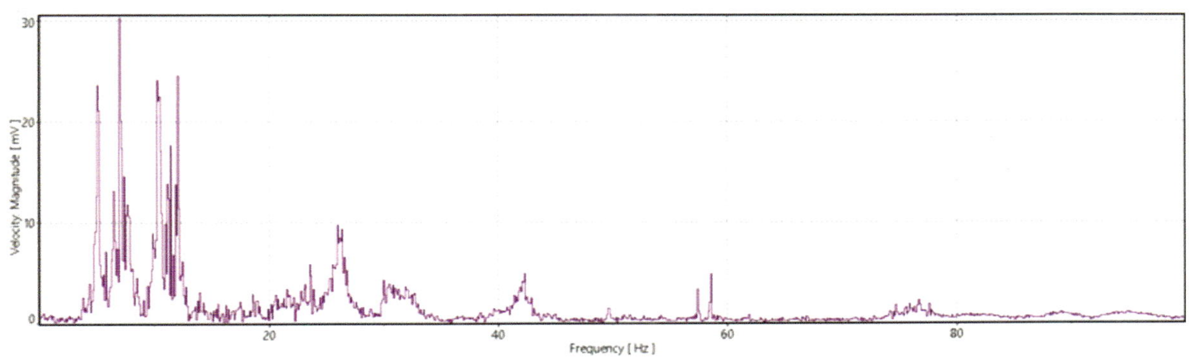

Fig. 17 Average frequency spectrum of *Fishhook from Hawaii* over all points measured. The sensitivity for conversion is 5 mm/s/V.

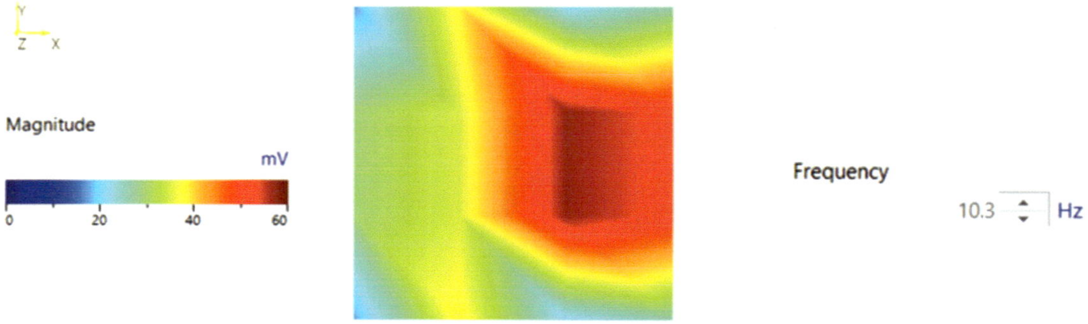

Fig. 18 First natural frequency. *Fishhook from Hawaii*, No. 2, 1939. The sensitivity for this deflection shape is 5 mm/s/V.

Fig. 19 Cracks definition resulting in preferential, non-linear displacement response, especially at the edges.

of the of the edge displacements along the viewer right edge preferentially extend upwards and to the right, directly into a series of vertical cracks through the paints that were discovered slightly below, and to the left of the fishing line swivel figure in the painting. These cracks can be seen when the painting is viewed in strongly contrasting raking light (Fig. 19).

Conclusion

The condition of a work of art is influenced by its materials, history, and environment. Conservators use various techniques to understand and protect these complex objects. Laser Doppler vibrometry offers a non-destructive, non-contact way to assess a painting's mechanical responses to vibration excitations, which can illustrate and document both general natural frequency responses and idiosyncratic characteristics that can detect the existence of isolated damage or reflect the ability of previous repairs to ensure structural continuity with the surrounding surfaces. Benefits of vibrometry include but are not limited to:

- Non-contact measurement: Avoids physical contact, preventing potential damage to delicate artworks.
- High sensitivity: Detects even subtle vibrations that could lead to long-term deterioration.
- Wide frequency range: Measures vibrations across a broad spectrum, capturing both low-frequency and high-frequency components.
- Real-time analysis: Provides immediate results, allowing for timely intervention if necessary.

For priceless artworks, vibrometry offers a valuable tool for the detection and monitoring of damage and deterioration. By identifying potential vulnerabilities, conservators can take steps to isolate works from damaging resonant frequencies in handling and transit. Moreover, detection of isolated, idiosyncratic vibrational responses across a work of art can help conservators detect and identify fatigue failures in materials and structures early in their development. Likewise, early detection provides greater opportunity for stabilizing conservation treatments. Following treatment, a work can be re-measured to both ensure that the repair has been effective in creating a more continuous vibrational response across the damaged area and makes possible periodic inspections of the work to monitor the life of the repair over time.

Acknowledgments Special thanks to Sarah Mastrangelo, the Cunningham Associate Conservator in Paintings at Museum of Fine Arts, Boston. Ms. Mastrangelo coordinated our visit, provided secure studio space for our testing equipment and provided safe transport, handling and access to the paintings. Thanks to the entire staff of the Paintings and Objects Conservation Departments for their participation and assistance.

Chapter 5
Expanding Single Frequency Excitations to Piecewise Bandwidth Excitations for Low Speed Cameras Vision-based vibration Measurements

Davide Mastrodicasa, Andrea Sabatini, Emilio Di Lorenzo, Bart Peeters, and Patrick Guillaume

Abstract Mechanical products in industries such as automotive and aerospace are tested for validation using vibration measurements and experimental modal analysis. These tests are usually carried out using pointwise sensors connected to the structure. Image processing techniques like Digital Image Correlation (DIC) are developed to perform modal analysis. One area where DIC is of interest is in the use of cheap, light, and low-speed cameras for detecting the high-frequency behavior of structures. This work focuses on the use of signal reconstruction algorithms for bandlimited signals, which makes it possible to go beyond the Nyquist Shannon frequency limit even without the use of high-speed cameras. The approach involves exciting a structure using a band-limited signal, reconstructing the output, and performing modal analysis. A numerical validation of the approach has been conducted on a simple simulated MDOF system. An experimental validation has been conducted on a simple cantilever beam.

Keywords DIC · Modal Analysis · Vibration Analysis · Low-speed cameras · Down-sampling

Introduction

Experimental Modal Analysis (EMA) is an important technology used to study structural dynamics. It breaks down complex phenomena into separate modes, which reveal natural frequencies, modal damping, and mode shapes [1]. These modes provide insights into the dynamic properties of structures and are derived from controlled forced vibrations and natural ambient vibrations [2–4]. Together, these modal parameters form a Modal Model that is crucial for various applications, from automotive to aerospace industries. It helps in the design, optimization, and validation of Finite Element (FE) models.

Traditionally, EMA requires measuring system responses through point-wise sensors like accelerometers. However, this can be labor-intensive and introduce electrical noise [5]. Vision-based methods, such as Digital Image Correlation (DIC), are becoming increasingly popular because they enable full-field 3D measurements with high accuracy. DIC was traditionally used for static measurements but has now expanded into dynamic analysis. Phase-based Motion Magnification (PMM) and Lucas–Kanade (LK) optical flow offer similar approaches. Additionally, deep learning techniques are employed to predict displacements and strains, which outperform traditional methods in terms of computational time. These non-contact vision-based methods are useful in Structural Health Monitoring (SHM) in various fields, from aeroelasticity to wind energy and aerospace.

However, measuring high-frequency displacements presents challenges due to camera speed and resolution limitations. Recent advancements extend the use of low-speed cameras and down-sampling methods to accommodate the use of low-speed cameras for high-frequency measurements [6–7]. This paper presents a novel data acquisition and post-processing

Davide Mastrodicasa · Andrea Sabatini · Emilio Di Lorenzo · Bart Peeters
Siemens Industry Software NV, Interleuvenlaan 68, Leuven, 3001, Belgium
e-mail: davide.mastrodicasa@siemens.com; andrea.sabatini.ext@siemens.com; emilio.dilorenzo@siemens.com; bart.peeters@siemens.com

Davide Mastrodicasa · Patrick Guillaume
Vrije Universiteit Brussel, Pleinlaan 2, Brussels, 1050, Belgium
e-mail: davide.mastrodicasa@siemens.com; patrick.guillaume@vub.be

Janko Slavic et al. (eds.), *Computer Vision & Laser Vibrometry, Vol. 6*, Conference Proceedings of the Society for Experimental Mechanics Series, https://doi.org/10.13052/9788743801511-5

strategy to reconstruct high-frequency measurement responses without the need for single harmonic excitation.

This approach is numerically validated on an MDOF system. After, the methodology is applied on a simple cantilever beam, and validated against accelerometer data.

Background

A bandlimited continuous-time signal $x(t)$ in the bandwidth $B = F_H - F_L$ Hertz can be uniquely recovered from its samples provided that the sampling rate $f_s \geq 2B$ Hz. This chapter illustrates two different methods of signal reconstruction for a bandlimited down-sampled signal.

- The first method (Interpolation method) exploits the reconstruction process in the time domain.
  A signal $x(t)$ can be recovered in the time domain from its down-sampled version $x(nT_s)$ using the equation:

$$x(t) \approx \sum_n x(nT_s) \bullet g(t - nT_s)$$

  where $g(t) = \frac{\sin(\pi Bt)}{\pi Bt}\cos(2\pi F_c t)$ and Fc is the central frequency of the band B calculated as $F_c = \frac{(F_L + F_H)}{2}$
- The second method (repMat method) instead makes use of the concept of frequency shift without reconstructing the signal $x(t)$ in the time domain. Any bandpass signal $x(t)$, having a Fourier transform $X(f)$, can be represented by an equivalent lowpass signal $x_l(t)$, having a Fourier transform $X_l(f)$ whose relation with $X(f)$ is given by the equation:

$$X_l(f) = X(f + F_c)$$

  which means that the spectrum of the bandpass signal $x(t)$ can be obtained from the spectrum of the complex signal $x_l(t)$ by a frequency translation of a known quantity. The concept behind this approach is shown in Figure 1.

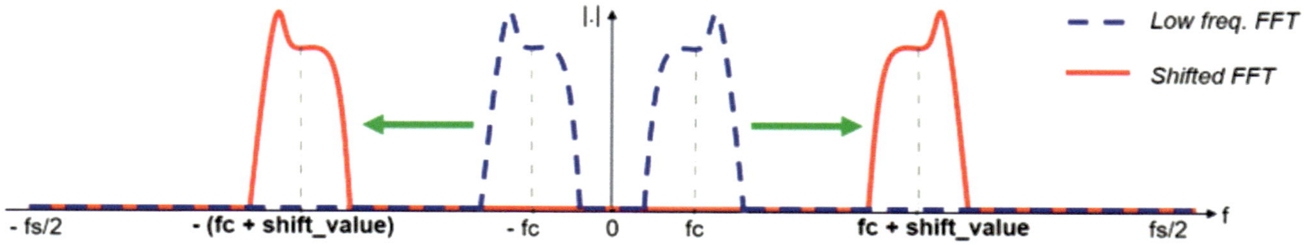

Fig. 1 Frequency Shift concept of a bandlimited spectrum.

These two methods are analyzed separately and a numerical validation over a generic bandlimited signal such as chirp or pseudorandom is performed.

Numerical Validation

Numerical validations have been conducted on a multi-degree of freedom system, shown in Figure 2. The model is built with the following parameters: $k = 900\ kN/m$; $m = 50\ kg$; $m_3 = 20\ kg$; $c = 10\ Ns/m$; $f_1 = f_2 = 0$; $f_3 = Chirp(50 - 75)\ Hz$.

To validate the methodology, the response x_2 is computed with a sampling frequency $f_{ori} = 800$ Hz (black signal). The x_2 signal is down-sampled reducing the sampling frequency to $f_{ds} = 50$ with respect to the constraint of $f_{ds} \geq 2B = 2(f_H - F_L) = 50$ Hz (red signal). Finally, the Interpolation method previously introduced is used to reconstruct the original signal and to compare the reconstruction with it (green signal).

This approach shows high fidelity of the reconstructed data, showing TRAC/FRAC values higher than 99% and a PAC value of 79%.

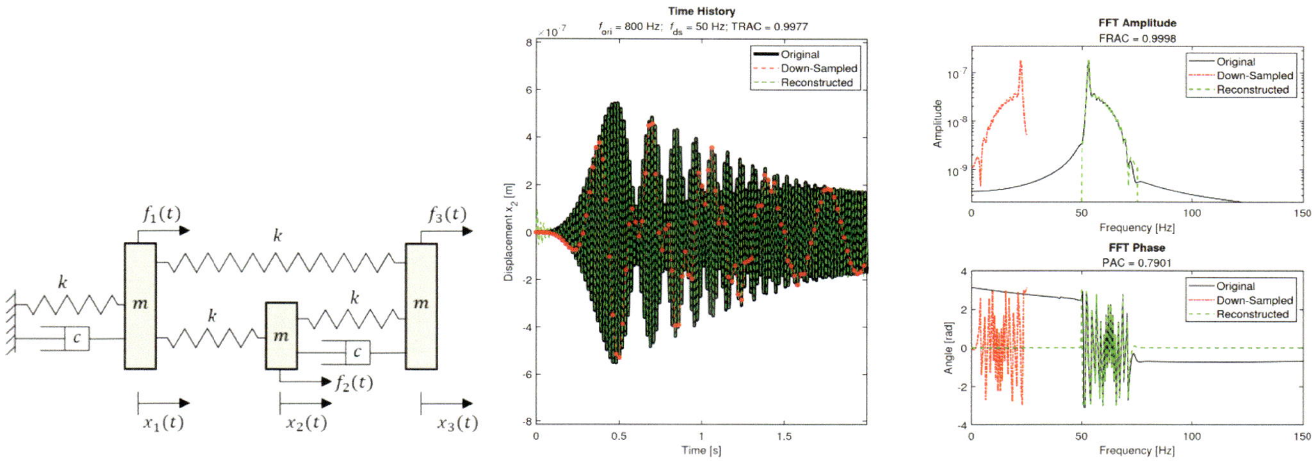

Fig. 2 MDOF Validation. left) MDOF model. right) Original-Down sampled-Reconstructed Signal using the Interpolation method.

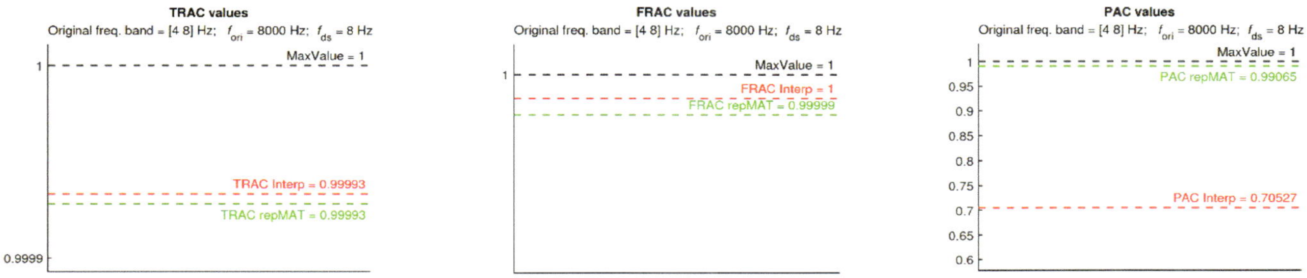

Fig. 3 TRAC/FRAC/PAC for Interpolation method and repMat method reconstruction Chirp Excitation F_L =4 Hz; F_H =8 Hz.

The same model has been used to make a comparison between the Interpolation method and the RepMat method in the frequency band $[4-8]$ Hz shows a good agreement between the two methodologies, as shown in Figure 3.

In Figure 3, it is clear that there is good agreement between the two methodologies regarding the TRAC and the FRAC. However, there is a difference in the PAC, favoring the RepMat approach. Therefore, this methodology has been used in the experimental validation.

Experimental Validation

The experimental validation was carried out on a cantilever beam (400×40×5 mm) excited by using an electrodynamic shaker in multiple sub-bands in the frequency range $[0-820]$ Hz. The different sub-bands are reconstructed using the RepMat approach. The full experimental setup, including two 5 MPx cameras, the lights, and the SCADAS acquisition systems, is shown in Figure 4.

The structure was also instrumented with 1 load cell to compute the FRFs and 10 accelerometers to validate the proposed camera-based approach.

In the test case, 40 sub-bands of around 40 Hz overlap with each other by 50%. The purpose is to cover the global range [0 820] Hz, including all five modes identified in a preliminary accelerometer test.

The measuring phase involves the acquisition of the input force via the force cell (useful to compute the FRF for the modal analysis in post-processing), the accelerations via the 10 accelerometers, and the down-sampled displacements via the cameras by using DIC. After the measuring phase where all the needed data are collected, in the post-processing phase all the down-sampled displacements are reconstructed in the frequency domain via the application of the RepMat algorithm.

Now that the displacement signals are reconstructed the computation of the FRF can be computed by using the H_1 estimator:

$$H_1 = \frac{G_{XF}}{G_{FF}} = \frac{X(f)F^*(f)}{F(f)F^*(f)}$$

where $X(f)$ the FFT of the reconstructed displacements and $F^*(f)$ the FFT complex conjugate of the input force.

Fig. 4 Experimental Setup.

As mentioned above, a preliminary test was conducted on the beam to investigate its dynamic behavior and understand the modes targeted by the reconstruction approach.

Figure 5 shows the FRF sum of the 10 accelerometers and the characterized mode shapes.

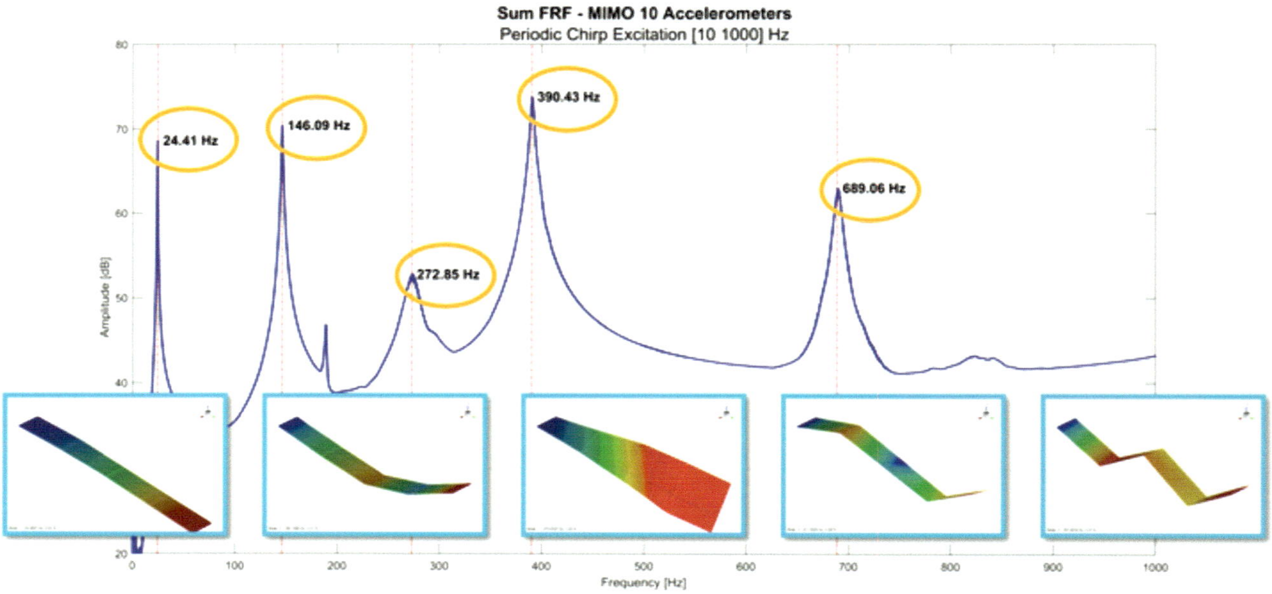

Fig. 5 FRF Sum of the preliminary test and the computed mode shapes.

Knowing the location of the modes helps in choosing some target bands to be analyzed for the reconstruction. Successively all the 40 sub-bands will be analysed for assembling all the tests in a unique FRF.

The results of the modal analysis procedure on 3 of the 40 sub-bands are shown in Figure 6 together with the estimated DIC mode-shapes.

For each frequency band, two FRFs are calculated: one representing the sum of the FRFs from the 10 accelerometers (denoted by the red line), and another representing the sum of the FRFs from the 2123 points of the DIC reconstructed displacements (denoted by the blue line). Additionally, an integrated FRF derived from the accelerometers is displayed (shown as the light blue line). The degree to which the light blue line (i.e., the integrated accelerometer FRF) aligns with the blue line (i.e., the FRF from reconstructed displacements) indicates the accuracy of the procedure. A high level of overlap

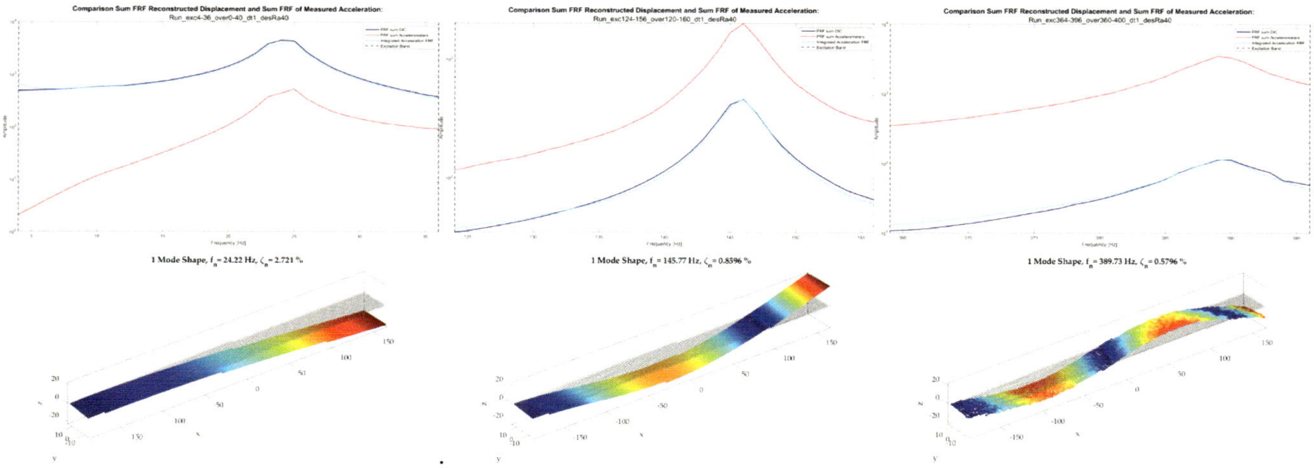

Fig. 6 Experimental Validation in the frequency bands left) 4-36 Hz center) 124-156 Hz right) 364-396 Hz.

suggests that comparable FRF data are obtained using both the accelerometer measurements and the low-speed cameras in conjunction with the reconstruction method. Furthermore, for each band, the mode shapes output from the PolyMax algorithm are also presented.

A general dynamic procedure for FRF branch assembly has been developed, consisting of the following steps:

1. Loading of FRFs and Coherences: Load the FRFs and Coherence functions for all frequency bands under consideration.
2. Segmentation of FRF and Coherence: Segment the FRF and Coherence data based on the excitation conditions.
3. Identification of Overlap Zones: Identify the overlapping regions between different frequency bands.
4. Selection of Optimal Coherence Branches: Within the overlap zones, select the branches that exhibit the best Coherence.
5. Concatenation of All Branches: Assemble the selected branches to form a continuous FRF across all bands.

Figure 7 illustrates the simultaneous loading of all the FRF and coherence results across the 40 frequency bands. It is important to note that the FRF data should be analyzed within the corresponding excitation band to ensure accuracy and relevance to the dynamic response under study.

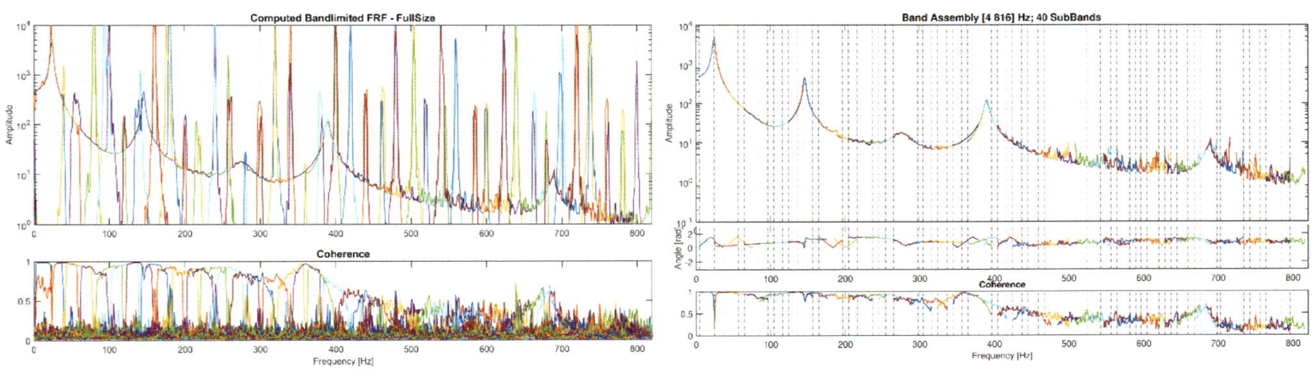

Fig. 7 left) Loaded FRFs and Coherences of the chosen bands right) FRFs and Coherences considered only in the excitation frequency range.

Figure 7 (right) presents the results obtained by selecting the specific FRF branches for each band within the excitation range, effectively "cleaning" the plot shown in Figure 7 (left). Additionally, Figure 7 (right) displays the phase of the FRF, where the phase trend reveals inversions corresponding to the resonance peaks. However, the phase does not always transition smoothly, unlike the amplitude, particularly at frequencies around 50 Hz, 200 Hz, and 370 Hz. These discontinuities in the phase may affect the accuracy of the PolyMax algorithm. The underlying causes of the imperfect phase transitions remain unclear and require further investigation. Moreover, the overall upward trend in the phase could indicate imperfect synchronization between the excitation force and the system's response. At higher frequencies, coherence decreases, leading to noisy FRF data in both amplitude and phase.

Conclusion

This study provided an in-depth analysis of the implementation and development of reconstruction algorithms for down-sampled displacements, using DIC for vibration measurements. The methodology enabled high-frequency measurements using low-speed cameras while achieving higher resolution.

Key accomplishments include:

- The development and numerical validation of two reconstruction algorithms, tested on simple signals and a simulated MDOF system.
- Validation of experimental results using accelerometer sensors.
- Creation of a dynamic and parametric FRF assembly algorithm for modal analysis after collecting data from band-limited runs.
- Successful PolyMax analysis, yielding accurate mode shapes comparable to those obtained with accelerometers.

Future improvements should focus on reducing post-processing time, particularly by developing systematic and iterative scripts to streamline the workflow. Further investigation is needed to address the noise at higher frequencies, including validation with high-speed cameras and testing on more complex structures. Overall, the approach demonstrated promising results in capturing the full dynamics of a structure with low-speed cameras, reducing the need for contact sensors and managing smaller datasets.

References

1. B. Peeters, H. Van der Auweraer, P. Guillaume, J. Leuridan, The polymax frequency-domain method: A new standard for modal parameter estimation?, Shock and Vibration 11 (2004) 395–409.
2. C. Pappalardo, D. Guida, System identification and experimental modal analysis of a frame structure, Engineering Letters 26 (2018) 56–68.
3. H. Van der Auweraer, Structural dynamics modeling using modal analysis: applications, trends and challenges, in: IMTC 2001. Proceedings of the 18th IEEE Instrumentation and Measurement Technology Conference. Rediscovering Measurement in the Age of Informatics (Cat. No.01CH 37188), volume 3, 2001, pp. 1502–1509 vol.3. doi:10.1109/IMTC.2001.929456.
4. A. Cusano, P. Capoluongo, S. Campopiano, A. Cutolo, M. Giordano, F. Felli, A. Paolozzi, M. Caponero, Experimental modal analysis of an aircraft model wing by embedded fiber bragg grating sensors, IEEE Sensors Journal 6 (2006) 67–77.
5. A. Zanarini, Full field optical measurements in experimental modal analysis and model updating, Journal of Sound and Vibration 442 (2019) 817–842.
6. J. Baqersad, P. Poozesh, C. Niezrecki, P. Avitabile, Photogrammetry and optical methods in structural dynamics – a review, Mechanical Systems and Signal Processing 86 (2017) 17–34. Full-field, non-contact vibration measurement methods: comparisons and applications.
7. Paolo Neri. Frequency-band down-sampled stereo-DIC: Beyond the limitation of single frequency excitation. Mechanical Systems and Signal Processing, 172:108980, June 2022.

Chapter 6
LiDAR-based Semi-Autonomous Inspection of a Coal Mine Tailing Impoundment Emergency Spillway

Feras Abla, Cengiz Kaydim, Paulo Simplicio, Onur Avci, Deniz Tuncay, Guilherme A. S. Pereira, and John Quaranta

Abstract Coal mine tailing impoundments are built primarily to dispose of coal refuse and have taken on the role of storing process water. In the Appalachian region of the eastern USA, impoundments are generally constructed as cross-valley embankment dams. These structures are regulated by the respective state and by the US Department of Labor, Mine Safety and Health Administration (MSHA). The tailings dams may be designed to have emergency spillways that passively maintain the required pool elevation level for structural stability and water discharge control. The emergency spillways are required to remain free of obstructions at all times. Emergency spillways of tailing dams are inspected to ensure that no objects are blocking the water discharge. To facilitate a more objective, comprehensive, and faster inspection process during sunny-day, excessive storm events, and night-time conditions, this paper presents a semi-autonomous inspection approach using Unmanned Aerial Vehicle (UAV) to identify, categorize, and geo-reference purposely placed obstructions as simulated hazards placed within the cross-section of a hydraulic spillway. The study focuses on the inspection of an emergency spillway for

Feras Abla
Graduate Student, Department of Civil and Environmental Engineering, Engineering Sciences Building, 1306 Evansdale Drive, West Virginia University, Morgantown, WV 26506
(Corresponding author)
e-mail: fa00060@mix.wvu.edu

Cengiz Kaydim
Graduate Student, Department of Mining Engineering, Mineral Resources Building, 1306 Evansdale Drive, West Virginia University, Morgantown, WV 26506
e-mail: ck00047@mix.wvu.edu

Paulo Simplicio
Graduate Student, Department of Mechanical, Materials and Aerospace Engineering, Engineering Sciences Building, 1306 Evansdale Drive, West Virginia University, Morgantown, WV 26506
e-mail: pvg00001@mix.wvu.edu

Onur Avci
Assistant Professor, Department of Civil and Environmental Engineering, Engineering Sciences Building, 1306 Evansdale Drive, West Virginia University, Morgantown, WV 26506
e-mail: onur.avci@mail.wvu.edu

Deniz Tuncay
Assistant Professor, Department of Mining Engineering, Mineral Resources Building, 1306 Evansdale Drive, West Virginia University, Morgantown, WV 26506
e-mail: deniz.tuncay@mail.wvu.edu

Guilherme A. S. Pereira
Professor, Department of Mechanical, Materials and Aerospace Engineering, Engineering Sciences Building, 1306 Evansdale Drive, West Virginia University, Morgantown, WV 26506
e-mail: guilherme.pereira@mail.wvu.edu

John Quaranta
Professor, Department of Civil and Environmental Engineering, Engineering Sciences Building, 1306 Evansdale Drive, West Virginia University, Morgantown, WV 26506
e-mail: jdquaranta@mail.wvu.edu

Janko Slavic et al. (eds.), *Computer Vision & Laser Vibrometry, Vol. 6*, Conference Proceedings of the Society for Experimental Mechanics Series, https://doi.org/10.13052/9788743801511-6

a tailing dam facility located at a permitted and operating coal mine located in Greene County, Pennsylvania, USA. Airborne Laser Scanning (ALS) of the spillway was carried out using the L1 Zenmuse LiDAR sensor mounted on a UAV. Two flight missions were planned: one to scan the spillway without anomalies (baseline), and another with simulated anomalies placed on the spillway. After collecting LiDAR datasets, three-dimensional (3D) point cloud models of the spillway were reconstructed for both missions. The point clouds of both models were registered and then the Cloud to Cloud (C2C) surface change detection algorithm was implemented to detect changes in the baseline model. It was found that the C2C algorithm successfully detected all simulated anomalies on the spillway, even the ones that were the size of a soccer ball. It is concluded that relying on such technology has the potential to minimize the dependence on subjectivity of physical inspections and reduce the safety concerns for inspectors.

Keywords UAV · Anomaly detection · Tailing dam · LiDAR

Introduction

Coal tailing dams are inspected by trained inspectors and Professional Engineers to comply with state and Federal mining regulations. The dam inspection consists of a comprehensive field review of critical aspects of the dam structure and hydraulic appurtenant structures [1]. Inspecting large dam and hydraulic control structures is time-consuming and poses hazards to the inspectors. Therefore, it is essential to investigate methods to employ the latest technological advances to facilitate this process for humans and contribute to increase the likelihood that anomalies are detected. The authors have been exploring incorporating UAVs and advanced algorithms in the inspection process of tailing dams. In this paper, the authors present a simple approach for semi-autonomous anomaly detection for an emergency open channel spillway of a tailing dam on the East Coast of the USA.

The use of UAVs in structural health monitoring has caught the attention of researchers in the past. A dam health monitoring model using UAV photogrammetry was proposed by [2]. The authors reconstructed tri-dimensional (3D) models of the dam under study from UAV photogrammetry and utilized the Multiscale Model-to-Model Cloud Comparison (M3C2) method for anomaly detection. The M3C2 method was also used in another study to explore the potential of leveraging multi-temporal 3D surveying data to monitor archaeological heritage sites [3]. A deep learning-based approach was proposed to detect rill erosion on tailing dam slopes from UAV images [4].

Matrice 350 RTK and L1 Zenmuse

The UAV platform used in this study is the DJI Matrice 350 RTK shown in Figure 1(a). This advanced drone platform has a wide range of state-of-the-art features such as a wide-range transmission system that enables triple channel 1080p HD live feeds, an efficient battery system, several safety features, and the ability to carry three payloads simultaneously [5]. These features together with the proper payload make this UAV platform a very powerful asset for flight operations such as high-precision mapping, air-to-ground coordination, and automated precision inspection.

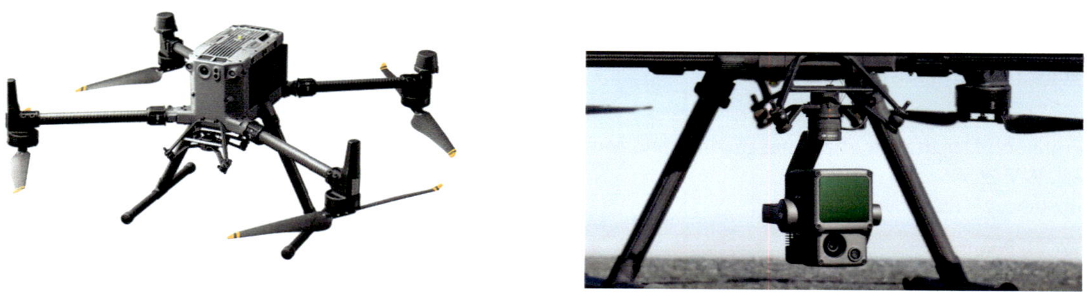

(a) (b)

Fig. 1 DJI Matrice 350 RTK (a), and L1 Zenmuse LiDAR sensor (b).

L1 Zenmuse is a LiDAR sensor that can be mounted to the DJI Matrice 350 RTK drone platform as shown in Figure 1(b). This LiDAR sensor incorporates a Livox Lidar module, a high-accuracy Inertial Measurement Unit (IMU), and a camera with a 1-inch CMOS on a 3-axis stabilized gimbal. It has a detection Range of 450 m with 80% reflectivity and 0 klx, whilst a lesser range of 190 m when the reflectivity is poor (10% reflectivity, and 100 klx). The effective point rate is 240,000 pts/s and it supports three different return modes: single, dual, and triple. It also offers two scanning pattern options: repetitive and non-repetitive. The RGB camera is 20 MP with 1-inch CMOS and a mechanical shutter. The payload has a vision sensor for positioning accuracy and is capable of GNSS, IMU, and RGB data fusion. According to the specs provided by the manufacturer, the system accuracy (RMS 1σ) is 10 cm at 50 m horizontal and 5 cm @ 50 m vertical, while the LiDAR ranging accuracy (RMS 1σ) is 3 cm at 100 m [6].

The open channel emergency spillway

The open channel emergency spillway shown in Figure 2 is for the tailing dam facility studied in the project. It plays an essential role in managing the disposal of excess water, particularly in times of emergency (excessive precipitation inflow). The spillway attributes are a trapezoidal cross-section with a base width of approximately 15 meters, top width of 32 meters, and side slopes of 32 degrees. The spillway length is approximately 155 meters, and discharges to an existing stream. It was constructed using concrete-filled fabric bags having a rough surface to reduce water velocity. Spillways are required always to remain free from obstructions that prevent water discharge during emergency events.

Fig. 2 Open channel emergency spillway.

Mission Planning

The DJI Matrice 350 RTK is used to execute the spillway surveying mission. The L1 Zenmuse sensor is mounted on the drone to collect LiDAR data points. The mission is initially created on the remote controller of the drone on the Pilot 2 app. The area of interest is selected on the map, and the mission and payload settings are set to build accurate 3D models of the spillway. The sampling rate is set to 240 kHz for the payload. Because the drone can do real-time kinematic (RTK) correction of the GPS signal while surveying the area of interest, no ground control points are needed to verify the drone's measurements and point cloud model accuracy. The collected LiDAR data points are stored on a MicroSD card and used later to reconstruct detailed three-dimensional spillway models.

To implement an anomaly detection method, the spillway had to have surface changes that represent potential anomalies blocking the spillway. Consequently, two surveying missions are planned and flown. For the first surveying flight, the open channel spillway was clear from any objects (simulated anomalies). Therefore, the dataset collected from the first mission served as a baseline for comparing with the datasets collected from the second mission. To create a scenario that simulates the existence of potential anomalies on the spillway, objects such as boxes and soccer balls are dispersed on the spillway, as well as some of the team members stood in different areas while the UAV was scanning to create the second dataset as shown in Figure 3.

After collecting the LiDAR datasets from the two missions, DJI Terra [7] software is used to reconstruct three-dimensional point cloud models of the spillway. These models are then exported in formats compatible with point-cloud software for

Fig. 3 Simulated anomaly locations of the spillway.

further analysis. The next step is to register the two 3D point cloud models to align and merge them into a common coordinate system. Over the years, various point cloud registration algorithms were introduced such as the classic ICP-algorithm, feature-based methods, learning-based methods, and others [8]. CouldCompare [9] software is used to register the reconstructed 3D point cloud models by manually picking at least 4 equivalent point pairs i.e. correspondences in each model. Then, the registration algorithm will find the correct translation and rotation between the two clouds.

The registered point clouds are then utilized as inputs for the Direct Cloud-to-Cloud Comparison algorithm (C2C). This is an efficient approach to compare 3D point clouds directly without gridding or meshing the data and calculation of surface normal. It compares two sets of point clouds simply by finding the closest point in the first cloud for each point in the second point cloud. Subsequently, the distances between the points indicate surface change. However, the surface roughness and point cloud density affect the closest point distance L_{C2C} between the point clouds. Therefore, this approach is well-suited for change detection for highly dense point clouds [10].

Results

Three-dimensional model reconstruction of the spillway

The reconstructed 3D point cloud models of the spillway are illustrated in Figures 4(a) and 4(b) representing the models without and with anomalies, respectively. Each point cloud has approximately 150 million points. These high-density point clouds show fine details of the open channel spillway with remarkable clarity. In addition, the models exhibit crucial characteristics such as surface features, which makes it possible to analyze both normal and hazardous situations in great detail. Moreover, the accuracy of these reconstructions provides insightful information about the state of the spillway which enables more effective monitoring plans.

Before performing further analysis, it is imperative to investigate how well the LiDAR scan captured the anomalies with varying sizes. A close-up view of the reconstructed spillway model with anomalies shows that even an object as small as a soccer ball is clearly visible and well represented in the point cloud model as shown by the color-coded elevation map of the reconstructed model with anomalies in Figure 5. It is worth mentioning that the flight altitude for this mission is 30 m above the bottom of the spillway which emphasizes the ability of the LiDAR sensor to capture impressive details from such high altitude.

Point cloud registration

To align both models of the spillway with and without anomalies, 4 pairs of points are selected from each model at identical locations. To minimize registration errors, these points are strategically picked at distinct locations of the spillway such as the spillway's corners and key features. Subsequently, the models are aligned and registered accordingly, and the software dis-

(a)

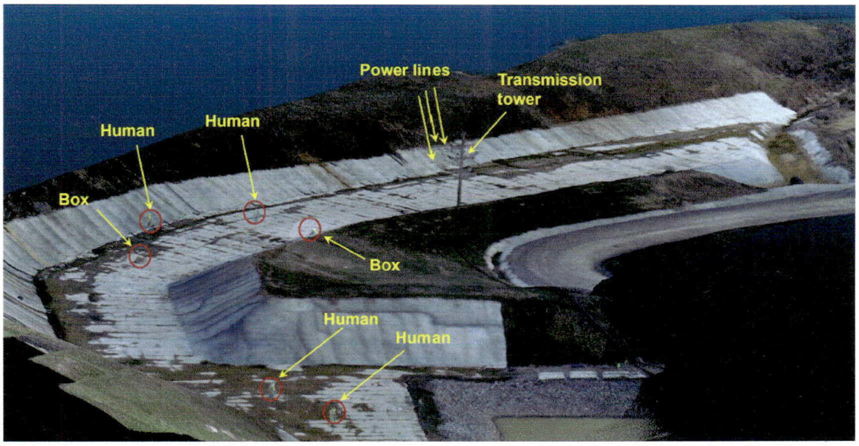

(b)

Fig. 4 Reconstructed 3D point cloud models of the spillway without anomalies (a), and with anomalies (b).

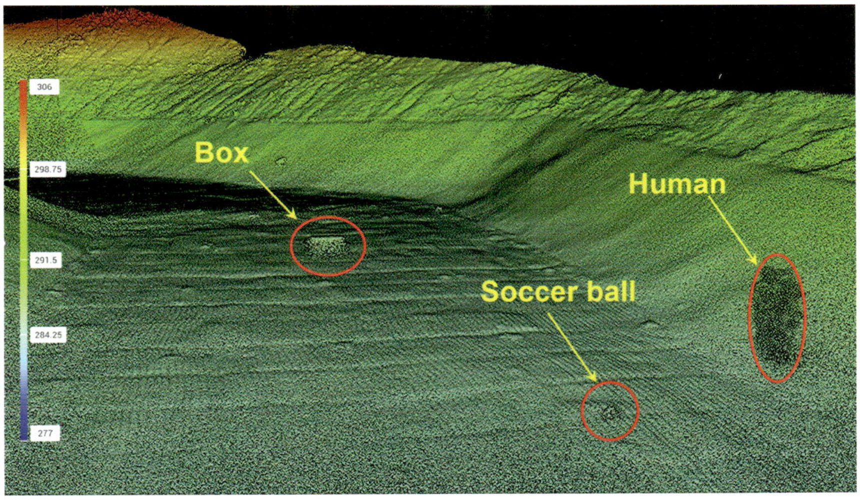

Fig. 5 Color-coded elevation map of the reconstructed 3d with anomalies model.

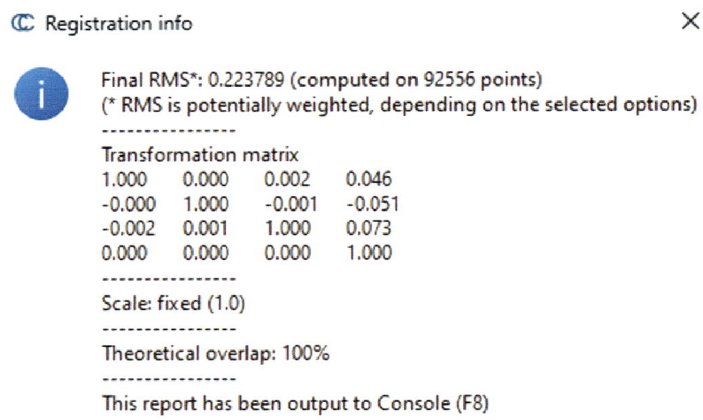

Fig. 6 Registration RMS error and transformation matrix.

plays the registration information as shown in Figure 6. The final root mean square error (RMS) for the registration process is computed based on nearly 46,000 points, yielding an RMS error of 22.37 cm. This degree of accuracy is considered adequate for identifying the existence of anomalies impeding the hydraulic discharge that are detected using the C2C algorithm, as it ensures the alignment is accurate enough for reliable comparison and assessment of the spillway models.

Implementing the C2C algorithm

The C2C algorithm is applied to the registered models to identify changes in the surface between the baseline model (without anomalies) and the model with anomalies. This algorithm highlights surface changes between point cloud models where these discrepancies represent anomalies in our case. Figure 7 illustrates a color-coded map of the compared models generated by the C2C algorithm. The map clearly shows distinct color variations corresponding to the locations of the simulated anomalies. Therefore, this indicates that the C2C algorithm successfully detected all surface changes including the smallest object such as the soccer ball. This robust detection ability is essential for evaluating the spillway's integrity and making sure that all pertinent anomalies are found and handled effectively.

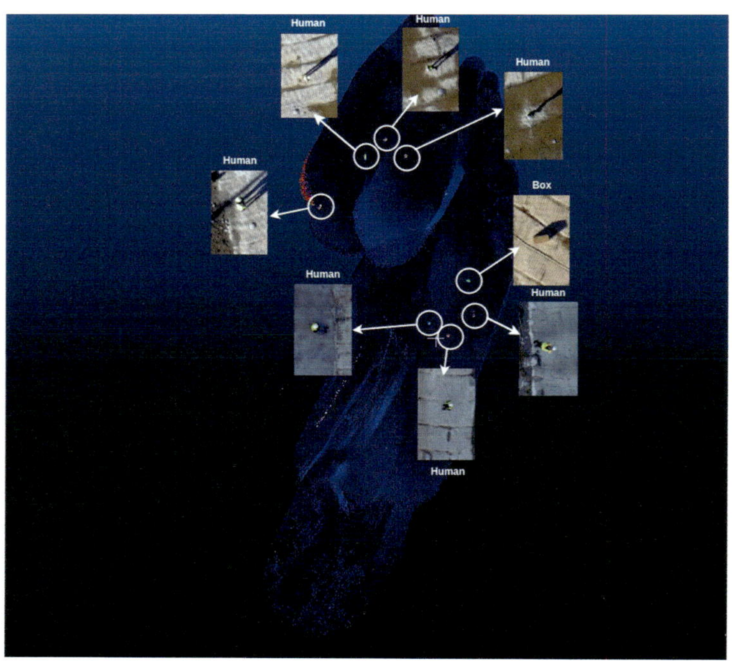

Fig. 7 Color-coded map generated by C2C algorithm.

Conclusions

In this paper, the authors introduce an approach to utilize an ALS LiDAR in the inspection of an open channel emergency spillway at an existing coal mine tailing impoundment. With this approach, accurate 3D point cloud models of the spillway are reconstructed and the C2C algorithm is utilized to compare them. In conclusion, this work provides insights into the efficiency and practicality of employing ALS LiDAR in structural health monitoring and damage detection applications. Also, the findings demonstrate that relying on such technology has the potential to minimize the dependence on the subjectivity of physical inspections and reduce the safety concerns for inspectors. In the future, the authors plan to investigate additional hazards associated with coal mine tailing impoundments such as leakage and seepage. Furthermore, they aim to develop a user-friendly software interface to facilitate hazard detection in these environments.

References

1. Quaranta, J. D., Stawovy, J., & Fourie, A. (2022). Study on mine tailings classification effected by co-disposal of drilling wastes with geochemical cations. Mining, Metallurgy & Exploration, 39(4), 1391–1402.
2. Zhao, Sizeng, et al. "Structural health monitoring and inspection of dams based on UAV photogrammetry with image 3D reconstruction." Automation in Construction 130 (2021): 103832.
3. Lercari, N. (2019). Monitoring earthen archaeological heritage using multi-temporal terrestrial laser scanning and surface change detection. Journal of Cultural Heritage, 39, 152–165.
4. Nasategay, F. F. U. (2020). Detection and monitoring of tailings dam surface erosion using UAV and machine learning (Doctoral dissertation, University of Nevada, Reno).
5. Matrice 350 RTK. DJI. (n.d.). https://enterprise.dji.com/matrice-350-rtk?site=enterprise&from=nav
6. Support for Zenmuse L1 - DJI. (2014). DJI. https://www.dji.com/support/product/zenmuse-l1
7. DJI Enterprise. (2024). (V 4.2.5). [DJI Terra]. https://enterprise.dji.com/dji-terra
8. Li, L., Wang, R., & Zhang, X. (2021). A tutorial review on point cloud registrations: principle, classification, comparison, and technology challenges. Mathematical Problems in Engineering, 2021(1), 9953910.
9. CloudCompare project. (2024). (V 2.14.alpha). [CloudCompare]. https://www.cloudcompare.org/
10. Lague, D., Brodu, N., & Leroux, J. (2013). Accurate 3D comparison of complex topography with terrestrial laser scanner: Application to the Rangitikei canyon (NZ). ISPRS journal of photogrammetry and remote sensing, 82, 10–26.

Chapter 7
DIC Test Planning with Blender – A Focus on Lighting

Marc A. Eitner, Jorge Alejandro Ricaurte, and Jayant Sirohi

Abstract The growing maturity of digital image correlation (DIC) capabilities lead to more applications outside academic and research environments, and thus need to be integrated into common workflows used by industry testing professionals. A major part of typical vibration or modal testing is the pretest planning analysis, where decisions about excitation and sensing hardware and their locations are made. For camera-based testing, this pretest planning is commonly neglected, though it could prove of significant advantage to reduce test uncertainty and limit the time spent at the test site. The open-source software Blender allows one to perform such pretest analysis and is used in this paper to showcase its application for a DIC test. An overview of the different lighting methods is given, along with the available camera setups. A variety of methods for filtering is shown and a method to simulate fluorescent light emission is provided, using Blender's emission shader.

Data from a previously performed test campaign on a model-scale helicopter rotor in hover (90 m/s tip speed) is used for this demonstration. The test involved stereoscopic, high-speed DIC and used a complex lighting method based on laser-induced fluorescence. The test setup is recreated in the software, which includes the rotor test stand geometry, the camera positions and settings, and the lighting. The discussion focuses on DIC lighting techniques and their implementations in Blender. A good match between the real-world images and the rendered images can be created, provided sufficient information about the light power and camera sensitivity is available, thereby enabling Blender's use for test planning of camera-based experiments such as DIC or photogrammetry.

Keywords Blender · Digital Image Correlation · Rotor · Structural Dynamics · Test Planning · Laser

Introduction

Camera-based test methods, such as photogrammetry, digital image correlation (DIC), or even pressure-sensitive paint (PSP) measurements, require a substantial amount of experimental effort. This starts with handling the transport of expensive equipment such as cameras, lenses, filters, lights, etc., and then requires a considerable effort in creating a test setup that achieves all the desired outcomes, such as proper focus, illumination, field of view, etc. Depending on the type of test, surface preparation of the test specimen is also commonly needed. During the test, camera vibrations and image noise influence test data and calibration, and camera positions need to be adjusted to account for glare and other unwanted lighting effects. All of these components need to be handled by the test engineers, commonly working under tight schedules in expensive facilities such as wind tunnels. This is especially true in big national laboratories or at customer sites. For this reason, this paper investigates the use of a simplified digital twin of the test setup, to enable some aspects of test planning a priori of the test. The test setup, including the test specimen, the light source(s), and the camera(s), are simulated in a virtual environment. Engineers can then use this digital setup to experiment and find optimal values for, for example, camera positions, lamp power, or camera settings, thus saving time during the actual test campaign.

The creation of such a virtual environment requires software that can handle all of these different aspects. Since the early days of rising computing power and 3D graphics, engineers have used software to render photorealistic images.

As early as 1982, Potmesil and Chakravarty added lens and aperture effects to simulate depth of field and improve the

Marc A. Eitner
ATA Engineering Inc.San Diego, CA
e-mail: marc.eitner@ata-e.com

Jorge Alejandro Ricaurte · Jayant Sirohi
The University of Texas at AustinDept. of Aerospace Engineering
e-mail: jorge.ricaurte@austin.utexas.edu; sirohi@utexas.edu

Janko Slavic et al. (eds.), *Computer Vision & Laser Vibrometry, Vol. 6*, Conference Proceedings of the Society for Experimental Mechanics Series, https://doi.org/10.13052/9788743801511-7

realism of their rendered images [1]. Forty years later, a variety of 3D computer graphics software is available with simple user interfaces to allow even beginners to render photorealistic images. One of these tools (among many others) is the open-source software Blender [2], which is the focus of this paper.

The open-source software Blender [2] is a tool intended for artists to use advanced 3D computer graphics technology to create photorealistic images and movies. It is packed with features that allow character rigging and animation, design, texturing, modeling, and even physical simulations, ranging from rigid body dynamics to fluid simulations. While the original intent was artistic, Blender's versatility and ease of access has led to applications far outside the scope of its original design.

Blender's first application to DIC was by Rohe in 2019 [3], which introduced the structural dynamics community to the advantages of the software, and showcased its strength as a virtual test simulator and test planning software. Blender's Python-based programming interface allows automation of mesh deformations and enables deformation reconstruction even at sub-pixel levels. In this paper, the focus will be placed on the available lighting methods in Blender, that can be used to recreate the illumination conditions commonly used in camera-based experiments such as DIC or photogrammetry.

Blender Overview

Blender's primary use is to render high-quality images of computer-generated environments. Models can be imported in a variety of formats (obj.,stl.,abc.,. . .) that describe surface meshes. Additionally, Blender contains a modeling tool, where surface meshes can be created. Textures can be applied to the surface meshes through a variety of methods. Those textures can be imported from online sources, such as Poliigon [4], or can be created within Blender using a combination of image textures and shaders. Objects can be animated on a frame-by-frame basis to allow rendering of moving scenes. Blender is based on the Python programming environment and conveniently allows automation of task through scripting interface. There is also a post-processing step for rendered images, the compositing feature, where filters and other effects can be applied to images. Blender uses a global illumination scheme, where a path tracing algorithm accounts for surface illumination from direct and indirect light. Such computations are expensive to render, but Blender supports graphics process unit (GPU) rendering, making the computations orders of magnitude faster than if central processing unit (CPU) rendering is used.

Blender'S Camera Model

Cameras are created in Blender as objects with user-defined attributes. Blender supports orthographic, panoramic, and perspective cameras, with the latter being of primary importance. The perspective cameras are based on the pinhole model, where the aperture is a single point and the relation between the image coordinates (u, v) and the real-world coordinates (x_w, y_w, z_w) of an object in the field of view are described by the linear relation:

$$\begin{pmatrix} u \\ v \\ 1 \end{pmatrix} = [K]\,[R|t] \begin{pmatrix} x_w \\ y_w \\ z_w \\ 1 \end{pmatrix} \tag{1}$$

The camera matrix $[K]$ describes the intrinsic parameters of the camera, such as focal length, pixel resolution, and sensor size. The matrix $[R|t]$ describes the extrinsic parameters and contains the rotation angles and translations of the camera with respect to the world coordinate system origin, i.e., the camera's position and orientation in space. In Blender, the user directly defines all the camera intrinsic and extrinsic parameters. In a real-world experiment, the camera parameters are obtained through a calibration routine, where typically images of a plate with a distinct pattern (such as a checkerboard pattern) are taken, while the calibration target is translated and rotated. Such a calibration routine can also be performed in Blender, by the user importing a calibration target and imaging it with cameras in multiple positions and orientations. Those calibration images can then be exported and analyzed with commercial or open-source software [5] to obtain the matrices $[K]$ and $[R|t]$. A primary use for this specific process is the development and testing of calibration algorithms.

The settings for the camera are shown in Fig. 1. Depth-of-field effects can be activated by the user defining an f-stop value, along with a focal distance, which is either an actual distance or an object in the scene that is to be in focus. The focal length of an ideal convex lens is set by the user, as well as the sensor size (in mm). The resolution of the camera is that of the rendered images, and is set separately in the rendering menu. The location and rotation of the camera are based on a camera coordinate system located at the lens aperture.

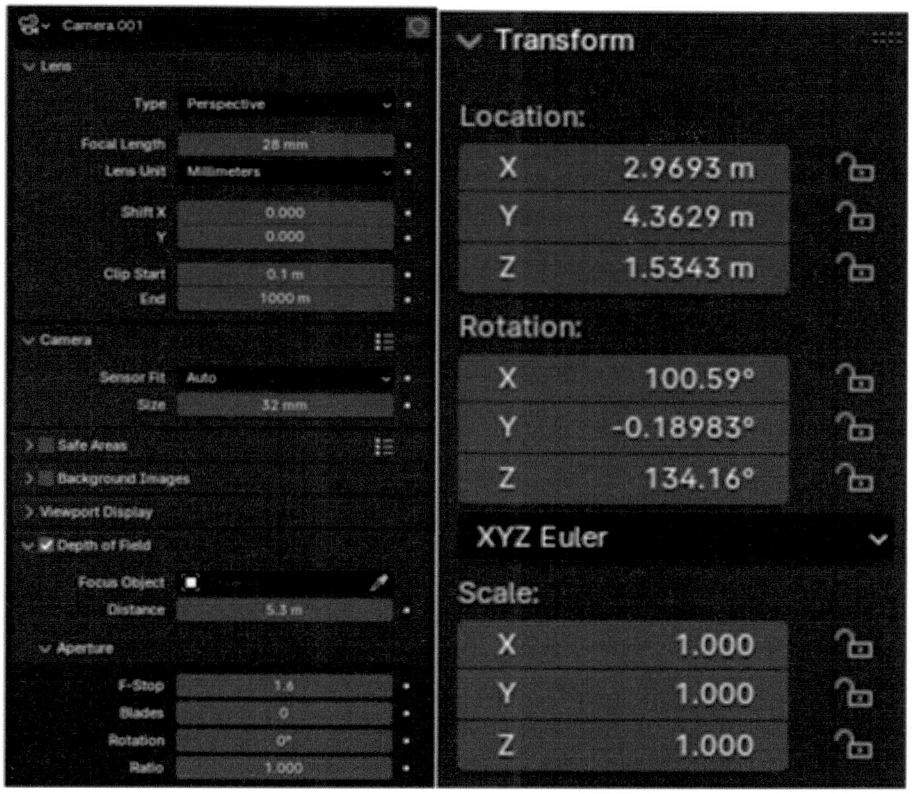

Fig. 1 Camera settings.

Note that the camera settings do not allow for effects from non-ideal lenses, such as barrel distortions or chromatic aberrations. Chromatic aberrations, if needed, can be added as a postprocessing step, using Blender's compositing feature, Lens Distortion Node. Radial distortions that are positive (pincushion distortion) or negative (barrel distortion), can be set using Movie Distortion Node, which allows the user to define the three radial distortion coefficients commonly used in most calibration routines. Alternative distortion models implemented in Blender are the Brown-Conrady model [6], which contains both radial and tangential distortions.

Lighting In Blender

From a test engineer's point of view, arguably the most difficult aspect of taking high-quality images is lighting. Setting the focal distance is typically straightforward; the test specimen needs to be in focus. Shutter speed, exposure length, ISO settings, and filters are in turn all dependent on the available light and are typically chosen in a time-consuming and trial-and-error manner. It is thus critical to have some idea of how much light is needed or available to choose appropriate settings and equipment.

Light sources can be added to a Blender scene as one of four objects:

1. Point: Omnidirectional point light source
2. Sun: Constant direction parallel rays light source
3. Spot: Directional cone light source
4. Area: Directional area light source

These objects can be placed in a scene or attached to a physical object. Common light sources used in experiments are LED panels. An example of such a panel is shown in Fig. 2. The square area light is attached to a simplified representation of the LED panel. A physical object attached to the light source helps to find appropriate space in the laboratory, considering the actual dimensions of the light.

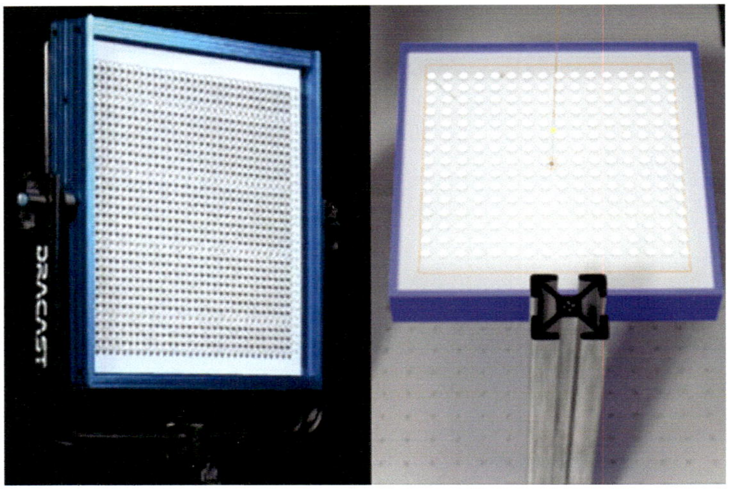

Fig. 2 Representation of area light in Blender. (Left) image of Dracast LED1000 Plus Series, 60 W power; (Right) simplified Blender representation.

Table 1 Suggested power values for typical lights [7]

Real-World Light	Power	Suggested Light Type
Candle	0.05 W	Point
800 lm LED bulb	2.1 W	Point
1000 lm light bulb	2.9 W	Point
1500 lm PAR38 floodlight	4 W	Area, disk
2500 lm fluorescent tube	4.5 W	Area, rectangle
5000 lm car headlight	22 W	Spot, size 125°

An important aspect to get right is the power setting of the light source, as this is not the electrical power needed by the light, but rather radiant power or radiant flux. The problem is that off-the-shelf lamps typically do not advertise the radiant power. Instead, consumer lamps often use the radiant flux in lumen, which accounts for the sensitivity of the human eye, as their measure of light power. The relation between radiant power and luminous flux depends on the type of lamp; a few examples for common light sources are provided on the Blender website and shown here in Table 1, which lists the appropriate radiant power settings and light type in Blender. The most common light sources for DIC experiments performed in laboratories are white LEDs. As a first approximation, the 800 lumen LED bulb listed in Table 1, can be scaled approximately to account for stronger lights, such as LED panels. The efficiency of LEDs varies from manufacturer to manufacturer and also depends on the quantity and type of single LEDs used in the lamp.

Exposure Settings

For a real camera image, the brightness of the image (the pixel value) is determined by four factors: (**??**) the light reflected and emitted from the imaged object, (2) the aperture, (3) the shutter speed, and (4) the sensor gain (ISO value). As discussed in the previous section, the light environment can be well simulated in Blender, by the user modeling the light sources with appropriate power settings as well as adjusting the surface texture of the imaged object. Blender's aperture and shutter speed settings have influence on the depth of field and motion blur, but they do not impact the image brightness (as opposed to reality). There is also no real ISO setting to adjust the image sensor gain. Instead, the image brightness can be controlled with a single Exposure setting (Render options $\rightarrow$ Color Management $\rightarrow$ Exposure). This exposure value A controls the relation between the initial rendered image brightness I_{render} and the output brightness I_{out} as follows:

$$I_{out} = I_{render} \times 2^{A} \tag{2}$$

The use of this setting is demonstrated with a simple setup shown in Fig. 3. The Dracast LED light was used to illuminate a flat panel showing a calibration pattern. A Phantom Miro M310 camera was used with a 28 mm focal lens and imaged the object with a variety of shutter speeds and aperture settings. Note that the ratio of light that reaches the sensor is approximately divided by 2 going from left to right in the figure. The ISO setting of the camera remained constant for all these images. Since Blender does not have an ISO setting, the brightness of the darkest image (f/8, 1/10 s) needs to be matched manually. This was achieved with an exposure value of $A = 1.15$. By consecutively increasing the exposure value by one, which doubles the image brightness, the rendered image brightness matches the real-world images. Note that some of the brighter images contain saturated pixels, which complicates the visualization in Fig. 3; however, the concept of brightness adjustment and matching real-world camera settings still holds.

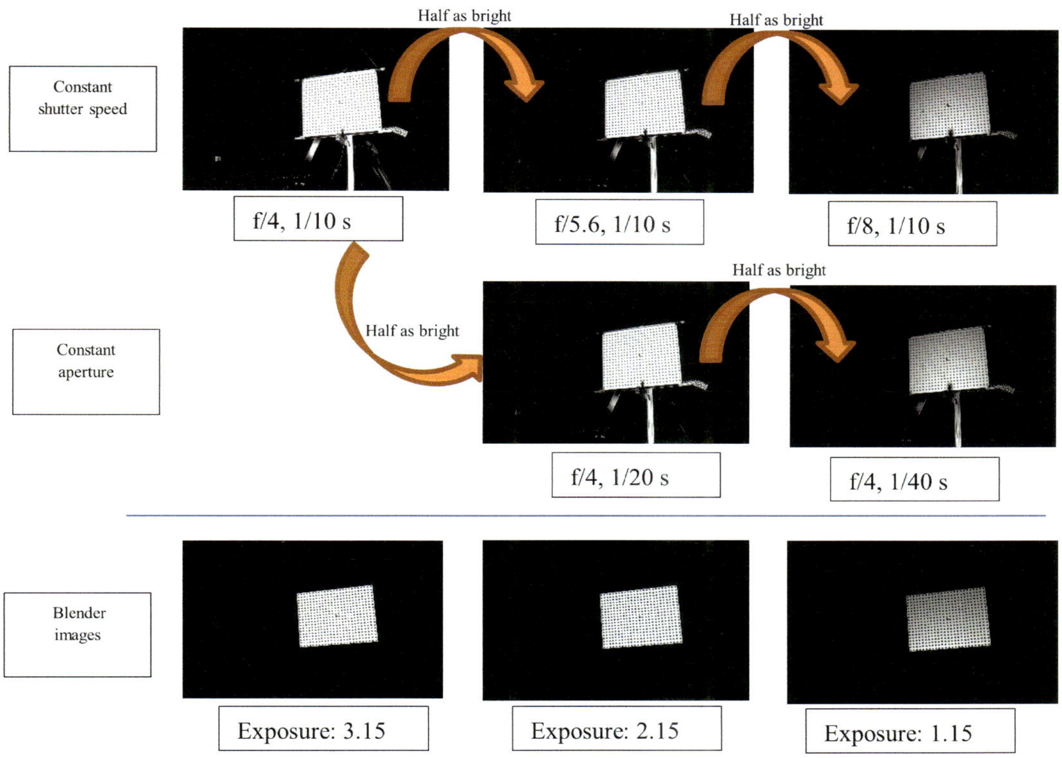

Fig. 3 Demonstration of Blender's Exposure value to match real-world images.

Filters in Blender

In many DIC/photogrammetry experiments, there is an insufficient amount of light available and it is undesirable to use lens filters. There are, however, multiple scenarios in which filter lenses might be needed for an experiment. One of these cases occurs when the surface of the imaged object is painted with fluorescent paint, which emits light in a specific narrow-band wavelength when excited. An example of such a test is shown in Fig. 4, where discrete markers are painted on the exit of a small-scale rocket model. The model is illuminated with a blacklight, which optically excites the paint, making it emit orange light. The contrast of the resulting images (marker to background) can then be greatly improved by using a filter lens on the camera. This can be done in two ways: (1) A low-pass filter lens can be used to remove the high-wavelength light (purple light emitted from the lamp), or (2) a band-pass filter lens can be used to only allow the fluorescent (orange) light emitted from the markers to reach the camera sensor. Either way, this method can lead to images with high contrast, which is desired to improve the DIC/photogrammetry process.

In Blender, many different methods of filtering specific light from an image exist. A simple method that will be demonstrated in the next section is to separate the initial render into its red, green, and blue channels. This is effective when dealing with light sources that are dominant in one of these channels, i.e., narrowband sources.

A second class of optical experiments that suffer from unwanted illumination occurs in hypersonic and high-temperature testing, when the model temperatures or that of the surrounding gas become so high, that their radiation becomes the dominant source of light. In such a case, a neutral-density filter needs to be employed, which is a broadband filter that cuts out a certain percentage of the total incoming light. These neutral density filters can most conveniently be simulated by adjusting the previously discussed Exposure value, which results in an overall reduction of brightness. For example, the addition of a neutral density filter with optical density $d = 3$ results in a 1000x smaller image brightness, based on the relation:

$$\frac{I_{transmitted}}{I_{incident}} = 10^{-d} \tag{3}$$

To achieve the same reduction in brightness, Blender's current exposure value A_0 needs to be decreased by a value of approximately 10, since

$$\frac{I_{out}}{I_{render}} = 2^{A_0-10} = 2^{A_0} \times 2^{-10} \approx 2^{A_0} \times 0.00 \tag{4}$$

For dealing with fluorescence or high-temperature light emission Blender's primary shader allows to assign an "Emission" value, which turns the material into a light source. Similar to the light sources discussed in the previous section, the radiant power of the material can be set in W/m^2.

Fig. 4 Use of fluorescent paint in point tracking of rocket nozzle [8].

Application to a Dic Experiment of a Rotor Blade

The techniques discussed up to now are demonstrated on a real-world experiment, in which the vibration of a rotating helicopter blade model is measured using stereoscopic DIC. This specific experimental campaign was performed prior to the Blender simulations, which means the simulations are adjusted to match the experiment. The objective of this section is to showcase in what ways the Blender simulated experiment can represent an actual real-world experiment, and where its limitations are.

The experimental setup and its virtual reconstruction in Blender are shown in Fig. 5. The setup shows two high-speed, grayscale cameras (Phantom Miro M310), with an f-stop of 2.8, a 28 mm Nikon focal lens, and an exposure time of 0.5 ms. The cameras image a rotating rotor blade with tip speed 90 m/s.

An ND:YLF solid state laser was used in pulse mode to illuminate the rotor during rotation. The emitted light has a wavelength of 527 nm (green), and the firing rate was synced to the rotor to obtain images at precise azimuthal locations across a rotation. This is then filtered out using a 540 nm cut-off high-pass filter lens. The surface of the blade is painted with an orange spray paint with fluorescent properties, which emits orange light when excited [9]. Note that this is the same paint

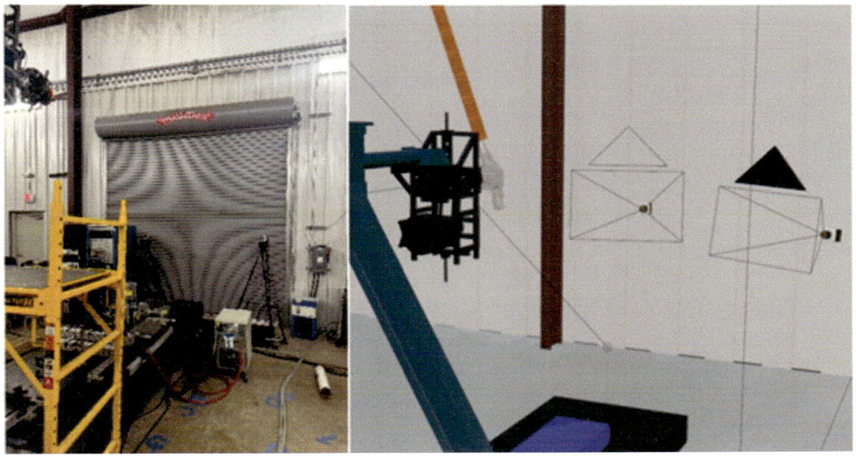

Fig. 5 View of laboratory (left) and recreation in Blender (right).

used on the rocket nozzle shown in Fig. 4, except this time it is excited by the high-power laser. Speckles of black paint are then added to the blade, which creates the random pattern needed for DIC.

In Blender, the CAD file of the rotor test stand is imported and a simplified laboratory environment is created. The camera parameters and positions are matched to the real world. The speckled orange pattern is mapped to the blade surface as an image texture, whereby the black dots are designated as the background (alpha) layer. An emission shader is then applied to the speckle pattern, resulting in an orange glow. This is to approximate the fluorescent emission. A cone light source that emits green light is added to the scene to represent the laser. The cone has a beam angle of 40° and a very strong "Blend" value, which means the intensity toward the edge of the cone is much weaker than along its centerline. This is done to achieve the effect that the blade is hardly illuminated near its tip. The rendered color image along with the real-world grayscale image are shown in Fig. 6. The rendered color image needs to be filtered appropriately to match the real-world image. Unfortunately, this is not as simple as removing all the green light from the scene (as is done in reality with the filter lens). The resulting image would show the entire blade emit light evenly. However, the fluorescent effect only acts on the parts of the blade that are illuminated (i.e., optically excited) by the laser light. This effect is achieved by first separating the red and green channels of the rendered image, and then multiplying them with each other. This effectively creates a spatial weighing of the red channel, which is mostly responsible for the orange emission. The entire process is summarized in Fig. 7.

Fig. 6 Rendered color image (left) and real-world grayscale image (right).

At this point several issues need to be highlighted. First of all it can be seen that the rendered image compares very well with the real-world image, both in perspective and intensity. This was however only achieved by manually adjusting the radiant power of the emission shader (which creates the orange fluorescent light), as well as the radiant power of the green laser source. While the laser power settings are known, there is no information on the paint, that would allow to quantify the fluorescent emission. If such experiments are to be repeated, the fluorescence should be quantified to allow this process

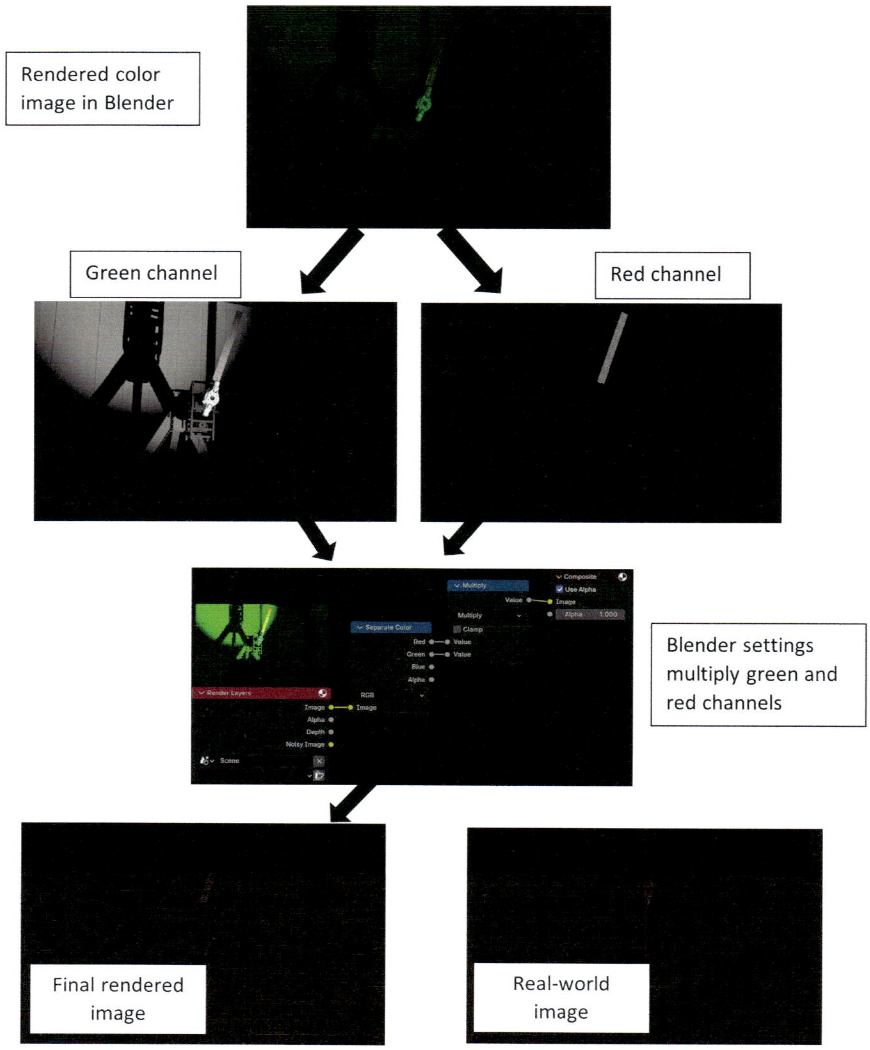

Fig. 7 Postprocessing of rendered color image to match real-world grayscale image.

to become more rigorous. An additional uncertainty comes from the laser cone intensity distribution, which was manually adjusted to match the real-world intensity fall off towards the blade tip. Again, this beam profile could be further quantified using a simple photo-diode in combination with the laser and diffusive lens.

Conclusion

Several techniques were presented that can be used to simulate DIC/photogrammetry experiments using the open-source software Blender. The focus was placed on lighting, and it was shown that Blender supports a variety of tools to model natural (sunlight) and artificial light sources (lamps). Blender's emission shader allows to even approximate effects such as fluorescence, which was demonstrated successfully on a real-world rotorcraft experiment. More advanced methods involving the emission of surfaces and even plasmas in high-temperature experiments, such as wind tunnels or plasma torches, could technically also be modeled using this same approach. By taking reference images of a given camera and light source, users can find appropriate exposure settings within Blender, which can then be used for reference in future simulated experiments. With a limited amount of effort in modeling the experimental setup in Blender, valuable time can be saved during experiments, which is often of primary importance when for test engineers performing tests outside of university laboratories, such as at national labs or customer sites. This can be achieved by optimizing the position of cameras in the scene, as well as camera settings, such as aperture, or picking a suitable lens. Furthermore, locations and power settings of light sources can be approximated, provided sufficient information is known. This step might require some initial experimentation with the available equipment, to find for example, the relation between the camera ISO and the exposure setting in Blender. The combination of all of these steps have the potential to save valuable time during experiments performed under tight schedules.

References

1. M. Potmesil and I. Chakravarty, "Synthetic image generation with a lens and aperture camera model," ACM Transactions on Graphics (TOG), vol. 1, no. 2, pp. 85–108, 1982. W309W10080
2. The Blender Foundation, "Blender," 2024. [Online]. Available: https://www.blender.org/.W309W10080
3. D. Rohe, "Rohe, Daniel Peter. An Optical Test Simulator Based on the Open-Source Blender Software," SAND2019-12756C. Sandia National Lab, Albuquerque, NM, 2019.W309W10080
4. A. Price, "poliigon," [Online]. Available: https://www.poliigon.com/.W309W10080
5. "OpenCV," Open Source Vision Foundation, [Online]. Available: https://opencv.org/.W309W10080
6. D. C. Brown, "Decentering distorion of lenses," Photogrammetric Engineering, vol. 32, no. 3, pp. 444–462, 1966. W309W10080
7. "Blender Manual - Power of Lights," The Blender Foundation, [Online]. Available: https://docs.blender.org/manual/en/latest/render/lights/light_object.html#power-of-lights. [Accessed 2024].W309W10080
8. M. Eitner and J. Sirohi, "Modal parameter estimation of a reduced-scale rocket nozzle using blind source separation," Measurement Science and Technology, vol. 30, p. 095401, 2019. W309W10080
9. D. Uehara and J. Sirohi, "Full-field optical deformation measurement and operational modal analysis of a flexible rotor blade," Mechanical Systems and Signal Processing, vol. 133, 2019. W309W10080

Chapter 8
Mitigating Speckle Noise while Preserving Impact Response in Vibration Measurements with Laser Doppler Vibrometer on a Moving Platform

Yuanchen Zeng, Alfredo Núñez, and Zili Li

Abstract Laser Doppler Vibrometer measures structural vibrations contactlessly. By using a laser Doppler Vibrometer on a moving platform (LDVom), the vibration of a structure can be measured continuously along a specific path, covering all points on the path in a single scan. Impact responses are prevalent in structural vibrations and useful for modal analysis and structural health monitoring. However, speckle noise can significantly degrade the quality of LDVom signals, so it is essential to remove speckle noise while preserving structural vibrations, including impact responses. However, both impact response and speckle noise are impulsive and transient in the time domain and broadband in the frequency domain, making it difficult to distinguish them and treat them differently in signal processing. To address this challenge, this paper develops a novel denoising method to mitigate speckle noise while preserving impact response in LDVom signals. Instead of directly detecting speckle noise, the proposed method is inspired by conventional wavelet denoising that works in an orthogonal domain but introduces key differences tailored to address the unique properties of speckle noise. Wavelets with shapes similar to speckle noise are utilized, and the strategy of manipulating approximation and detail coefficients based on local noise intensity is proposed. Then, the developed method is validated through vibration measurements in two cases – one with artificial excitations and the other with operational excitations. In the former case, a full-scale slab track sample subjected to manual impact excitations is measured by an LDVom. In the latter case, the rail of the TU Delft V-Track test rig subjected to a moving wheel load over a rail joint is measured by an LDVom. Different scanning speeds are tested in both cases, and the impact responses extracted from the LDVom signals are compared with signals measured by accelerometers mounted on the target structures. The results demonstrate the effectiveness and robustness of the developed method in mitigating speckle noise and preserving impact responses in the LDVom signals measured under different conditions and at different speeds.

Keywords Laser Doppler Vibrometer · Speckle Noise · Impact Response · Wavelet Denoising · Vibration Measurement

Introduction

Laser Doppler Vibrometer (LDV) is widely used for vibration measurements across various fields due to its non-contact nature, broad frequency band, and high sensitivity [1]. In most applications, an LDV remains static with respect to its target structure, such as fixed-point LDV [2], stepped-scanning LDV [3], and continuous-scanning LDV [4]. In contrast, a novel setup is to mount an LDV on a moving platform that moves with respect to the target structure, referred to as LDVom. This setup enables the vibration of a structure to be measured continuously along open paths. A typical example of LDVom is scanning transportation infrastructures using LDVs on a moving vehicle [5]~[8].

In structural vibration measurements, capturing responses to impact loads is of significant interest because they carry the structural dynamics over a wide frequency range and are useful for modal analysis [9], fault diagnosis [10], and structural health monitoring [11]. Impact responses are typically transient in the time domain and broadband in the frequency domain. However, utilizing LDVom to capture such responses introduces a significant challenge – speckle noise. Speckle noise, caused by the change of interference patterns as a laser spot scans a rough surface [12], can significantly degrade the quality

Yuanchen Zeng · Alfredo Núñez · Zili Li
Department of Engineering Structures, Faculty of Civil Engineering and Geosciences, Delft University of Technology, Stevinweg 1, 2628 CN, Delft, The Netherlands
e-mail: Y.Zeng-2@tudelft.nl; A.A.NunezVicencio@tudelft.nl; Z.Li@tudelft.nl

Janko Slavic et al. (eds.), *Computer Vision & Laser Vibrometry, Vol. 6*, Conference Proceedings of the Society for Experimental Mechanics Series, https://doi.org/10.13052/9788743801511-8

of LDVom signals. This challenge is usually addressed through specialized post-processing denoising methods. In [13], speckle noise is detected using the segment kurtosis ratio and then replaced with linear prediction. In [6] and [7], speckle noise is detected by a combination of Hampel identifier and stationary wavelet packet transformation clustering and then replaced through interpolation and smoothing. In [8] and [14], speckle noise is detected based on discrete wavelet transform, replaced with time series models, and then smoothed.

In summary, most existing denoising methods rely on the impulsive nature of speckle noise to detect and replace it in LDVom signals. However, impact responses also exhibit impulsive, transient, and broadband behaviors, making it difficult to distinguish between speckle noise and impact responses. This can lead to denoising errors, including impact responses being mistakenly treated as speckle noise or speckle noise not being fully detected and mitigated.

This paper aims to develop a novel denoising method that can effectively mitigate speckle noise in LDVom signals containing impact responses. Instead of directly detecting speckle noise, the proposed method decomposes an LDVom signal, manipulates the decomposition coefficients, and reconstructs the signal. The motivations behind and details of this method are presented in the Background and Methodology sections, respectively. The developed method is then validated experimentally through vibration measurements using LDVom in two cases – one with artificial excitations and the other with operational excitations. Comparisons are made with accelerometer measurements as well as the denoising results of two existing methods. These results are presented in the Validation Case 1 and Validation Case 2 sections.

Background

The developed method utilizes discrete wavelet transform (DWT), which allows for the separation and manipulation of different signal components in an orthogonal domain [15]. DWT also enables specific dynamic behaviors to be captured using wavelets that have similar shapes to these behaviors [16]. Conventional wavelet denoising generally consists of three steps: **1) decomposing** an original signal using DWT, **2) manipulating** the wavelet coefficients through thresholding, and **3) reconstructing** the signal by performing inverse DWT to the manipulated wavelet coefficients.

In cascade DWT, the raw signal is first decomposed (called level 1), and the approximation coefficient of the last level is decomposed from the second decomposition (called level 2). At each decomposition level, the approximation coefficients carry the main trend (low-frequency components) of the decomposed signal, while the detail coefficients carry the local feature (high-frequency components). Conventional wavelet denoising mainly aims to remove Gaussian white noise [17], so both approximation coefficients and large-magnitude detail coefficients are retained to preserve key features, whereas small-magnitude detail coefficients are removed or suppressed to filter out Gaussian white noise [17][18]. A variety of thresholding strategies have been developed, such as hard thresholding, soft thresholding, and universal threshold selection [19].

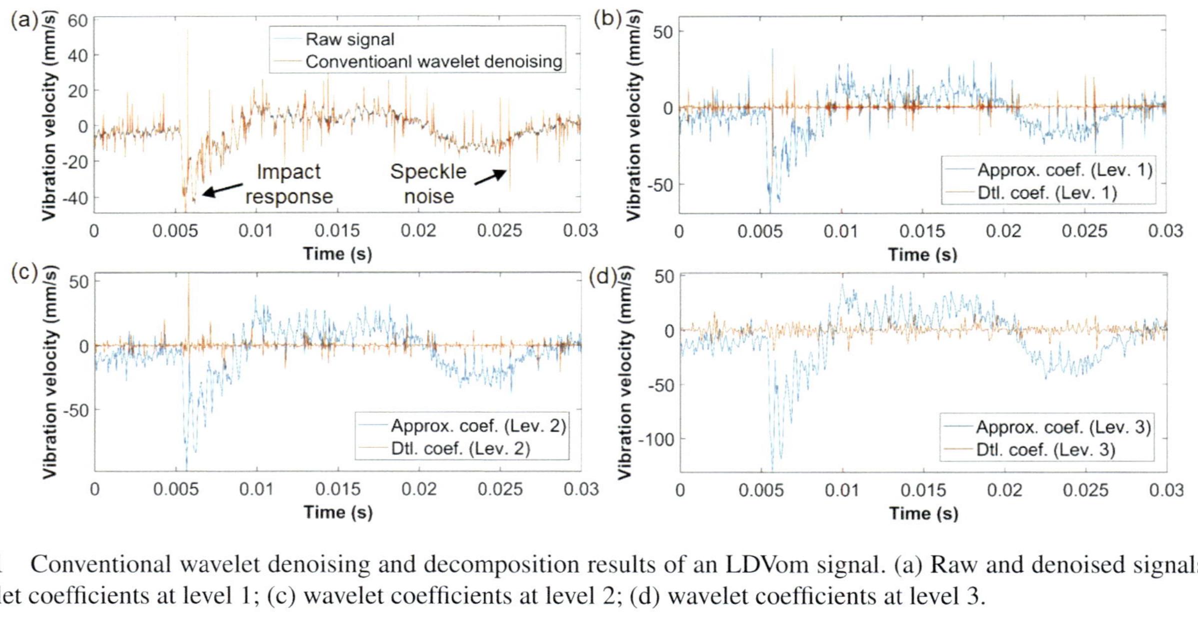

Fig. 1 Conventional wavelet denoising and decomposition results of an LDVom signal. (a) Raw and denoised signals; (b) wavelet coefficients at level 1; (c) wavelet coefficients at level 2; (d) wavelet coefficients at level 3.

However, conventional wavelet denoising has been found to be ineffective in preserving sharp features [17][19]. It has also been reported in [20] that the wavelet denoising fails to remove spikes. These problems reduce the effectiveness of wavelet denoising in mitigating speckle noise while preserving impact responses in LDVom signals. For example, Fig. 1 (a) shows a measured LDVom signal and the result of conventional wavelet denoising. The impact response starts from approximately 5 ms, and speckle noise can be observed throughout the signal. Speckle noise still persists despite using a denoising level of 12 with a Symlet wavelet. The wavelet decomposition results in Fig. 1 (b)~(d) reveal that at each level, the detail coefficient peaks at locations of speckle noise. However, conventional wavelet denoising removes only detail coefficients that are small, leaving speckle noise mostly unaddressed. Moreover, though the detail coefficients indicate the presence of the speckle noise, DWT cannot directly isolate the speckle noise and impact response into different wavelet coefficients: The approximation coefficients carry the impact response, but they are also affected by speckle noise; The detail coefficients at level 3 are also affected by the impact response, and the amplitude of the approximation coefficients is enlarged accordingly.

Methodology

The above analysis shows that speckle noise affects both detail and approximation coefficients in cascade DWT and conventional wavelet denoising is unsuitable for separating speckle noise in LDVom signals. To overcome this limitation, we develop a new wavelet denoising method that manipulates both detail and approximation coefficients at each decomposition level, which can mitigate speckle noise while preserving impact responses in LDVom signals. The steps are as follows.

Step 1. Apply DWT to the raw signal $x(n)$ and obtain the detail coefficients $cD_i(n)$ and the approximation coefficients $cA_i(n)$ for decomposition level i (starting from 1), where n denotes the index of the time series.

$$x(n) \underset{i=1}{\overset{\text{DWT}}{\longrightarrow}} (cD_i(n),\ cA_i(n)) \tag{1}$$

Step 2. Map the detail coefficients $cD_i(n)$ to weights $w_i(n)$ through a weight function $W(\bullet)$ as follows. In general, larger detail coefficients are more affected by speckle noise locally and should be assigned lower weights.

$$w_i(n) = W(cD_i(n)) \tag{2}$$

Step 3. Perform a moving weighted average to both the detail coefficients $cD_i(n)$ and the approximation coefficients $cA_i(n)$ using the weights $w_i(n)$ as follows,

$$cD_i{}^*(n) = \frac{\sum_{j=n-N_{\mathrm{D}i}}^{n+N_{\mathrm{D}i}} cD_i(j)\cdot w_i(j)}{(2N_{\mathrm{D}i}+1)\sum_{j=n-N_{\mathrm{D}i}}^{n+N_{\mathrm{D}i}} w_i(j)} \qquad cA_i{}^*(n) = \frac{\sum_{j=n-N_{\mathrm{A}i}}^{n+N_{\mathrm{A}i}} cA_i(j)\cdot w_i(j)}{(2N_{\mathrm{A}i}+1)\sum_{j=n-N_{\mathrm{A}i}}^{n+N_{\mathrm{A}i}} w_i(j)} \tag{3}$$

where $cD^*{}_i(n)$ and $cA^*{}_i(n)$ denote the manipulated wavelet coefficients and N_{Di} and N_{Ai} represent the half window size for smoothing $cD_i(n)$ and $cA_i(n)$, respectively.

Step 4. Apply DWT to the manipulated approximation coefficient $cA^*{}_i(n)$ and obtain the wavelet coefficients at the next level, denoted again as $cD_i(n)$ and $cA_i(n)$ with i increased by 1.

$$cA^*{}_i(n) \underset{i\leftarrow i+1}{\overset{\text{DWT}}{\longrightarrow}} (cD_i(n),\ cA_i(n)) \tag{4}$$

Step 5. Repeat Steps 2~4 until i exceeds the desired decomposition level I.

Step 6. Reconstruct the signal by applying cascade inverse DWT to the manipulated wavelet coefficients $cA^*{}_I(n)$, $cD^*{}_I(n)$, $cD^*{}_{I-1}(n)$, …, $cD^*{}_1(n)$. Denote the reconstructed signal as $x^*(n)$.

$$cA^*{}_I(n), cD^*{}_I(n), cD^*{}_{I-1}(n), \ldots, cD^*{}_1(n) \overset{\text{Cascade inverse DWT}}{\longrightarrow} x^*(n) \tag{5}$$

This procedure is illustrated in Fig. 2, which follows a similar structure to conventional wavelet denoising but introduces key differences tailored to address the unique properties of speckle noise. These distinctions are essential for more effectively mitigating speckle noise and preserving impact responses in LDVom signals. In the proposed method,

- Both detail and approximation coefficients are manipulated, whereas, in conventional wavelet denoising, only detail coefficients are manipulated.

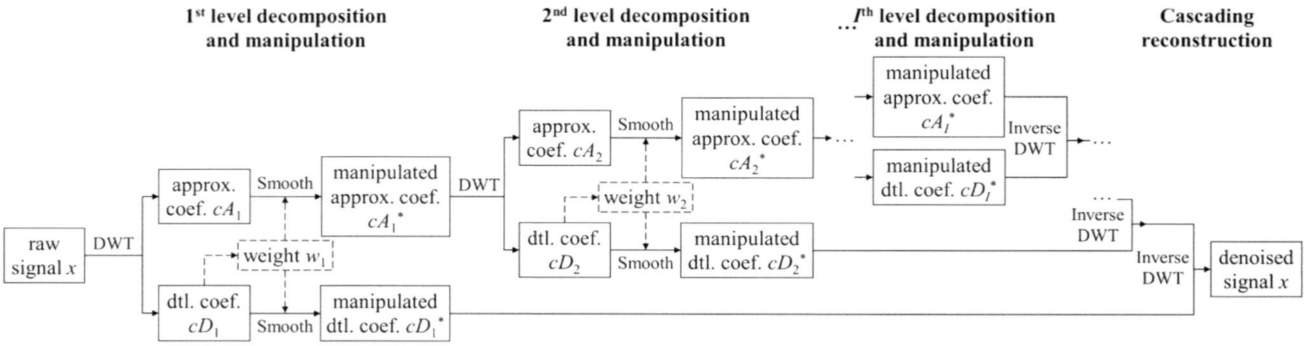

Fig. 2 Illustration of the proposed wavelet denoising method.

- The entire series of wavelet coefficients are manipulated based on the local noise intensity, whereas, in conventional wavelet denoising, only detail coefficients below certain thresholds are manipulated.
- Wavelet coefficients at locations of larger detail coefficients are assigned lower weights and are less influential to the manipulated wavelet coefficients, whereas, in conventional wavelet denoising, smaller detail coefficients are made less influential to the manipulated detail coefficients.

Several key parameters need to be selected properly in the proposed method:

- **Wavelet type**: Different wavelets are suited to capture different dynamic patterns [16]. In the proposed method, the wavelet should provide large detail coefficients exactly at locations of speckle noise. Therefore, wavelets with shapes similar to speckle noise are preferred. Considering the shape of speckle noise, as exemplified in Fig. 3 (a), the Coiflets or Symlets families [21] are recommended. Examples of these wavelets are shown in Fig. 3 (b)~(c).
- **Weighting function**: The weighting function $W(\bullet)$ is designed to assign higher weights $w_i(n)$ to locations with less speckle noise. While detail coefficients $cD_i(n)$ are effective indicators of speckle noise, they cannot perfectly localize speckle noise due to scaling effects, phase shifts, and autocorrelation among neighboring coefficients [19][22]. Therefore, we recommend that the weight at each location should consider the detail coefficients of the location and its neighbors. An example of such a weighting function is given as follows.

$$w_i(n) = \frac{1}{\sum_{j=n-N_{w_i}}^{n+N_{w_i}} cD_i(j)^2} \tag{6}$$

  where N_{wi} represents the number of neighbors considered on each side and it should be no smaller than the decorrelation lag of the detail coefficient as follows.

$$N_{w_i} \geq \min\{k\colon |\mathrm{ACF}_i(k)| < \varepsilon\} \tag{7}$$

  where $\mathrm{ACF}_i(k)$ represents the autocorrelation of $cD_i(n)$ at lag k and ε is a small number.

- **Window sizes**: To effectively smooth the wavelet coefficients without overly compromising local signal behaviors, such as impact responses, we recommend tuning the window sizes based on the following relationship. This relationship suggests that N_{Di} and N_{Ai} should be larger than N_{wi} to incorporate more neighboring points in the moving average process. N_{Di} is recommended to be larger than N_{Ai} as the detail coefficients primarily represent speckle noise and require more aggressive smoothing, whereas the approximation coefficients may have broadband responses and should not be smoothed excessively.

$$N_{w_i} < N_{\mathrm{A}i} < N_{\mathrm{D}i} \tag{8}$$

- **Decomposition level**: The decomposition level I determines the maximum level at which the wavelet coefficients can be manipulated. We recommend increasing it incrementally from level 1 until the denoising result begins to deteriorate, e.g., the impact response gets overly smoothed.

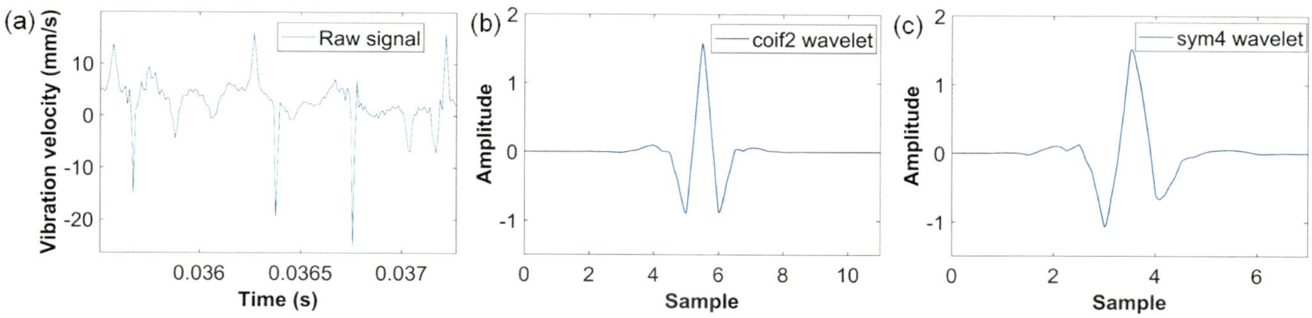

Fig. 3 Selecting wavelet to capture speckle noise into detail coefficients. (a) Examples of speckle noise in a raw LDVom signal; (b) a 2nd-order Coiflet wavelet; (c) a 4th-order Symlet wavelet.

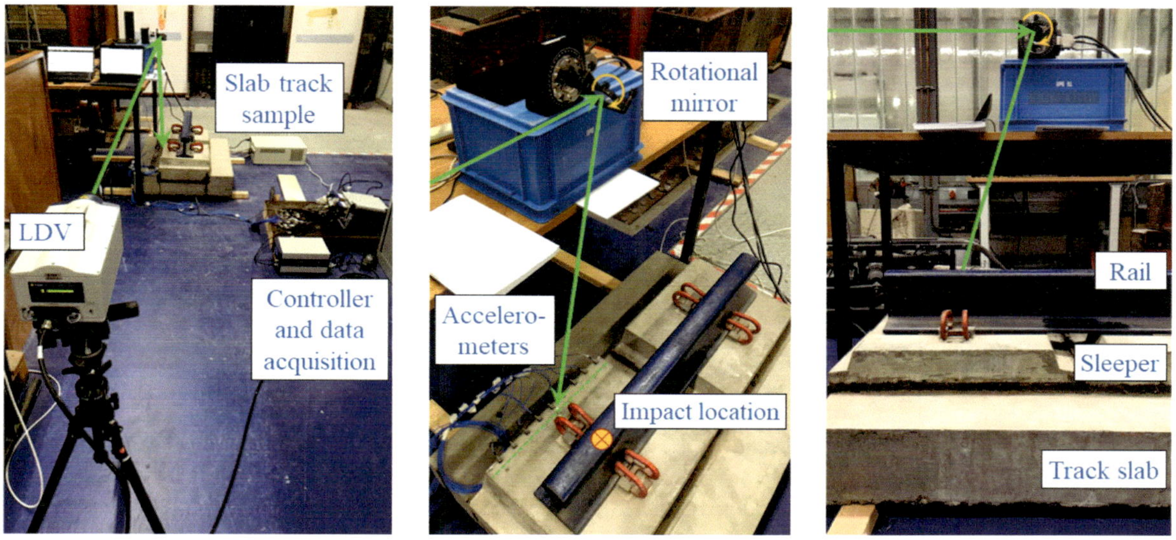

Fig. 4 Experimental setup of Case 1.

Validation Case 1

The developed denoising method is validated through two experimental case studies. In Case 1, a full-scale slab track sample subjected to manual impact excitations is measured by an LDVom. The experimental setup is shown in Fig. 4. An LDV is placed at a distance from the slab track sample, and its laser is directed onto one of the concrete sleepers by a rotational mirror. By controlling the rotation of the mirror, the laser scans the sleeper along its longitudinal direction, resembling LDV scanning from a moving train. The vibration of the sleeper is generated by manually hitting the rail head, exciting the dynamics of the slab track over a wide range of frequencies. At the same time, an array of accelerometers placed along the scanning path measures the sleeper vibration for comparison. Both the LDV and accelerometer signals are acquired through a data acquisition system at a sampling rate of 102.4 kHz.

The measurement and denoising results at two different scanning speeds are shown in Fig. 5. The impact response begins with a sudden drop and then gradually decays with both high-frequency and low-frequency oscillations. Speckle noise with random occurrence and varying amplitude can be observed, which is more severe at the higher scanning speed. The proposed denoising method is applied to the raw signals using the 2nd-order Coiflet wavelet and the parameters N_{wi}=2, N_{Ai}=3, N_{Di}=5, and I=1. A low-pass (LP) filter is applied at the final stage to smooth the signals. It can be seen from the upper plots that the speckle noise is effectively eliminated. The accelerometer closest to the laser spot during the impact is selected, and its signal is converted to velocity through integration. It can be seen that the denoised results closely match the accelerometer response.

Then, the proposed method is compared with two other denoising methods in the lower plots of Fig. 5. The first comparison is with the detection-based method adapted from [14], which detects speckle noise and then replaces it with pre-

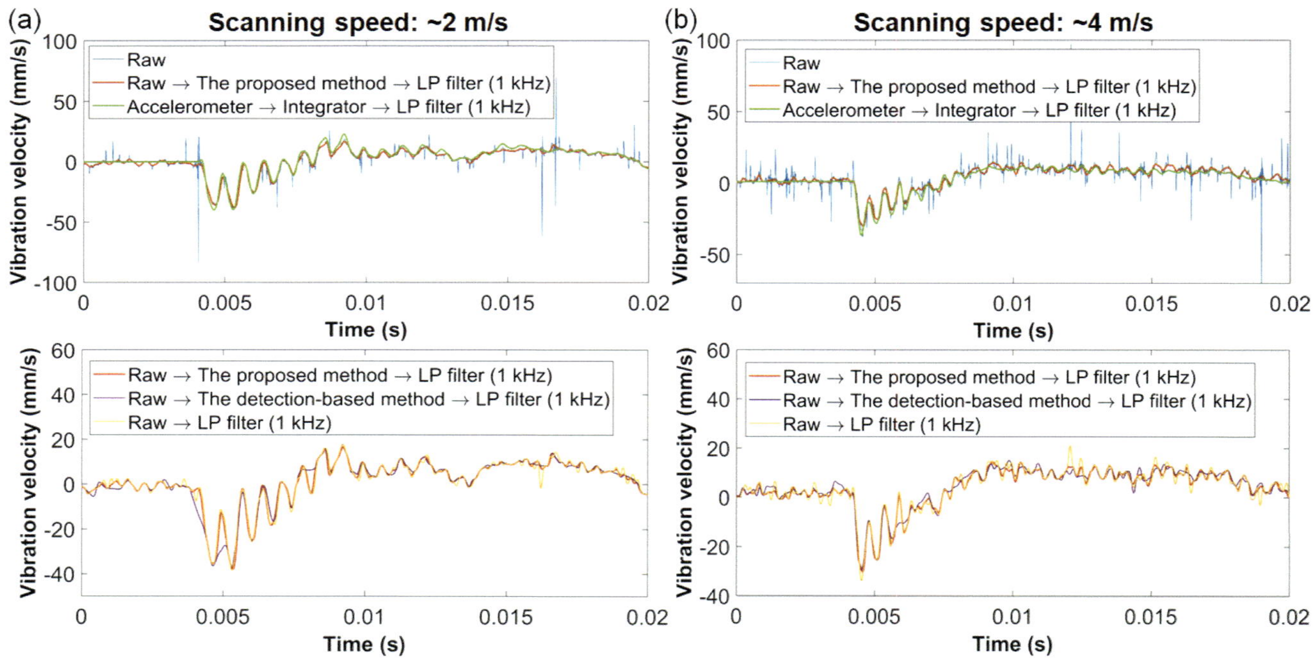

Fig. 5 Measured signals and denoising results in Case 1 (Upper plots: comparison with accelerometer signals; Lower plots: comparison with the other denoising methods). (a) At the scanning speed of around 2 m/s; (b) At the scanning speed of around 4 m/s.

dictions from time series models. In both examples, the impact response is distorted since some points are mistakenly detected as speckle noise and replaced. Meanwhile, the prediction error during the impact response is large due to the strong nonstationarity. For instance, significant speckle noise occurs at the instant of the impact in Fig. 5 (a), but the time series model cannot anticipate the sudden impact, thus producing large prediction errors. The second comparison is with the direct use of the LP filter. While the filter covers the frequency range of the impact response, it also retains some speckle noise components. This smooths large-amplitude speckle noise into vibration-like patterns, inducing additional errors in the denoised signals. In contrast to the two methods, the proposed method reduces speckle noise without distorting the impact response or inducing additional disturbances, demonstrating its effectiveness at different scanning speeds.

Validation Case 2

The 2nd case study extends the validation by testing the denoising method using impact responses measured under operational conditions. It is conducted on the TU Delft V-Track test rig [23], as shown in Fig. 6. The test rig replicates wheel-rail interaction with scaled vehicle and track structures. An LDV is mounted on the test rig, and its laser beam is reflected by a mirror and pointed onto the rail. As the test rig operates, the wheel runs along the rail, and the LDV measures the rail vibration caused by the wheel-rail contact. Impacts occur at local irregularities, such as the rail joint in Fig. 6. The LDVom signal near this rail joint is used to test the developed method. Accelerometers are mounted on the rail near the joint for comparisons. Moreover, an accelerometer is mounted on the mirror to compensate for the influence of the mirror vibration on the LDVom signal. Both the LDV and accelerometer signals are acquired at a sampling rate of 102.4 kHz.

The measurement and denoising results at two different running speeds are shown in Fig. 7, in which the influence of the mirror vibration has been removed. The dynamic response over the joint first begins with a gradual increase, followed by the sudden and decaying impact response. As the running speed increases, the impact response becomes larger, and the speckle noise becomes more severe. The proposed denoising method, using the same wavelet type and parameters as Case 1, is applied to the raw signals. It can be seen that the speckle noise is effectively removed and the denoised signals are close to the vibration measured by the accelerometer on the rail.

The results are further compared with the other two denoising methods in Fig. 7. The detection-based method again distorts the impact responses due to inaccurate detection and prediction. The LP filter retains some components of the speckle noise. The results demonstrate the effectiveness of the proposed method in removing the speckle noise and capturing the impact response under operational conditions. It is noteworthy that the same wavelet type and parameters are used for the two validation cases at different speeds, highlighting the generalizability and robustness of the proposed methods.

Fig. 6 Experimental setup of Case 2.

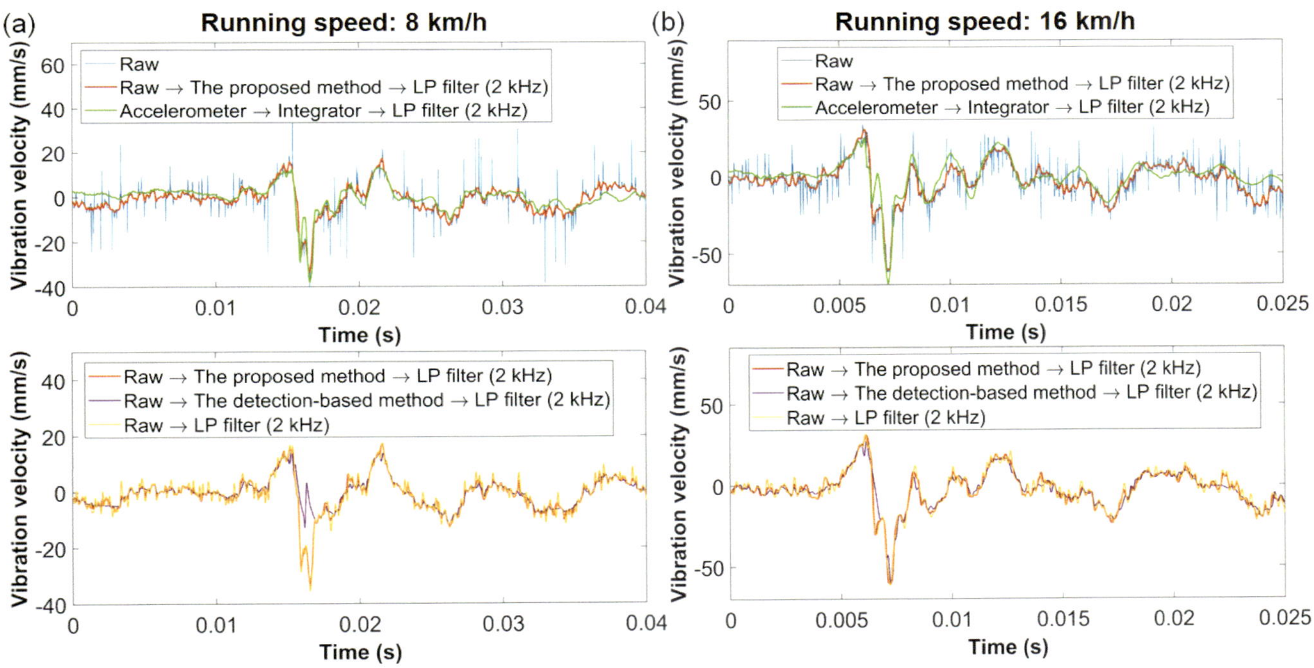

Fig. 7 Measured signals and denoising results in Case 2 (Upper plots: comparison with accelerometer signals; Lower plots: comparison with the other denoising methods). (a) At the running speed of 8 km/h; (b) At the running speed of 16 km/h.

Conclusion

This paper presents a novel wavelet denoising method for mitigating speckle noise while preserving impact responses in LDVom signals. The method adapts conventional wavelet denoising by manipulating both approximation and detail coefficients at each decomposition level to address the unique characteristics of speckle noise. Experimental validation is conducted under varying conditions and scanning speeds. The conclusions are summarized as follows.

- Conventional wavelet denoising is unsuitable since it removes only small detail coefficients, leaving speckle noise mostly unaddressed. DWT cannot effectively separate speckle noise from impact responses.
- Using wavelets with shapes similar to speckle noise, such as the Coiflets or Symlets families, can effectively indicate speckle noise in the time series. Manipulating both approximation and detail coefficients based on the local intensity of speckle noise in the time series can effectively mitigate speckle noise.
- The proposed method can effectively capture the impact responses in the LDVom signals with reduced speckle noise. The results match closely with the accelerometer measurements, and the method outperforms the two denoising methods under comparison.

The proposed method enables impact responses to be accurately captured by LDVom, which can potentially be used in various fields, including defect detection and modal identification. Future research will focus on designing wavelet functions that better resemble speckle noise characteristics and incorporating a smoothing step within the denoising framework.

Acknowledgments This research was partly supported by ProRail and the project Reliable Embankments for Safe Expansion in Rail Traffic (RESET). The authors would like to thank Dr. Pan Zhang for his support in the test on the TU Delft V-Track test rig.

References

1. Tomasini, E. P., Castellini, P. Laser Doppler Vibrometry. Springer Berlin Heidelberg. 2020
2. Lasen, M., Di Maio, D., De Bono, D., & Peluzzo, M. An investigation of fatigue damage growth in composites materials using the vibration response phase decay. arXiv Preprint, 2404.14834. 2024.
3. Chen, Y., & Griffith, D. T. Experimental and numerical full-field displacement and strain characterization of wind turbine blade using a 3D Scanning Laser Doppler Vibrometer. Optics & Laser Technology, 158, 108869. 2023.
4. Yuan, K., & Zhu, W. D. A novel mirror-assisted method for full-field vibration measurement of a hollow cylinder using a three-dimensional continuously scanning laser Doppler vibrometer system. Mechanical Systems and Signal Processing, 216, 111428. 2024.
5. Jiang, L. A., Albota, M. A., Haupt, R. W., Chen, J. G., & Marino, R. M. Laser vibrometry from a moving ground vehicle. Applied Optics, 50(15), 2263–2273. 2011.
6. Yang, C., Kaynardag, K., & Salamone, S. Missing rail fastener detection based on laser Doppler vibrometer measurements. Journal of Nondestructive Evaluation, 42(3), 68. 2023.
7. Kaynardag, K., Yang, C., & Salamone, S. A rail defect detection system based on laser Doppler vibrometer measurements. NDT & E International, 137, 102858. 2023.
8. Zeng, Y., Nunez, A., & Li, Z. Railway sleeper vibration measurement by train-borne laser Doppler vibrometer and its speed-dependent characteristics. Computer-Aided Civil and Infrastructure Engineering, 39(16), 2408–2426. 2024.
9. Zeng, Y., Núñez, A., & Li, Z. Measuring transfer functions of track structures in a test rig with laser Doppler vibrometer and accelerometers on a moving vehicle. Mechanical Systems and Signal Processing, 214, 111392. 2024.
10. Zhang, G., Huang, X., Xu, J., & Li, Z. Adaptive detection of impact signals with two-dimensional piecewise tri-stable stochastic resonance and its application in bearing fault diagnosis. Applied Acoustics, 214, 109702. 2023.
11. Jing, L., Wang, K., & Zhai, W. Impact vibration behavior of railway vehicles: a state-of-the-art overview. Acta Mechanica Sinica, 1–29. 2021.
12. Sracic, M. W., & Allen, M. S. Experimental investigation of the effect of speckle noise on continuous scan laser Doppler vibrometer measurements. In 27th International Modal Analysis Conference (IMAC XXVII). 2009.
13. Lv, T., Han, X., Wu, S., & Li, Y. The effect of speckles noise on the Laser Doppler Vibrometry for remote speech detection. Optics Communications, 440, 117–125. 2019.
14. Zeng, Y., Nunez, A., & Li, Z. Speckle noise reduction for structural vibration measurement with laser Doppler vibrometer on moving platform. Mechanical Systems and Signal Processing, 178, 109196. 2022.
15. Berkner, K., & Wells Jr, R. O. Smoothness estimates for soft-threshold denoising via translation-invariant wavelet transforms. Applied and computational harmonic analysis, 12(1), 1–24. 2002.
16. Junsheng, C., Dejie, Y., & Yu, Y. Application of an impulse response wavelet to fault diagnosis of rolling bearings. Mechanical Systems and Signal Processing, 21(2), 920–929. 2007.
17. Bui, T. D., & Chen, G. Translation-invariant denoising using multiwavelets. IEEE Transactions on Signal Processing, 46(12), 3414–3420. 1998.
18. Berkner, K., & Wells Jr, R. O. Smoothness estimates for soft-threshold denoising via translation-invariant wavelet transforms. Applied and Computational Harmonic Analysis, 12(1), 1–24. 2002.

19. Halidou, A., Mohamadou, Y., Ari, A. A. A., & Zacko, E. J. G. Review of wavelet denoising algorithms. Multimedia Tools and Applications, 82(27), 41539–41569. 2023.
20. Sharan, T. S., Sharma, S., & Sharma, N. Denoising and spike removal from Raman spectra using double density dual-tree complex wavelet transform. Journal of Applied Spectroscopy, 88, 117–124. 2021.
21. Zaeni, A., Kasnalestari, T., & Khayam, U. Application of wavelet transformation symlet type and coiflet type for partial discharge signals denoising. In 2018 5th International Conference on Electric Vehicular Technology (ICEVT), IEEE, 78-82. 2018.
22. Sudha, S., Suresh, G. R., & Sukanesh, R. Speckle noise reduction in ultrasound images by wavelet thresholding based on weighted variance. International journal of computer theory and engineering, 1(1), 2009.
23. Naeimi, M., Li, Z., Petrov, R. H., Sietsma, J., & Dollevoet, R. Development of a new downscale setup for wheel-rail contact experiments under impact loading conditions. Experimental Techniques, 42, 1–17. 2018.

Chapter 9
Neural Radiance Fields for Low-cost, High-resolution 3D Scanning of Critical Infrastructures

Thiago F. Ribeiro, Felipe S. Brandão, Victor H. R. Cardoso, João C. W. A. Costa, Túlio N. Bittencourt, and Moisés F. Silva

Abstract 3D reconstruction of civil and mechanical structures is vital for structural assessment and health management. Often, 3D scan data are integrated into finite element analysis (FEA) to refine geometry, evaluate models, and measure displacement, stress, and loading effects. This process offers a cost-effective way to capture surface geometry and create a digital structure representation, with accuracy heavily dependent on the 3D scanning technology used. Terrestrial laser scanning is frequently employed for non-contact estimation of accurate, high-fidelity geometry data to predict a structure's remaining service life. However, high costs and budget limitations restrict its use in many applications. Multi-View Stereo (MVS) based photogrammetry has many advantages, such as high resolution and the ability to capture complex surfaces. However, it also has drawbacks such as heavy computational loads, high sensitivity to conditions and image quality, and the need for specialized software. The photogrammetric process can be time-consuming, especially with many images. Moreover, these methods generate large data volumes, which can be prohibitive for continuous data acquisition in real-time applications. In counterpart, Neural Radiance Fields (NeRF) based 3D reconstruction models are capable of producing photo-realistic renderings of complex geometries with significantly reduced memory overhead and storage requirements, as it creates an implicit 3D representation of the structure (i.e., the geometry is implicitly embedded into the weights of a neural network). This can be achieved by combining recent advances in inverse neural rendering and volumetric representation of scenes to create a generative model inherently capable of representing the 3D light field from a scene from a small and sparse set of images. In this work, we demonstrate the capabilities of using NeRF-based point cloud extraction to evaluate critical elements of civil infrastructures from sparse sets of smartphone images. We compare this to traditional laser scanning and photogrammetry techniques. We propose a non-contact workflow aimed at providing rapid, extremely low-cost, high-fidelity, and accurate 3D scanning for fast assessment. Our findings show the potential of NeRF-estimated dense point clouds as a viable alternative to MVS photogrammetry and terrestrial laser scanning.This approach offers an efficient, cost-effective method for precise measurement and analysis in engineering and scientific contexts.

Keywords Neural radiance fields · 3D scanning · 3D Point clouds · Volumetric rendering · Building information modeling

Introduction

Extracting geometric data from civil and mechanical structures is fundamental for accurate structural assessment and modeling. Whether it is applied to design, numerical simulations, or operational inspections, 3D scanning and measurement technologies are indispensable tools across the architecture, engineering, and construction (AEC) sectors. With the ongoing

Thiago F. Ribeiro · Victor H. R. Cardoso · João C. W. A. Costa
Applied Electromagnetism Laboratory, Federal University of Pará, Belém, PA, Brazil
e-mail: thiago.figueiro.ribeiro@gmail.com; victorcard@ufpa.br; jweyl@ufpa.br

Felipe S. Brandão · Túlio N. Bittencourt
Department of Structural and Geotechnical Engineering, University of São Paulo, São Paulo 05508-900, Brazil
e-mail: fbrandao2k@usp.br; tbitten@usp.br

Moisés F. Silv
Los Alamos National Laboratory, Los Alamos, New Mexico 87544, USA
e-mail: cien.felipe@gmail.com

Janko Slavic et al. (eds.), *Computer Vision & Laser Vibrometry, Vol. 6*, Conference Proceedings of the Society for Experimental Mechanics Series, https://doi.org/10.13052/9788743801511_9

deterioration of key infrastructures due to the escalating effects of climate change, there is an increasing need for structural assessments that are rapid, efficient, and cost-effective [1]. These assessments are vital not only to ensure safety but also to optimize maintenance efforts and prolong the lifespan of infrastructure assets.

The rapid evolution of non-contact sensing technologies for structural health monitoring (SHM) has expanded the use of cameras, smartphones, unmanned aerial vehicles (UAVs), satellites, ultrasonic devices, and other imaging sensors for conducting structural assessments [2]. A significant number of these technologies rely on the generation of 3D point clouds, which are collections of spatial data points that describe the shape and surface characteristics of structures [3, 4]. The versatility of point cloud data extends far beyond visual representation; it allows engineers to derive 3D models that highlight discrepancies between projected models and as-built structures, facilitating critical evaluations of construction quality, structural deformation, and deviations from original design specifications [5]. Additionally, point cloud data has become essential for monitoring construction progress, assessing geometry quality, and detecting early-stage damage, such as cracks, in buildings and civil infrastructures.

Among the numerous techniques for non-contact geometry extraction, terrestrial laser scanning (TLS) stands out as one of the most widely used. It is predominantly based on Light Detection and Ranging (LiDAR) technology, which uses pulses of laser light to measure distances between the scanner and surrounding objects. LiDAR operates similarly to Radar (Radio Detection and Ranging), but instead of using radio waves, it relies on discrete pulses of laser light [6]. This allows for high-precision distance measurements, which are crucial for producing accurate 3D models of complex structures. However, the advanced technological complexity of LiDAR systems, which includes high-quality optical components and precise rotational mechanisms, results in considerable costs. The Ouster OS1-64 used by AARON et al. (2023) [7] retails at €17,424.00 as of September 2024. Consumer-grade LiDARs such as the Livox Mid-40, which retails for $599 as of September 2024, trade higher accuracy for lower prices as noted by ARTEAGA et al. (2019) [8].

Despite their effectiveness, the high cost and technical challenges associated with LiDAR limit its accessibility for routine inspections, particularly for smaller projects or organizations with limited budgets. To address these cost-related barriers, recent research has explored alternative methods for 3D geometry extraction. One promising approach involves the use of Neural Radiance Fields (NeRF), a machine learning-based technique that generates high-fidelity 3D models from sparse sets of images, such as those captured by standard consumer-grade cameras, including smartphones. REMONDINO et al. (2023) [9] has shown a generalized capacity of 3D reconstruction using NeRF. HUANG et al. (2024) [10] recently demonstrated the successful application of NeRF in reconstructing individual trees in 3D with remarkable accuracy. Although NeRF has primarily been applied in fields like computer graphics and environmental modeling, its potential for civil and mechanical infrastructure assessment remains largely unexplored. As far as the authors knowledge goes this is the first study and application of NeRFs with a focus on civil infrastructures. We aim to demonstrate that NeRF, trained on a limited set of photos taken with inexpensive cameras, can generate 3D models that rival those produced by more expensive and complex systems like LiDAR and photogrammetry. By conducting a detailed comparison with alternative methods, such as stabilished LiDAR systems and traditional photogrammetry, we seek to provide a comprehensive evaluation of NeRF's viability for large-scale structural assessments. The ultimate goal is to offer a more accessible, scalable, and accurate approach to infrastructure monitoring, one that addresses the growing demands of the AEC industry while reducing costs and technological barriers.

3d Reconstruction Techniques

Lidar

Laser-based ranging, profiling, and scanning have been subjects of research since the 1960s. Two primary techniques are employed to measure the distance to an object (Figure 1). The first method involves precisely measuring the Time-of-Flight (ToF) of a short but intense laser pulse emitted from the instrument, which is then reflected back from the object's surface. Given the precise knowledge of the speed of light, the accuracy or resolution of the distance measurement is determined by the precision of the time measurement [11]. In the second method, distance is determined by accurately measuring the phase difference between the emitted signal and the signal received after reflecting off the object. Modern LiDAR systems scan objects by sampling a series of measured distances, capturing both vertical and azimuthal angles. Each sampled point is digitally recorded and stored, creating a detailed 3D representation of the scene.

Photogrammetry

Photogrammetry is a technique that involves creating a three-dimensional model of an object, structure, or environment by analyzing two-dimensional images taken from multiple angles. The process starts with detecting and matching features

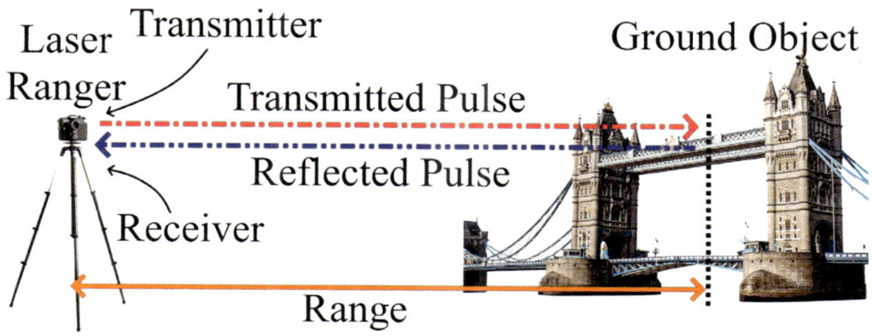

Fig. 1 Operation of a laser rangefinder that is using the timed pulse or TOF method. Adapted from SHAN et al. (2018) [11]

between the images using algorithms like Scale Invariant Feature Transform (SIFT) [12] (Figure 2). Next, Structure-from-Motion (SfM) techniques [13] are applied to estimate the camera's position and orientation for each image. Using these camera poses, the 3D coordinates of the matched feature points are determined through triangulation, producing an initial sparse point cloud. To improve the alignment of point clouds, the Iterative Closest Point (ICP) algorithm [14] is often used. Afterward, Multi-View Stereo (MVS) techniques [15] are applied to produce a dense point cloud. For each photo, a depth map is generated by triangulating matching features from multiple images, which represents the distance of each pixel from the camera. These depth maps are subsequently aligned and merged. Since every pixel in the depth map corresponds to a 3D point, and its distance to the camera is known, this approach effectively generates a dense 3D point cloud.

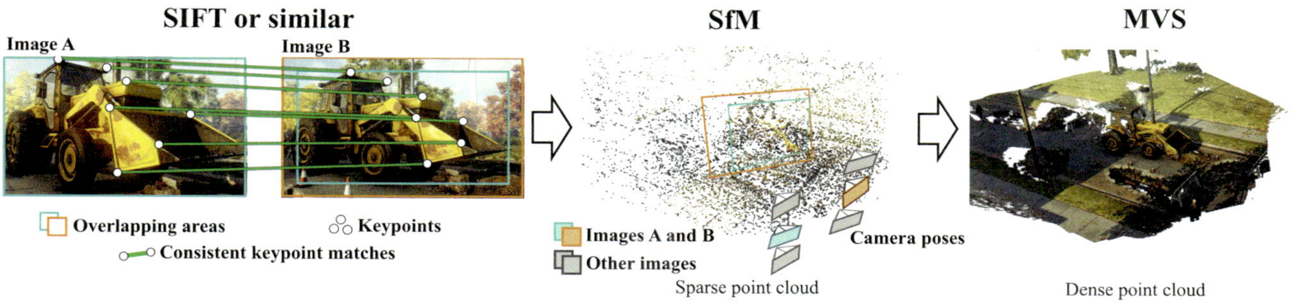

Fig. 2 The three key stages in a SfM-MVS workflow illustrated on two hypothetical images of a construction truck: (1) keypoint identification and matching (e.g. SIFT), (2) SfM with camera parameters and a sparse point cloud as output and (3) the densified point cloud following MVS. Adapted from IGLHAUT et al. (2018) [16]

NeRF

In contrast to the conventional LiDAR-based approaches, NeRFs are considered an implicit representation, as it stores the 3D lightfield of a given scene into the weights of a neural network [17, 18]. In this method, a static scene is represented as a continuous 5D function that outputs the radiance emitted in each direction, (θ, ϕ), and each point in the 3D coordinate system, (x, y, z). Ray samples are created from each pixel using a classical pinhole camera model using conventional ray tracing. Thus, the overall goal is, for each ray sample, to learn an RGB color and a density value σ at the 3D location, acting as a differential opacity, controlling how much radiance is accumulated by a ray passing through that point (x, y, z). Later, volume rendering algorithms can be applied over all samples to reconstruct the final pixel color of an image rendered from a given camera angle [19, 20]. The NeRF architecture first processes the input with a multilayer perceptron (MLP) with 9 fully connected layers (ReLU activated, 256 neurons each) with skip connections implemented in the fifth and ninth layers where positional information [21] is integrated (Figure 4). The ninth layer outputs σ. A tenth, 128 neurons-wide, ReLU layer processes the remaining 256-dimensional feature vector. A final Sigmoid activated layer outputs the emitted RGB radiance at position x, as viewed by a ray with direction d [19].

NeRFs perform the reconstruction of color and opacity of ray samples using an MLP to represent the lightfield of the scene, which regresses a 5D coordinate-space (x, y, z, θ, ϕ) to the appropriate σ and view-dependent color $C = (r, g, b)$,

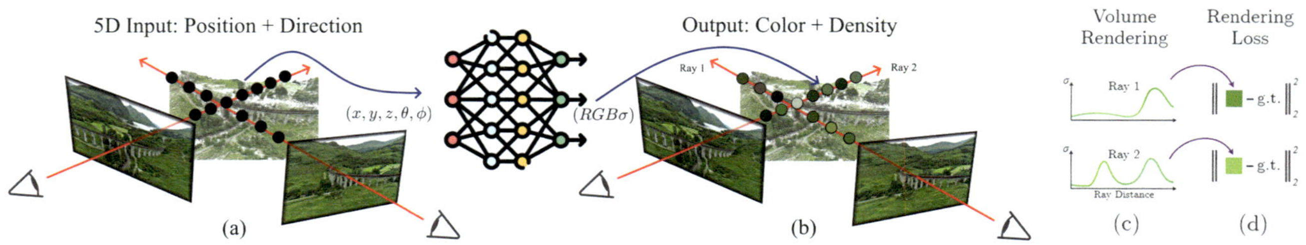

Fig. 3 An overview of the neural radiance field scene representation and differentiable rendering procedure. Images are synthesized by sampling 5D coordinates (location and viewing direction) along camera rays (a), feeding those locations into an MLP to produce a color and volume density (b), and using volume rendering techniques to composite these values into an image (c). This rendering function is differentiable, so optimization of the scene can be done by representation by minimizing the residual between synthesized and ground truth observed images (d). Adapted from MILDENHALL et al. (2020) [19]

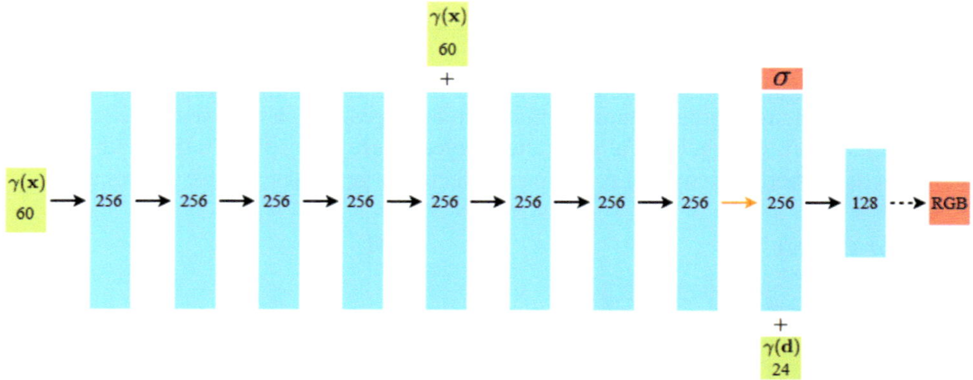

Fig. 4 Visualization of the NeRF's fully-connected network architecture. Adapted from MILDENHALL et al. (2020) [19]

$$C \approx \sum_{i=1}^{N} T_i \alpha_i c_i, \tag{1}$$

where N represents the number of samples along a given ray, T_i is the accumulated transmittance from the camera to the i-th sample point, α_i represents the opacity at the i-th point, which is related to the volume density σ by $\alpha_i = 1 - \exp(-\sigma_i \delta_i)$, and c_i represents the color of the i-th point, expressed as an RGB vector. In this context, a high volume density σ_i will increase the opacity α_i, making the color c_i of that point more significant in the final rendered color C.

The formula states that the accumulated density between the near bound t_n and t will decrease the importance of whatever lies between t and the far bound. The function $T(t)$ denotes the accumulated transmittance along the ray from t_n to t, i.e., the probability that the ray travels from t_n to t without hitting any other particle [19]. In other words, the opaquer surfaces between the near bound and position t, the less light will go through that space, and the lower the accumulated transmittance be at position t are measured. Therefore, the resulting color for this ray on the image, C, is a linear integral over the scene's domain of the color function weighted by the density and accumulated transmittance [22]. The loss function of the NeRF is the total squared error between the rendered and true pixel colors for both the coarse ($N = 64$) and fine ($N = 192$) renderings:

$$L = \sum_{r \in R} \left[\|\hat{C}_c(r) - C(r)\|_2^2 + \|\hat{C}_f(r) - C(r)\|_2^2 \right]. \tag{2}$$

NeRFs are trained on batches of rays rather than entire images. R represents the set of rays in each training batch. For each ray r in a batch, $C(r)$ denotes the ground truth RGB color, while $\hat{C}_c(r)$ and $\hat{C}_f(r)$ represent the predicted RGB colors from the coarse and fine models, respectively. The training aims minimizing the difference between these predicted colors and the ground truth, ensuring that both coarse and fine models learn to accurately reconstruct the scene. The input data for the training of a NeRF must include the spatial location of each camera, a 5D coordinate (x, y, z, θ, ϕ) tensor. In computer-generated images, this process is simpler since you already have this information in order to generate the image. In contrast, to generate the input data for real-world photos, we use Structure from Motion (SfM) algorithms [13]. They use triangulation techniques to estimate the three-dimensional coordinates and angles from each input photo.

Performance Metrics

Cloud to Cloud Comparison

When comparing distances between cloud points, different techniques can be used, such as nearest neighbor and nearest neighbor with local modeling. The first measures the distance between a 3d point to the nearest euclidean point of the other cloud. However, since point clouds are discrete models, this distance can be inaccurate when compared to the true continuous reality. This problem can be mitigated when using local modeling. Instead of using a single point and measuring distance, the K closest points (K Nearest Neighbors or KNN) are selected and a spatial plane is calculated from these points using linear regression. Then, the height of this plane and the initial point is measured. This height will be the distance using KNN. This technique gives a continuity to the compared model, but introduces errors if the model is indeed discontinuous in a region. Similarly to Elkhrachy (2020) [23], this paper considers the Mean value and Root Mean Square (RMS) of all differences to assess the accuracy of the model, while the Standard Deviation (Std. Dev.) was used as an indicator of data noise.

Level of Accuracy (LOA)

The Level of Accuracy (LOA) is a scale that is used to determine how precise a model should be to the point cloud [24]. The LOA scale was developed by the United States Institute of Building Documentation (USIBD) to standardize specifications industry-wide. The LOA standard is structured in five increments of ten, beginning with LOA10 (lower precision) and extending through LOA50 (higher precision). We will examine LOA30 (with an error margin of up to 15 mm), LOA40 (up to 5 mm), and LOA50 (up to 1 mm) for the smaller structure (USP Bridge). Since the distance from the camera to the Octavio Frias Bridge is approximately 100 times greater than that to the USP Bridge, we anticipate that the LoA will be affected by the same factor. Therefore, we will use LOA30 x 100 (with an error margin of up to 1500 mm), LOA40 x 100 (up to 500 mm), and LOA50 x 100 (up to 100 mm) to evaluate the errors between the reconstruction models of this bridge and the ground truth.

Objects of Study

The USP Footbridge

The University of São Paulo (USP) Bridge is a footbridge over the Tejo stream, located between the Mining Engineering and Civil Engineering buildings on the USP-SP campus, São Paulo, Brazil. The bridge has a deck with a span of approximately 12m made up of cross-section timber pieces measuring 7 cm x 16 cm, transversely prestressed, covered with reinforced concrete, and a prestressing system consisting of a metal diverter that rests under the deck in the middle of the span and tendons that in turn fix the diverter and are anchored at the ends of the deck. The superstructure is supported by reinforced concrete blocks that were recovered from a pre-existing structure on the site, and at the access on the Minas side there is a precast reinforced concrete slab approximately 2 meters long allocated to span the gap between the support pillars and the abutment [25].

The Cable-stayed Octavio Frias Bridge

The Octavio Frias de Oliveira Cable-stayed Bridge is in the Rela Parque Roadway Complex and cross the Pinheiros River at the end of Jornalist Robert Marinho Avenue, in São Paulo, Brazil [26]. It is constituted by two curved ramps giving direct access between the Journalist Roberto Marinho Avenue and the expressway of the Nações Unidas Avenue. The cable-stayed decks present a curvature in plan with a constant radius, with 140-meter spans to cross the Nacoes Unidas Avenue and the CPTM trains, and 150-meter spans across the Pinheiros River. Inaugurated in May 2008, it stands out for its unique "X" shape, with two curved decks that intersect around a single concrete tower standing 138 meters high. The bridge is named after Octavio Frias de Oliveira, a prominent Brazilian businessman and former owner of the newspaper Folha de S.Paulo. This structure stands out for its innovative design, which serves not only as a crucial connection between Marginal Pinheiros

Fig. 5 The USP Bridge over the Tejo stream, São Paulo, Brazil

Fig. 6 The Octavio Frias de Oliveira Cable-stayed Bridge crossing the Pinheiros River

and the surrounding regions but also as a symbol of modernity and architectural excellence in São Paulo. At night, the bridge is often illuminated in vibrant colors, making it a key landmark in the city's skyline.

Experiments

We captured 348 photos of the USP Bridge from various azimuths and used them as input for photogrammetry and NeRF-based point cloud generation. All photos were taken at a resolution of 960 x 1280 px. Figure 7 shows the camera angles for each image. The resulting point clouds were compared to a LiDAR-based point cloud generated by a Leica RTC360, a high-performance 3D laser scanner widely employed in industries such as construction and architecture. The RTC360 point cloud was considered the ground truth, providing a reliable benchmark for accuracy and precision in the comparison. This comparison was performed using K-Nearest Neighbors (K = 6) local modeling. To investigate the behavior of the errors between our models and the ground truth, we repeated the experiment with progressively fewer photos, generating point

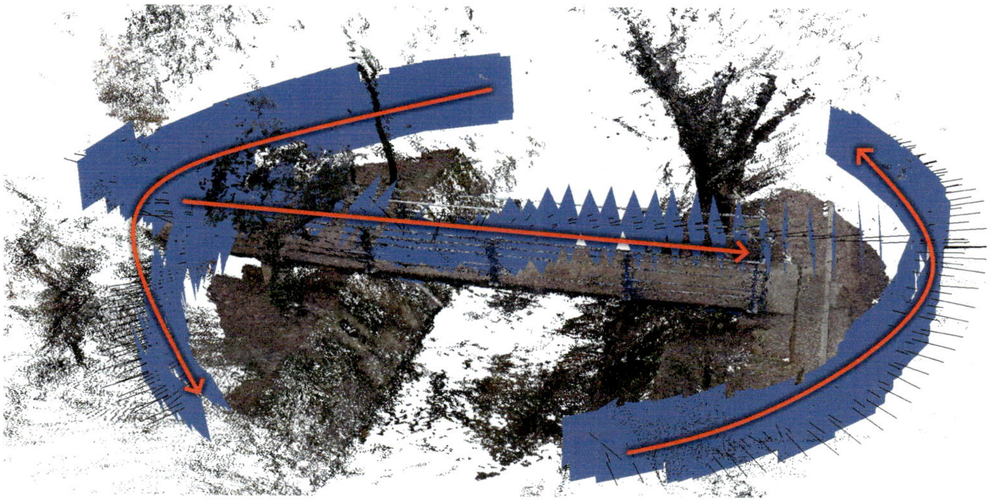

Fig. 7 Camera angles and positions of the photos taken over the USP Bridge.

Fig. 8 Camera angles of the photos taken by the drone over the Octavio Frias Bridge. Source: Google Maps

clouds from 75% (261 photos), 50% (174 photos), 25% (87 photos), 10% (35 photos), and 5% (17 photos) of the original dataset.

For the Octavio Frias Bridge, we extracted 500 equally spaced frames from a YouTube video. Figure 8 shows the camera angles for each frame. The frames, originally in UHD 4K resolution, were downsampled to Full HD (1080p) and used as input for photogrammetry and NeRF-based point cloud generation. These models were then compared to the LiDAR-based point cloud, which was used as the ground truth. Additionally, we created point cloud models using progressively fewer frames, ranging from 100% of the original 500 frames down to 2.5%, to compare their accuracy against the ground truth.

The distance-errors between models and the LiDAR-Ground Truth are presented as color-coded point clouds, where the color bar indicates the magnitude of error from blue (lower error) to red (higher error).

Results

The USP Footbridge

Figure 9 (B) shows a comparison between NeRF and photogrammetry models of the USP Bridge. NeRF manages to reconstruct the bridge with high accuracy from 100% down to 25%, preserving structural and visual details. 25% some degradation starts occurring, especially in the handrails. The 10% model shows reduction on the density of the model, with gaps forming

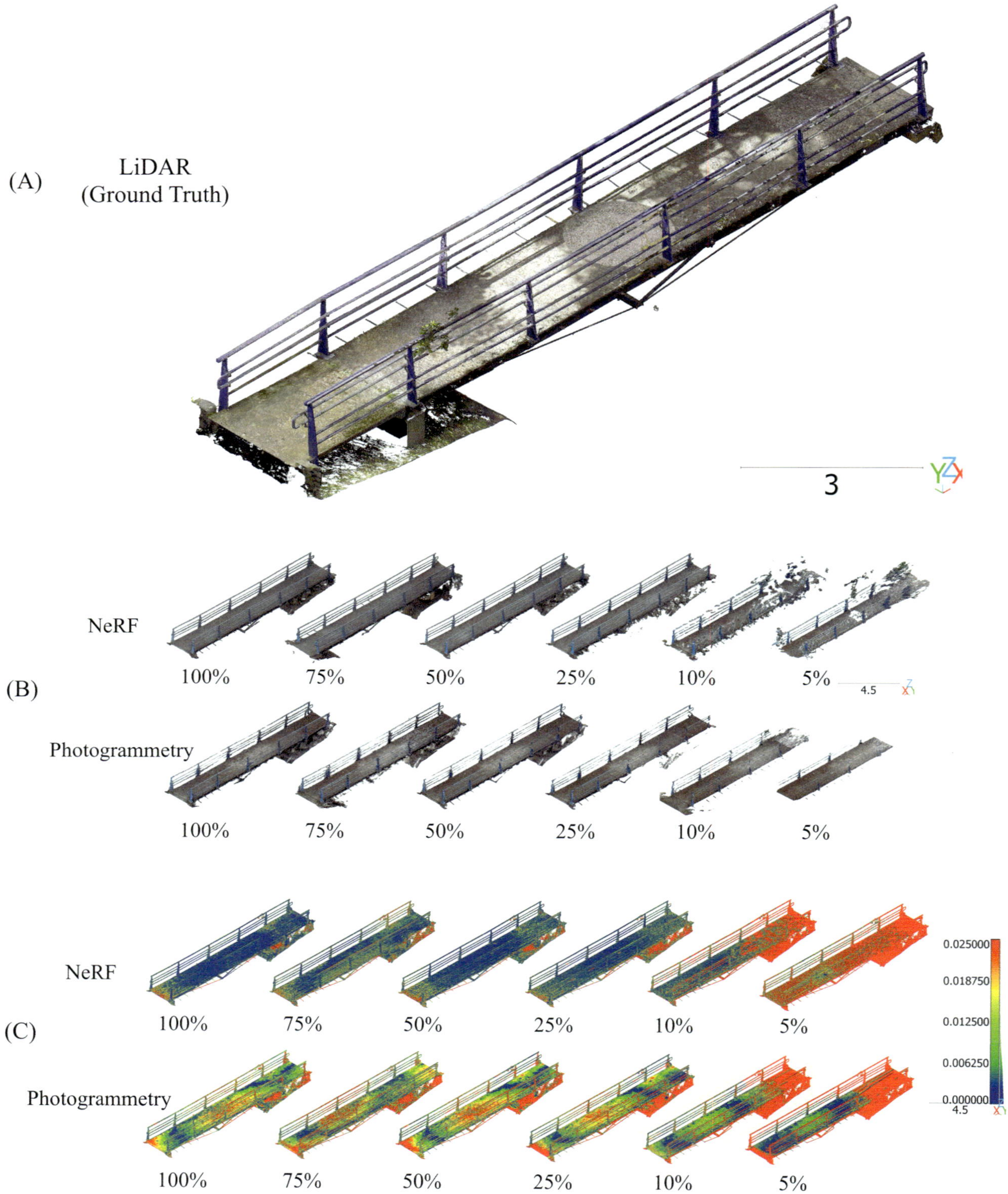

Fig. 9 (A) LiDAR based point cloud from the USP Bridge. (B) RGB NeRF-based and Photogrammetry-based point clouds created utilizing progressively fewer pictures of the USP Bridge. (C) Models compared to the LiDAR-based ground truth using Nearest Neighbors (KNN = 6) local modeling using progressively fewer pictures of the USP Bridge.

on the bridge deck. Finally the 5% model shows large portions of the structure missing. The photogrammetry model provides accurate reconstruction up to 50%, where some parts of the structure start disappearing. 25% shows gaps in the rails. 10% shows parts of the structure missing and 5% shows greatly reduced details on the rails, with only part of the bridge deck maintaining a coherent shape.

Figure 9 (C) shows photogrammetry and NeRF reconstruction models errors when compared to the LiDAR ground truth as the amount of available data is progressively reduced, ranging from 100% down to 5%. NeRF has a mostly blue representation up to 25%, showing a close proximity with the ground truth model. On the other hand, the photogrammetry demonstrates noticeable higher errors on the same range, with greener and red colors on the bridge deck. At 10% and especially 5% data availability, the photogrammetry reconstruction becomes significantly fragmented and incomplete. Despite this, unlike NeRF, at 5% the photogrammetry model maintains a more coherent representation.

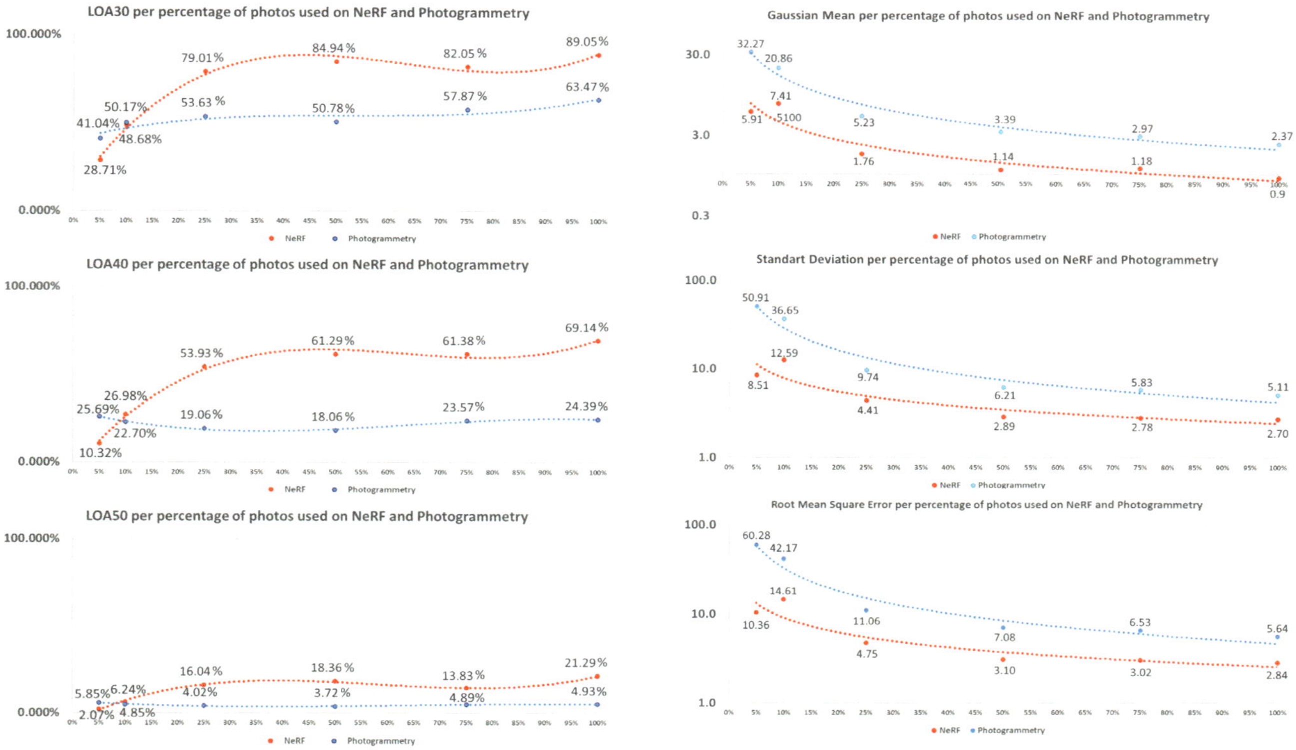

Fig. 10 Left column: LOA30, 40 and 50 by percentage of used photos when creating NeRF and Photogrammetry models (Higher is better). Right Column: Gaussian mean, Standart Deviation and Root Mean Square Error by percentage of used photos when creating NeRF and Photogrammetry models of the USP Bridge (Lower is better).

When comparing Levels of Accuracy, NeRF consistently outperforms photogrammetry at all percentages of photos, except for 5% (Figure 10). When using 100% of the photos, NeRF achieves an accuracy of 89.05% for LOA30 ($\leq$ 1.5 cm), whereas photogrammetry only reaches 63.47%. As the number of photos decreases, NeRF maintains a higher level of accuracy relative to photogrammetry. However, with as little as 5% of photos, photogrammetry performs better accuracy for LOA30, LOA40 and LOA50, compared to NeRF, but the gap narrows as the photo count decreases.

Moreover, NeRF maintains a lower Gaussian mean and RMSE compared to photogrammetry at higher percentages of photos, but both methods exhibit a sharp rise in error at lower photo percentages. With 100% of photos, NeRF's RMSE is 2.85 cm, significantly lower than photogrammetry's 5.64 cm. However, with only 5% of photos, NeRF's RMSE rises to 10.37 cm, while photogrammetry's skyrockets to 60.28 cm, indicating a much larger degradation in performance for photogrammetry.

The Cable-stayed Octavio Frias Bridge

Figure 11 (B) shows a completeness comparison between NeRF and photogrammetry reconstructions as the amount of data is progressively reduced. Both methods display reconstruction models of the Octavio Frias Bridge at different levels of data

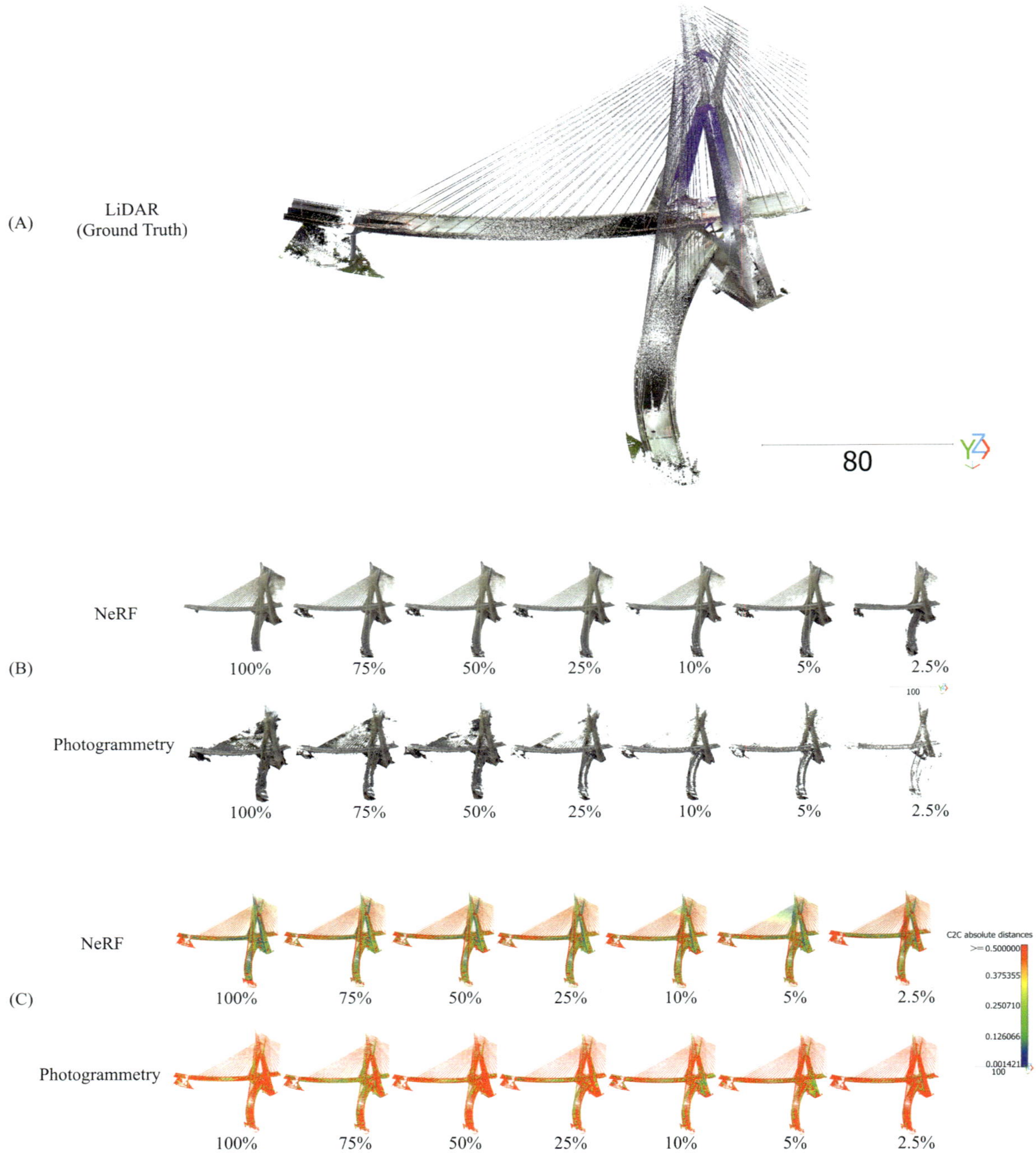

Fig. 11 (A) LiDAR based point cloud from the Octavio Frias Bridge. (B) RGB NeRF-based and Photogrammetry-based point clouds created utilizing progressively fewer pictures of the Octavio Frias Bridge. (C) Models compared to the LiDAR-based ground truth using Nearest Neighbors (KNN = 6) local modeling using progressively fewer pictures of the Octavio Frias Bridge

availability, ranging from 100% down to 5%. For NeRF, even at low percentages (2.5% and 5%), the model retains its structural integrity with minimal detail loss. However, photogrammetry exhibits significant degradation as the data percentage decreases. By the time only 10% or less of the data is available, the photogrammetry reconstruction becomes notably fragmented and incomplete, whereas NeRF maintains a more coherent representation of the bridge structure.

Figure 11 (C) presents a comparison between NeRF and photogrammetry based on error measurements across varying data availability levels. In the NeRF row, the reconstructions show relatively lower error, with most points in green and yellow, indicating smaller deviations from the reference model. Even at reduced data levels (2.5%), the NeRF model maintains lower errors, with fewer regions appearing in red. In contrast, the photogrammetry results exhibit a higher error distribution as the available data decreases. While the 100% and 75% models still maintain some accuracy (green and yellow regions), the 50%, 10%, 5%, and 2.5% models show a significant increase in red regions, indicating larger errors.

In terms of $LOA \times 100$ (Figure 12), NeRF scores fluctuate depending on the percentage of photos used, with higher percentages generally leading to higher $LOA \times 100$, except for the lowest error threshold ($\leq 10cm$), where the $LOA \times 100$ remains lower. The highest $LOA \times 100$ of 88.62% is observed when 5% of photos are used at $LOA30 \times 100$, but lower thresholds ($\leq 50cm$ and $\leq 10cm$) tend to perform better with a higher percentage of photos. For Photogrammetry, the trend is more consistent, with a drop in $LOA \times 100$ as fewer photos are used. Notably, Photogrammetry performs worse than NeRF across all $LOA \times 100$ thresholds, particularly for $LOA50 \times 100$ ($\leq 10cm$), where values remain below 3%, highlighting its difficulty in achieving low-error reconstructions.

NeRF displays more stability in terms of mean and error metrics as the percentage of photos decreases. Its lowest RMSE (1.14) occurs with 5% of photos, demonstrating better error tolerance than Photogrammetry at this level. Photogrammetry, however, shows higher error measures, particularly at lower photo percentages. When only 25% or 2.5% of photos are used, the RMSE increases sharply, reaching 4.55 and 4.66, respectively. This indicates that Photogrammetry struggles with fewer photos, becoming significantly less accurate compared to NeRF.

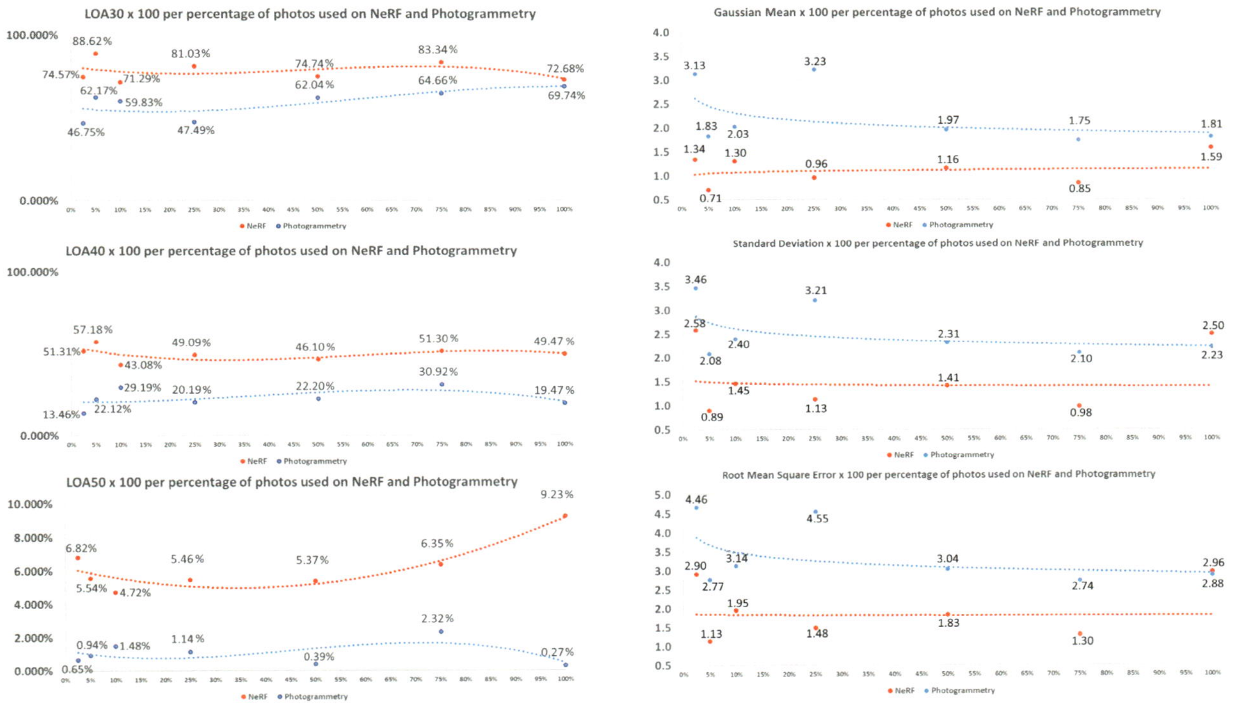

Fig. 12 Left column: LOA30 x 100, LOA40 x 100 and LOA50 x 100 by percentage of used photos when creating NeRF and Photogrammetry models (Higher is better). Right Column: Gaussian mean x 100, Standart Deviation x 100 and Root Mean Square Error x 100 by percentage of used photos when creating NeRF and Photogrammetry models of the Octavio Frias Bridge (Lower is better).

Conclusion

We demonstrate clear differences in the performance of NeRF and photogrammetry as the percentage of input photos decreases, impacting both LOA (Level of Accuracy) metrics and error measurements. NeRF maintains a higher level of detail and exhibits lower errors as data decreases, showing greater resilience to data reduction compared to photogrammetry.

Although photogrammetry retains some structural coherence at very low data levels (such as 5%), this comes at the cost of significantly higher errors and loss of detail, making it less reliable for high-fidelity reconstruction. Overall, NeRF proves to be the more robust and accurate method, especially when balancing data availability with reconstruction quality, positioning it as the preferred choice for 3D modeling in data-limited scenarios.

Acknowledgments This paper was made with the support of the Fundação Amazônia de Amparo a Estudos e Pesquisas (Fapespa), Process 2023/592598, Grant agreement 050/2023, the Fundação de Amparo à Pesquisa do Estado de São Paulo (Fapesp), Process 2022/10105-3 and CNPq.

References

1. Chalmers, P. "Climate change: Implications for buildings. key findings from the intergovernmental panel on climate change". Technical report, University of Cambridge (2014) Fifth Assessment Report.
2. Kaartinen, E., Dunphy, K., and Sadhu, A. "LiDAR-based structural health monitoring: Applications in civil infrastructure systems". *Sensors (Basel)*, 22(12):4610 (2022)
3. Lee, J. and Kim, R.E. "Noncontact dynamic displacements measurements for structural identification using a multi-channel lidar". *Struct. Contr. Health Monit.*, 29(11) (2022)
4. Jo, H.C., Sohn, H.G., and Lim, Y.M. "A LiDAR point cloud data-based method for evaluating strain on a curved steel plate subjected to lateral pressure". *Sensors (Basel)*, 20(3):721 (2020)
5. Wang, Q. and Kim, M.K. "Applications of 3D point cloud data in the construction industry: A fifteen-year review from 2004 to 2018". *Adv. Eng. Inform.*, 39:306–319 (2019)
6. Carter, J., Schmid, K., Waters, K., Betzhold, L., Hardley, B., Mataosky, R., et al. *LiDAR 101: An Introduction to LiDAR Technology, Data, and Applications*. National Oceanic and Atmospheric Administration (NOAA) Costal Services Center, Charleston, Silver Spring, MD (2012)
7. Aaron, J., Spielmann, R., McArdell, B.W., and Graf, C. "High-frequency 3D LiDAR measurements of a debris flow: A novel method to investigate the dynamics of full-scale events in the field". *Geophys. Res. Lett.*, 50(5) (2023)
8. Ortiz Arteaga, A., Scott, D., and Boehm, J. "Initial investigation of a low-cost automotive lidar system". *ISPRS - Int. Arch. Photogramm. Remote Sens. Spat. Inf. Sci.*, XLII-2/W17:233–240 (2019)
9. Remondino, F., Karami, A., Yan, Z., Mazzacca, G., Rigon, S., and Qin, R. "A critical analysis of NeRF-based 3D reconstruction". *Remote Sens. (Basel)*, 15(14):3585 (2023)
10. Huang, H., Tian, G., and Chen, C. "Evaluating the point cloud of individual trees generated from images based on neural radiance fields (NeRF) method". *Remote Sens. (Basel)*, 16(6):967 (2024)
11. Shan, J. *Topographic laser ranging and scanning*. CRC Press, Second edition. — Boca Raton : Taylor & Francis, CRC Press, 2018. (2018)
12. Lowe, D.G. "Object recognition from local scale-invariant features". In *Proceedings of the Seventh IEEE International Conference on Computer Vision*. IEEE (1999)
13. Schonberger, J.L. and Frahm, J.M. "Structure-from-Motion revisited". In *2016 IEEE Conference on Computer Vision and Pattern Recognition (CVPR)*. IEEE (2016)
14. Besl, P.J. and McKay, N.D. "A method for registration of 3-D shapes". *IEEE Trans. Pattern Anal. Mach. Intell.*, 14(2):239–256 (1992)
15. Goesele, M., Curless, B., and Seitz, S.M. "Multi-View stereo revisited". In *2006 IEEE Computer Society Conference on Computer Vision and Pattern Recognition - Volume 2 (CVPR'06)*. IEEE (2006)
16. Iglhaut, J., Cabo, C., Puliti, S., Piermattei, L., O'Connor, J., and Rosette, J. "Structure from motion photogrammetry in forestry: A review". *Curr. For. Rep.*, 5(3):155–168 (2019)
17. Yu, A., Li, R., Tancik, M., Li, H., Ng, R., and Kanazawa, A. "PlenOctrees for real-time rendering of neural radiance fields" (2021)
18. Barron, J.T., Mildenhall, B., Verbin, D., Srinivasan, P.P., and Hedman, P. "Zip-NeRF: Anti-aliased grid-based neural radiance fields" (2023)
19. Mildenhall, B., Srinivasan, P.P., Tancik, M., Barron, J.T., Ramamoorthi, R., and Ng, R. "NeRF: Representing scenes as neural radiance fields for view synthesis" (2020)
20. Niemeyer, M., Mescheder, L., Oechsle, M., and Geiger, A. "Differentiable volumetric rendering: Learning implicit 3D representations without 3D supervision" (2019)
21. Vaswani, A., Shazeer, N., Parmar, N., Uszkoreit, J., Jones, L., Gomez, A.N., Kaiser, L., and Polosukhin, I. "Attention is all you need" (2023)
22. Tagliasacchi, A. and Mildenhall, B. "Volume rendering digest (for nerf)" (2022)
23. Elkhrachy, I. "Modeling and visualization of three dimensional objects using low-cost terrestrial photogrammetry". *Int. J. Archit. Heritage: Conserv. Anal. Restor.*, 14(10):1456–1467 (2020)
24. USIBD. "Guide for usibd document c220tm: Level of accuracy (loa) specification for building documentation". Technical report (2016) Document C120TM, Version 2.0, URL: https://cdn.ymaws.com/www.nysapls.org/resource/resmgr/2019_conference/handouts/hale-g_bim_loa_guide_c120_v2.pdf.
25. Laboratório de Sistemas Estruturais Ltda. "Memorial descritivo da passarela de pedestres sobre o Córrego que separa os prédios da civil e da minas, no campus da USP-SP". Technical report (2006) Contrato DEE 77.
26. Pedro Manuel Calas and Sainz de Murieta, E., editors. *Multi-span large bridges*. CRC Press, London, England (2015)

Chapter 10
Turkiye Earthquake to Maui Wildfire: Generalizability of Aerial Rapid Post-Hazard Assessment Tools

Mohammad Hesam Soleimani-Babakamali, Rojiar Soleimani, Onur Avci, and Ertugrul Taciroglu

Abstract One well-known challenge in deploying parametric data-driven models is the generalizability to test data, which is also true for rapid post-hazard damage assessment methods. Generalizability to test data—refers to how well a parametric data-driven model, trained on specific sets of data, can adapt to new, unseen data. In the context of rapid post-hazard damage assessment methods, this problem arises because the model is typically trained on data from past disasters, which may not always reflect the full range of conditions or damage patterns seen in new events. The rapid post-hazard damage assessment tools are pre-trained on one or more "source" domains and are applied to "target" regions that can cover vast urban textures and different hazard-based damage manifestations, *i.e.,* the challenge of generalizability. To further explore and better understand such challenges, we study two recent major events: the 2023 Turkiye earthquake (source) and the Maui wildfire (target) incidents. The difference in hazard type and region highlights the challenges of generalizability and enables us to evaluate the effectiveness of several devised data and Machine Learning measures in addressing them. The proposed strategies include augmenting image channels with damage proxy maps, data fusion, deep ensemble learning, and Test Time Augmentation, which are needless for fine-tuning or labels from the target. The developed measures significantly improved damage detection with and without severity classification, with F1 scores rising from 0.71 to 0.87 and from 0.40 to 0.82 compared to the typical damage assessment pipelines, respectively

Keywords Natural hazards · Data fusion · Aerial damage assessment · Ensemble learning · Disaster response and recovery

Introduction

The increased urbanization and climate change have expanded human settlement exposures to natural incidents [1], causing considerable socio-economic consequences globally [2], [3], [4]. Following such incidents, response and recovery stages can alleviate these impacts and pave the road for quick socio-economic recovery [5]. The recent emergence of Artificial Intelligence (AI) and the broad application and ongoing research in the field of computer vision have allowed the employment of sources, such as satellite imagery, which become available post-hazard rapidly to accommodate these measures, offering great scalability and cost-effectiveness compared to traditional in-situ evaluations, either human-based or computer vision-

Mohammad Hesam Soleimani-Babakamali
Postdoctoral Research Associate, Civil & Environmental Eng Department, University of California, Los Angeles, CA 90095
e-mail: soleimanisam92@g.ucla.edu

Rojiar Soleimani
Ph.D. Student, Institute of Urban Design and Landscape Architecture, Vienna University of Technology, Vienna 1040, Austria
e-mail: rojiar.babakamali@tuwien.ac.at

Onur Avci
Assistant Professor, Department of Civil and Environmental Engineering, Engineering Sciences Building, 1306 Evansdale Drive, West Virginia University, Morgantown, WV 26506
(Corresponding author)
e-mail: onur.avci@mail.wvu.edu

Ertugrul Taciroglu
Professor, Civil & Environmental Eng Department, University of California, Los Angeles, CA 90095
e-mail: etacir@g.ucla.edu

Janko Slavic et al. (eds.), *Computer Vision & Laser Vibrometry, Vol. 6*, Conference Proceedings of the Society for Experimental Mechanics Series, https://doi.org/10.13052/9788743801511_10

based but with in-situ data acquisition, *e.g.,* 3D point clouds or Unmanned Aerial Vehicle (UAV)-based imagery. A rapid initial damage assessment also helps the response phase by guiding assessment efforts to more damage-prone areas, saving time and other resources. Satellite imagery offers fast availability, *e.g.,* public sources such as the MAXAR Open Data Program[1], which, for instance, publicly published the Maui wildfire [6] pre- and post-hazard satellite images one day after the incident. Satellite imagery coverage also offers unprecedented completeness, *i.e.,* coverage of the earth's surface, which becomes necessary as natural hazard events may occur at any point globally. Integrating AI-powered computer vision with satellite imagery can facilitate the automation of damage assessment, enabling a more efficient and accurate response to natural disasters. Still, computer vision tools for post-hazard damage assessment come with several limitations, including lower data resolution (details), which impacts the generalizability in downstream post-hazard tasks, compared to data sources such as field survey data with UAVs [7], street view images [8], and 3D point clouds [9].

Novel post-hazard rapid damage assessment methodologies have been proposed in the literature regarding two concepts: image segmentation [10] and object detection [11]. Image segmentation performs pixel-wise classification, *i.e.,* outputs patches of pixels representing a category (*e.g.,* building damage states). Object detection models detect and localize target objects on images through bounding boxes (*e.g.,* a damaged building). Both models are typically empowered with convolutional neural networks (CNNs) or other Deep Learning architectures, learning to identify patterns and features associated with the provided (damage-related) labels. One such challenge is the generalizability to unseen damage manifestations and urban textures. That is, when a model is trained on data from one type of hazard, it may not generalize well to another hazard with different characteristics. This issue is critical because post-hazard damage assessments often need to be quick and accurate in order to effectively allocate resources and prioritize rescue efforts. If a model overfits to its training data, it may fail to provide reliable assessments in future disaster scenarios, leading to inaccurate conclusions and suboptimal responses.

Although some studies have touched on generalizability issues with satellite images as the only visual input, they have focused mainly on small-scale regions, such as neighborhoods (*e.g.,* [12], [13], [14]) that do not provide significant variation in urban texture necessary to understand potential real-world generalizability challenges. Given the availability of state-of-the-art deep learning (DL) segmentation architectures, this study aims to enhance the generalizability of rapid (satellite image-based) damage assessment frameworks built upon these architectures. Specifically, the objective is to identify challenges associated with those frameworks' applications to unseen urban textures and diverse hazards and then develop data-oriented and Machine Learning-based strategies to overcome those challenges. To that end, a dataset comprising the 2023 Kahramanmaraş earthquake sequences in Turkiye [15] and the Lahaina wildfire on Hawaiian island of Maui [6] is constructed. The earthquake data serves as training (**source**) data, while the wildfire data simulates a real-world rapid damage assessment test area (**target**) without labeled data. These two locations exhibit significantly different urban textures and are affected by distinct predominant natural hazards, pushing state-of-the-art models to their limits and demonstrating real-world generalizability challenges. Various methods are designed and implemented to enhance generalizability during the training and testing. Devised measures include data fusion, *i.e.,* augmenting RGB channels with different damage proxy maps as additional data channels, deep pixel-wise ensemble learning across models with diverse architectures and inputs, and Test Time Augmentation (TTA) strategies.

Methodology

Considering a typical post-hazard damage assessment methodology using image segmentation models, a typical pipeline for the procedure is shown in Figure 1. This pipeline requires a pre-trained damage assessment model, a target region from which input data, namely satellite images, are captured, and the building footprints to overlay the damage maps and quantify damaged structures. A "measure" is introduced for each module to aid generalizability to the target area. To explore the pipeline and these measures, we first elaborate on assembling the Maui wildfire and Kahramanmaraş earthquake data, including the extraction and pre-processing of the images. The following subsection revolves around the Deep Learning (DL) network architectures, image data fusion, and training of these models using Kahramanmaraş (Turkiye) earthquake sequence data.

Satellite Imagery and Channel Augmentation

The first incident studied here is the 2023 Turkiye earthquake that struck southern and southeast Turkiye on February 6, 2023, near the city of Gaziantep [15]; this dataset serves the "source" role. These events affected at least 11 provinces and caused an estimated economic loss of around 110 billion dollars [16]. The "target" incident, namely the Maui (Hawaii)

[1]MAXAR, Maxar open data program, https://www.maxar.com/open-data, accessed: 2023-09-10.

wildfire, began on the morning of August 8, 2023, followed by another outbreak later that afternoon, causing significant damage to the city of Lahaina [6]. For both events, MAXAR released pre- and post-hazard satellite images for the affected regions within one day, demonstrating the practicality of satellite images for rapid post-hazard damage assessment purposes. These images became available shortly on Google Maps API and are downloaded for this study. For the Turkiye incident, the city of Antakya, located in the heavily affected Hatay province [17], offers abundant satellite images for training damage assessment models and is thus chosen to train the source models. Lahaina city is chosen from the Maui incident as the major impacted region. The satellite images come in a single large raster, and to make them suitable for input to the downstream Machine Learning models, given the maximum resolution of public MAXAR open data, images are taken with a width and height of 640 pixels, almost equal to 150-by-150-meter blocks. Accordingly, the ($\sim$35 km^2, Figure 2a) Lahaina city region is divided into 1551 blocks, and the Antakya region ($\sim$128 km^2, Figure 2b) is divided into 5,679 blocks with their associated pre- and post-hazard images stored. The images are downloaded from Google and Bing Maps (see Figure 3), which are MAXAR-based yet more conveniently retrievable through their application programming interfaces.

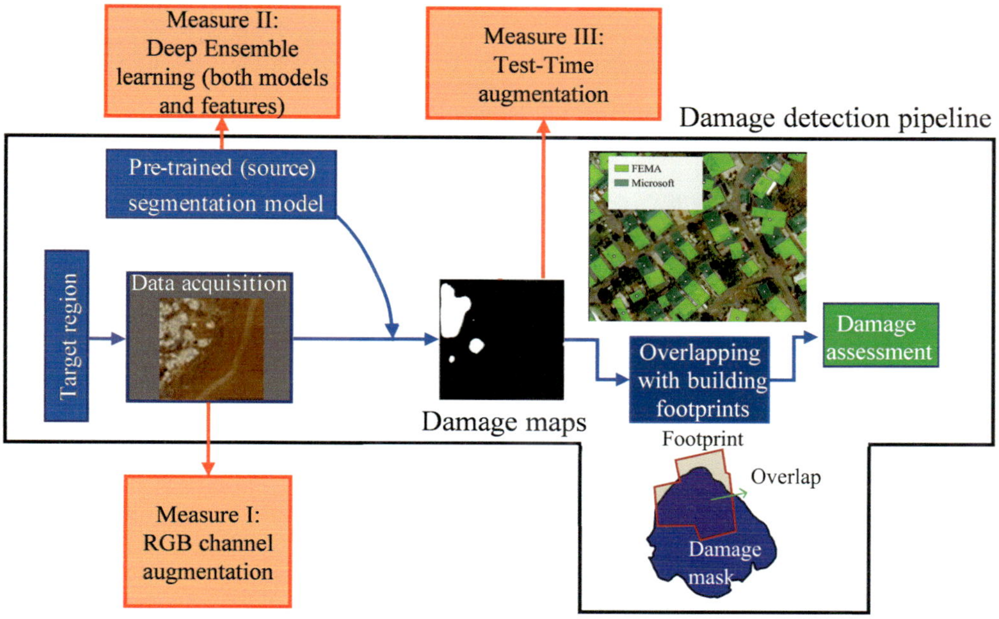

Fig. 1 Typical post-hazard damage detection pipeline from satellite imagery; the "Measure" boxes are the additional modules added in this research for possible generalizability enhancement.

Antakya image blocks are used to train the "source" damage detection model and must be labeled accordingly. The segmentation mechanism is chosen here over object detection, as building footprints are available from various public sources (*e.g.,* FEMA[2] and Microsoft[3]). The associated uncertainty can thus be removed from the study on the proposed models' generalizability. The segmentation model analyzes the visual features through Convolutional Neural Networks (CNNs) or other Deep Learning architectures and decides the label for each pixel. Hence, inside the training data, each pixel must be labeled (herein damage status) to train segmentation models. During inference, the model processes new images and generates segmentation masks, *i.e.,* pixel-wise probabilities for label categories, highlighting the areas likely to be damaged in the case of damage assessment problems (see Figure 1). These masks can then be used for further analysis or decision-making processes. For instance, they can be compared with building footprint boundaries (see Figure 1) to count damaged structures. Considering the pre-hazard image as the reference condition, any structural visual alterations can be interpreted as damages while labeling the Antakya image blocks.

Accordingly, two groups of labels are defined. For buildings with visible changes yet no collapsed parts, we use the label "D". For buildings that collapse entirely or partially, we use the label "C". Although C labels are relatively more straightforward, the D labels may require the pre-hazard images to be compared with post-hazard images. C and D aid in distinguishing damaged and collapsed buildings, which convey distinct consequences in loss estimation and affected population. Across 5,679 block images, 1,975 blocks include damaged buildings, with 879 C and 1,096 D segmentation patches labeled to train

[2]FEMA, Fema USA structures, https://gis-fema.hub.arcgis.com/pages/usa-structure, accessed: 2023-09-10.
[3]Microsoft US building footprints, https://github.com/microsoft/USBuildingFootprints, accessed:2023-09-10.

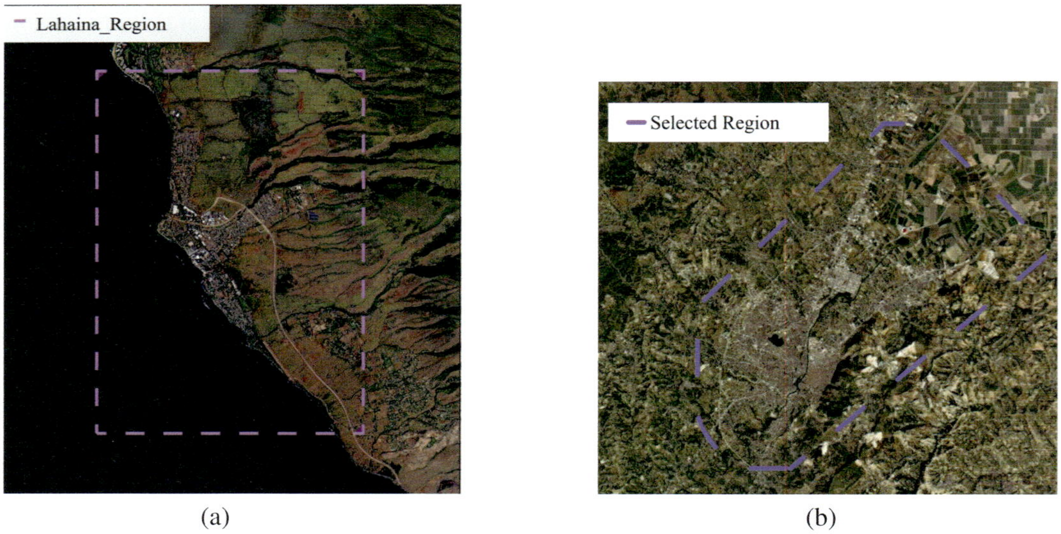

Fig. 2 Selected regions: (a) Lahaina, Maui County, Hawaii, USA and (b) the city of Antakya, Turkiye.

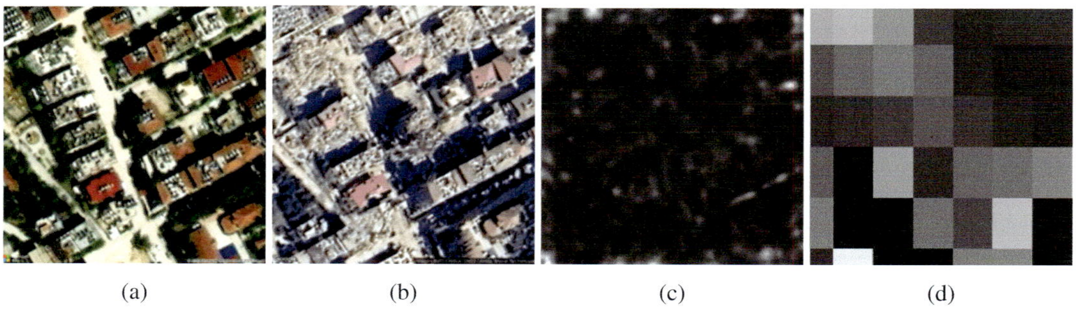

Fig. 3 Antakya block No. 1458 (a) pre-hazard satellite image, (b) post-hazard satellite image, (c) change detection (CD) maps between pre- and post-hazard images, (d) NASA ARIA map.

deep segmentation models. It is worth mentioning that even though two labels, C and D, are defined, the ultimate damage assessment will have three labels, D, C, and non-damaged (undamaged), which are the building footprints that did not match any damage masks (see Figure 1).

As noted earlier, regular post-hazard damage assessment techniques employ satellite images as inputs (only post-hazard images or combined pre- and post-hazard images). The first measure we implemented here is to augment the RGB satellite images with new channels (see Figure 1). The goal is to provide the network with salient visual features while minimizing visual intricacies caused by the varied urban textures between source and target regions, such as the predominant roof styles and building geometries, surrounding land cover, and differences in visual characteristics of training and testing images, *e.g.*, angle or luminosity. Two sets of such features are used, one by employing a **change detection** (**CD**) model and the other by employing a publicly available "damage proxy map" by NASA, namely the NASA ARIA[4] project. NASA ARIA publishes these maps after major incidents and thus can be included seamlessly in the rapid post-hazard damage assessment pipelines. In this study, the "general-purpose" CD model proposed by El Amin et al. [18] is used, with an example shown in Figure 3d. These ARIA and resulting CD maps are stored for all blocks in Antakya and Lahaina, with the former used to train the Deep-Ensemble models, explained in the following section, and the latter used to test the proposed strategies in promoting generalizability.

Deep Learning Mechanisms

The image segmentation task involves learning the mapping between satellite images, CD, NASA ARIA maps, and the C and D masks. C and D are segmentation mask labels, and later footprints with over-threshold overlapping with them are

[4]NASA, Nasa Aria program, https://aria.jpl.nasa.gov/, accessed: 2023-09-10.

termed as C and D buildings. Various architectures exist for this task, from which the U-Net [19] and U-Net++ [20] are employed. For the U-Net's encoder layers, two models are employed: EfficientNet-b7 (CNN-based [21]) and Segformer mit_b5 (Visual Transformer-based [22]). The input data includes satellite images with and without CD and ARIA proxy map augmentations. This combination leads to 6 models, summarized in Table 1. The models are optimized for 20 epochs, and training time augmentation strategies are included to enhance generalizability. The pixel accuracy of the selected models is calculated and found to be similar, indicating that variations in channel augmentations or transitioning from CNN-based to Transformer-based models do not consistently affect accuracy. The minor fluctuations in accuracy are expected due to the similarity between hazard damage features and urban textures in the training and validation data.

Table 1 Semantic segmentation models and input layers.

Name	Model	Input channels	Architecture	Train pixel accuracy
Eff-Net	Efficientnet-b7	3:(RGB)	U-Net++	95.7
Eff-CD-Net	Efficientnet-b7	4:3(RGB)+1(CD)	U-Net++	95.9
Eff-ARIA-Net	Efficientnet-b7	4:3(RGB)+1(ARIA)	U-Net++	95.6
Tra-Net	Segformer-mit-b5	3 (RGB)	U-Net	95.6
Tra-CD-Net	Segformer-mit-b5	4:3(RGB)+1(CD)	U-Net	95.7
Tra-ARIA-Net	Segformer-mit-b5	4:3(RGB)+1(ARIA)	U-Net	95.6

Some models may overfit the data and focus on image intricacies rather than damaged features during optimization. To mitigate this, an ensemble strategy is devised across different models. The damage masks produced by each model (total of six) are combined to prioritize blocks with high agreement on damage presence and reduce masks in areas with low agreement. The segmentation mask for each damage category is represented as a 640-by-640 matrix mask with elements of 0 (non-damaged) or 1 (damaged). The following ruleset constructs the pixel-wise ensemble strategy: a) a block, *i.e.,* 640-by-640 pixel mask, is termed as damaged for a given model (total of six) if more than 1% of its pixels are labeled as damaged inside it, b) if a block is termed as damaged by all six models, the masks are then overlaid, c) if 4 to 5 models designate a block as damaged, then the majority vote ensemble is carried out for each pixel, and d) if less than four models agree on a block being damaged, then the whole block pixels' are labeled as non-damaged. The resulting "inter-model" ensemble is termed the Ensemble model henceforward.

Aside from using CD and ARIA maps, which are designed to provide the networks with salient damage features for both training and testing, the RGB satellite images are still used as the main input features, which may suffer from visual intricacies (*i.e.,* non-damage-related features, to which models may overfit). Accordingly, an "intra-model" ensemble strategy is employed using Test Time Augmentation (TTA) techniques. Although computer vision tools are mostly trained with image augmentation inside their training process, models may still lack generalizability to unseen areas. This problem, for instance, is shown in Figure 4 for Lahaina Block No. 408's C mask output. This figure reveals that the trained (with Antakya data) Eff-Net model suffers from generalizability issues for the Lahaina region, and the positioning of the damaged features inside the input image impacts its outcome. Accordingly, for each of the six models in Table 1, the intra-model ensemble is carried out with the main input image, 90°, 180°, and 270° clockwise (CW) rotated images, and the resulting masks are overlaid. The inter-model ensemble model with the masks produced by the intra-model ensemble is termed as the Ensemble-R model hereafter.

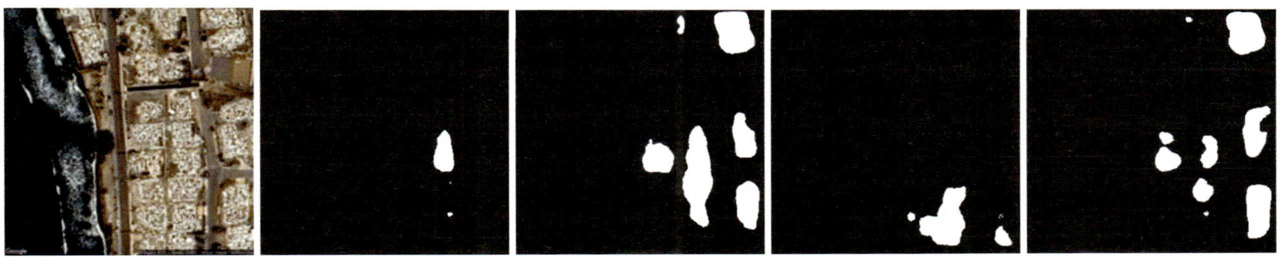

(a) (b) (c) (d) (e)

Fig. 4 Lahaina block No. 408's Eff-Net C mask output (a) regular image input, (b) 90° (CW) rotated image, (c) 180° (CW) rotated image, (d) 270° (CW) rotated image.

Results and Discussion

This section evaluates the damage estimation results using each deep segmentation model and assesses the proposed design's effectiveness in promoting generalizability. The results of the damage detection model are evaluated in two ways. First, damage maps produced by the proposed framework are compared to the Maui damage assessment by ESRI[5]. ESRI (Environmental Systems Research Institute) is a geographic information system (GIS) software providing tools and platforms for mapping, spatial analysis, and location intelligence. The Maui damage assessment results are produced using a Deep Learning model that was trained with UAV images from the Woolsey fire[6], and then fine-tuned for the Lahaina incident. The second evaluation is done by first labeling each building in categories D, C, and non-damaged (undamaged) to compare different models and strategies in terms of performance. Multi-class accuracy metrics, including precision, recall, and the F1 score, are used to assess damage detection outcomes and determine how different measures affect them.

For two individual Eff-Net and Tra-Net and the Ensemble and Ensemble-R model, the resulting damage map with the overlap threshold of 25% is plotted against the Maui damage assessment by ESRI in Figure 5 (if the building footprint overlapping with the damage mask is 25% or more of its area, then that structure is labeled as "damaged", see Figure 1 for mask/footprint matching). Eff-Net missed most of the damaged buildings compared to the damage map obtained by ESRI and the Ensemble approach. On the other hand, Tra-Net offers a better distribution and quantity of damaged buildings, yet it has numerous misclassifications between C and D masks. The Ensemble model tends to improve both the quantity and distribution of damaged buildings and offer a better damage severity assessment than the individual models. Ensemble-R reports a comparable ratio of damaged buildings, 61.73% compared to 64.68% of ESRI, with the damage distribution also in good agreement with that of ESRI. Furthermore, Ensemble-R can also classify the damage's severity into two classes of D and C damage, a capability that the ESRI model lacks. According to accuracy metrics.

Table 2 Lahaina wildfire damage detection performance with and without damage severity classification.

Model	Damage detection task (two categories)			Damage detection + Severity tasks (three categories)		
	Precision (macro)	Recall (macro)	F1 Score (macro)	Precision (macro)	Recall (macro)	F1 Score (macro)
Eff-Net	0.68	0.61	0.45	0.53	0.42	0.33
Eff-CD-Net	0.70	0.69	0.58	0.50	0.48	0.41
Eff-ARIA-Net	0.69	0.65	0.52	0.56	0.45	0.37
Tra-Net	0.73	0.77	0.71	0.55	0.51	0.40
Tra-CD-Net	0.76	0.80	0.73	0.59	0.58	0.55
Tra-ARIA-Net	0.67	0.66	0.54	0.61	0.50	0.45
Ensemble	0.76	0.80	0.73	0.63	0.58	0.55
Ensemble-R	0.85	0.89	0.87	0.80	0.85	0.82

To compare the six individual models and evaluate the effectiveness of the proposed measures, Table 2 recaps their damage detection performance. Table 2 includes two classification categories: the first refers to the binary damage/non-damage classification (*i.e.,* two-label damage detection) to evaluate overall damage detection performance, while in the second assessment, damage severity is also measured by having three labels of non-damage, and D and C damages. According to the earlier inventory labeling presented at the beginning of the Results section, 59.6% of the buildings were labeled as C, 9.8% as D, and 30.6% as no-damage buildings. Consequently, damage detection data has 69.4% damaged buildings, while the D class constitutes only 9.8% of data in the damage severity task, implying a severely imbalanced classification task. Due to the imbalance in both problems, macro-estimations are employed in Table 2. In the damage detection task, Eff-Net, a CNN-based model, falls short of even a random guess, achieving an F1 score below 50%. This poor performance is evident in Figure 5b also, where Eff-Net misses a significant number of damaged buildings. In contrast, Tra-Net achieves a respectable F1 score of 0.71. However, in the more challenging task of damage detection with severity measurement, Eff-Net again performs no better than chance, while Tra-Net, despite its strong performance in the damage detection task, struggles with misclassifying between (**fully or partially collapsed**) C and (**damaged but not collapsed**) D categories, as seen in Figure 5c. The incorporation of CD and ARIA maps as channel augmentations has been shown to improve damage detection

[5]ESRI, Maui damage assessment, https://www.arcgis.com/apps/dashboards/c6ba275c2762470bbaf0c660a89a46e7, accessed: 2023-09-10.

[6]USGS, Woolsey fire, https://www.usgs.gov/media/before-after/woolsey-fire, accessed: 2023-09-10.

and severity classifications. However, the RGB channels still carry environmental features that can affect model decision-making. This is exemplified by the generalizability issues observed in individual models, such as the Eff-Net C-mask model, which fails to capture damaged structures in certain images. Rotating the images by 90, 180, and 270 degrees counterclockwise and re-feeding them into the model improves segmentation outcomes, highlighting the impact of urban textures and building placement on model detection ability. The application of Test-Time Augmentation (TTA) methods, such as rotation, can enhance model performance and provide more accurate segmentation results. Moreover, the ensemble strategy, which combines the outputs of six models, can be further refined by applying TTA to each model, resulting in the Ensemble-R model. The performance of the Ensemble and Ensemble-R models is compared, demonstrating improved damage detection and severity measurement capabilities. The resulting masks offer greater confidence in detecting damage and measuring severity.

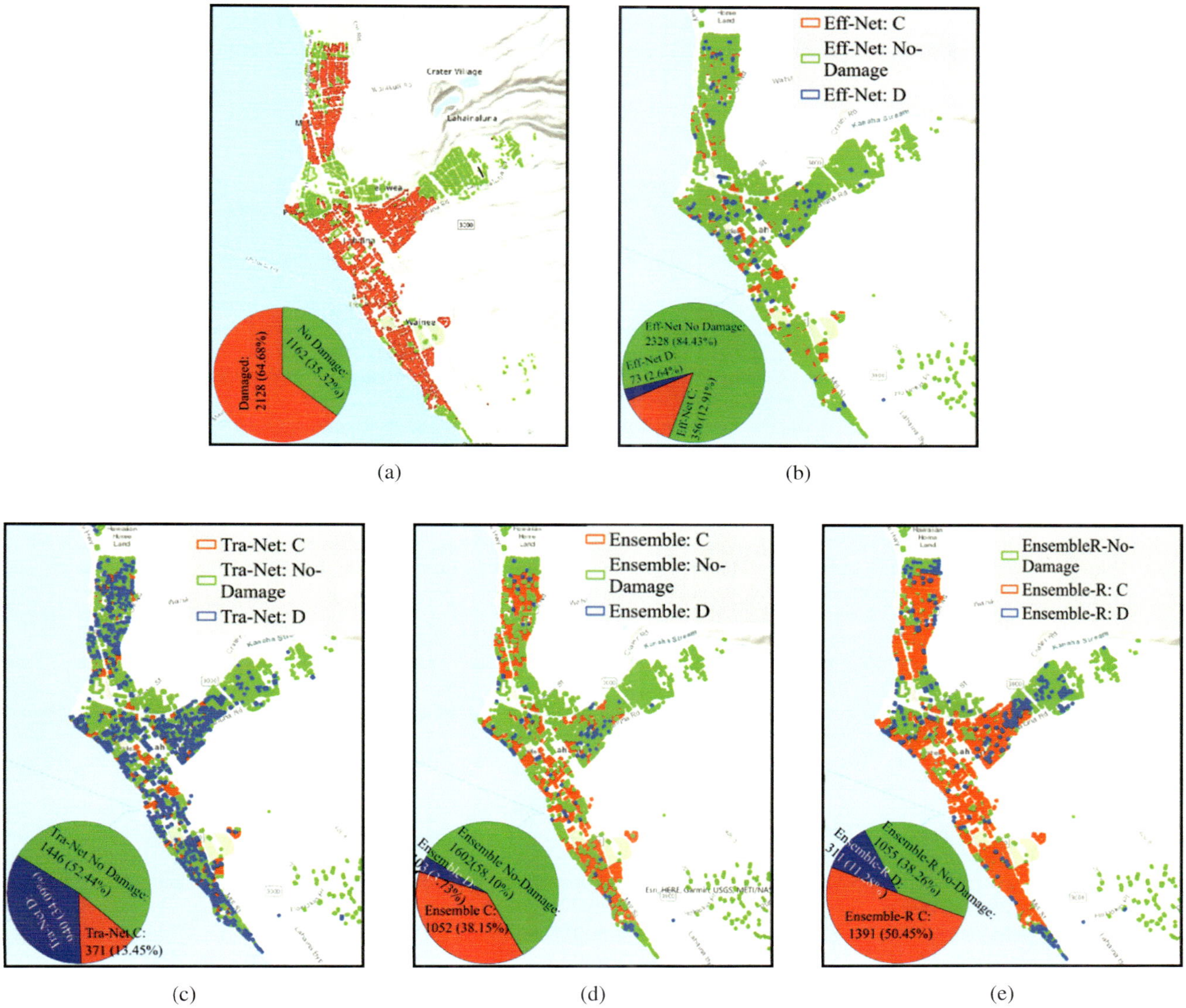

Fig. 5 Lahaina damage assessment: (a) ESRI estimation (ESRI, 2023), (b) Eff-Net estimation, (c) Tra-Net estimation, (d) Ensemble estimation, (e) Ensemble estimation.

Conclusions

Rapid first-hand damage assessments after natural incidents substantially aid response measures. The generalizability concerns in such assessments are vital as pre-trained models must be applied to the impacted region without any fine-tuning (due to lack of time due to the time-consuming labeling procedures) or verified inspection reports from the field. Thus, data and Machine Learning solutions must be devised to promote the generalizability of a rapid post-hazard assessment tool training performed in one or more "source" areas to the desired target regions. Based on this, various measures are developed and introduced in this manuscript that entail no fine-tuning on the target site and rely solely on public data sources. The aerial data from the 2023 Turkiye earthquake (source) and the Maui wildfire incident (target) are used first to investigate the generalizability concerns and then evaluate how well the developed Machine Learning strategies overcome those. To promote generalizability, various measures were designed and evaluated. Satellite images were fused with two damage proxy maps, a change detection model and NASA ARIA maps, to alleviate differences in urban texture and visual artifacts. Six deep segmentation models (CNN- and Visual Transformer-based) were trained with normal and augmented satellite images. A two-level ensemble strategy was developed: (1) an intra-model ensemble using Test Time Augmentation (TTA) strategies and (2) a pixel-based deep ensemble strategy across the six models. These measures were applied to the test area of Lahaina city, resulting in improved generalizability and high accuracies: macro F1 scores of 0.87 for damage detection and 0.91 for damage severity categorization, outperforming a single Visual Transformer-based model without interventions (0.71 and 0.50, respectively).

References

1. United Nations. Department of Economic and Social Affairs, *The World's Cities in 2016*. UN, 2016.
2. N. N. Ridder, A. M. Ukkola, A. J. Pitman, and S. E. Perkins-Kirkpatrick, "Increased occurrence of high impact compound events under climate change", *NPJ Clim Atmos Sci*, vol. 5, no. 1, p. 3, 2022.
3. K. Do, M. Mahish, A. K. Yeganeh, Z. Gao, C. L. Blanchard, and C. E. Ivey, "Emerging investigator series: a machine learning approach to quantify the impact of meteorology on tropospheric ozone in the inland southern California", *Environmental Science: Atmospheres*, 2023.
4. D. Nelson, Y. Choi, B. Sadeghi, A. K. Yeganeh, M. Ghahremanloo, and J. Park, "A comprehensive approach combining positive matrix factorization modeling, meteorology, and machine learning for source apportionment of surface ozone precursors: Underlying factors contributing to ozone formation in Houston, Texas", *Environmental Pollution*, vol. 334, p. 122223, 2023.
5. W. Sun, P. Bocchini, and B. D. Davison, "Applications of artificial intelligence for disaster management", *Natural Hazards*, vol. 103, no. 3, pp. 2631–2689, 2020.
6. T. W. Juliano *et al.*, "Brief communication: The Lahaina Fire disaster: How models can be used to understand and predict wildfires", *Natural Hazards and Earth System Sciences Discussions*, vol. 2023, pp. 1–7, 2023.
7. M. M. L. Ramadhan, G. Jati, M. R. Alhamidi, M. H. Hilman, W. Jatmiko, and others, "Collapsed Building Detection Using Residual Siamese Neural Network On LiDAR Data", in *2021 6th International Workshop on Big Data and Information Security (IWBIS)*, 2021, pp. 29–34.
8. D. Duarte, F. Nex, N. Kerle, and G. Vosselman, "Multi-resolution feature fusion for image classification of building damages with convolutional neural networks", *Remote Sens (Basel)*, vol. 10, no. 10, p. 1636, 2018.
9. F. Yamazaki, W. Liu, and K. Horie, "Use of Multi-Temporal LiDAR Data to Extract Collapsed Buildings and to Monitor Their Removal Process after the 2016 Kumamoto Earthquake", *Remote Sens (Basel)*, vol. 14, no. 23, 2022, doi: 10.3390/rs14235970.
10. Y. Wang, A. W. Z. Chew, and L. Zhang, "Building damage detection from satellite images after natural disasters on extremely imbalanced datasets", *Autom Constr*, vol. 140, p. 104328, 2022.
11. Z. Liu *et al.*, "Remote sensing-enhanced transfer learning approach for agricultural damage and change detection: A deep learning perspective", *Big Data Research*, vol. 36, p. 100449, 2024.
12. W. Yang, X. Zhang, and P. Luo, "Transferability of convolutional neural network models for identifying damaged buildings due to earthquake", *Remote Sens (Basel)*, vol. 13, no. 3, p. 504, 2021.
13. J. Ge, H. Tang, N. Yang, and Y. Hu, "Rapid identification of damaged buildings using incremental learning with transferred data from historical natural disaster cases", *ISPRS Journal of Photogrammetry and Remote Sensing*, vol. 195, pp. 105–128, 2023, doi: https://doi.org/10.1016/j.isprsjprs.2022.11.010.
14. M. Janalipour and A. Mohammadzadeh, "A fuzzy-ga based decision making system for detecting damaged buildings from high-spatial resolution optical images", *Remote Sens (Basel)*, vol. 9, no. 4, p. 349, 2017.
15. USGS, "M 7.8 - Pazarcik earthquake, Kahramanmaras earthquake sequence", 2023.
16. M. Ozturk, M. H. Arslan, and H. H. Korkmaz, "Effect on RC buildings of 6 February 2023 Turkey earthquake doublets and new doctrines for seismic design", *Eng Fail Anal*, vol. 153, p. 107521, 2023.
17. T. government, "2023 Kahramanmaras damage assessment (in Turkish)", 2023.
18. A. M. El Amin, Q. Liu, and Y. Wang, "Convolutional neural network features based change detection in satellite images", in *First International Workshop on Pattern Recognition*, 2016, pp. 181–186.
19. O. Ronneberger, P. Fischer, and T. Brox, "U-net: Convolutional networks for biomedical image segmentation", in *Medical Image Computing and Computer-Assisted Intervention–MICCAI 2015: 18th International Conference, Munich, Germany, October 5-9, 2015, Proceedings, Part III 18*, 2015, pp. 234–241.
20. Z. Zhou, M. M. R. Siddiquee, N. Tajbakhsh, and J. Liang, "Unet++: Redesigning skip connections to exploit multiscale features in image segmentation", *IEEE Trans Med Imaging*, vol. 39, no. 6, pp. 1856–1867, 2019.
21. M. Tan and Q. Le, "Efficientnet: Rethinking model scaling for convolutional neural networks", in *International conference on machine learning*, 2019, pp. 6105–6114.
22. E. Xie, W. Wang, Z. Yu, A. Anandkumar, J. M. Alvarez, and P. Luo, "SegFormer: Simple and efficient design for semantic segmentation with transformers", *Adv Neural Inf Process Syst*, vol. 34, pp. 12077–12090, 2021.

Chapter 11
Analyzing Rotating Bladed Disk Dynamics with Digital Image Correlation and Downsampling

Serena Occhipinti, Alessandra Cesaretti, Paolo Neri, Christian M. Firrone, and Daniele Botto

Abstract During operating conditions, turbomachinery blades are exposed to excitation forces that increase the risk of crack nucleation and reduce their fatigue life if the frequency of the forces matches the natural frequency of the blade. Therefore, it is essential to accurately measure the dynamics, in terms of vibration amplitude and resonance frequency, of rotating bladed discs. Conventional contact-based measurement methods have several drawbacks: they alter the dynamic behavior of the bladed disk by adding mass, have a limited lifetime, and face complex data transmission issues. This has led to a growing interest in non-contact measurement methods. In this context, this study presents an affordably, accurate, full-field, and high-resolution measurement technique. The proposed method employs low-speed, high-resolution cameras combined with down-sampling to capture images of a rotating bladed disk. These images are analyzed with the Digital Image Correlation (DIC) technique to identify the operational deflection shape of the bladed disk. This method was tested on a rotating bladed disk with simplified geometry. The method has been shown to work very well with only just one image per revolution, thus reducing the problems associated with large rigid body motions.

Keywords Turbomachinery blades · Dynamic measurement · Optical technique · DIC, · Down-sampling.

Introduction

Vibrations in turbomachinery can lead to crack formation, significantly shortening the fatigue life of blades. Consequently, effective blade health monitoring is essential to prevent unexpected failures. In synchronous vibrations, the frequency of the external force is an integer multiple of the rotational speed. When the rotational speed of the disk is such that the corresponding frequency of the exciting force approaches a natural frequency, the risk of dangerous resonance phenomena increases, making it crucial for designers to have an accurate knowledge of the disc dynamics.

Historically, strain gauges have been used to monitor rotor blade vibrations and remain the most reliable method of measurement. However, they require complex wireless systems or slip rings to transmit data, and their mass and stiffness interfere with the system dynamics. Furthermore, strain gauges have a limited service life and provide only localized data, related to discrete points on the bladed disk.

As a result, non-contact measurement techniques are becoming increasingly popular. One promising approach is the Scanning Laser Doppler Vibrometer (SLDV) combined with an optical derotator. This system offers a non-invasive way to accurately capture vibration data from specific points with high temporal resolution. The optical derotator ensures the laser beam accurately follows the rigid rotation of the disk by using a controlled optical rotation unit synchronized with the rotation frequency, effectively eliminating the effects of rigid body motion [1]. Despite its advantages, such as precise and non-contact measurements, the high cost of these devices is a significant limitation. In [2], an experimental vibration analysis of a fan was conducted using a custom-built derotator paired with an SLDV. Similarly, [3] describes a method for tracking specific points on rotating structures using a Laser Doppler Vibrometer and a system of plane mirrors—one mounted on the

Serena Occhipinti · Alessandra Cesaretti · Christian M. Firrone · Daniele Botto
Department of Mechanical and Aerospace Engineering, Politecnico di Torino, Torino, Corso Duca degli Abruzzi 24
e-mail: serena.occhipinti@polito.it; alessandra.cesaretti@polito.it; christian.firrone@polito.it; daniele.botto@polito.it

Paolo Neri
Department of Civil and Industrial Engineering, Università di Pisa, Pisa, Largo L. Lazzarno 1
e-mail: paolo.neri@unipi.it

Janko Slavic et al. (eds.), *Computer Vision & Laser Vibrometry, Vol. 6*, Conference Proceedings of the Society for Experimental Mechanics Series, https://doi.org/10.13052/9788743801511_11

rotating structure and the other stationary in front of it. However, these methods require complex hardware configurations, where even slight misalignments can lead to measurement inaccuracies, and the data collected is still limited to a small number of points.

In this context, Blade Tip Timing (BTT) technology has emerged as a valid alternative. BTT measures blade vibrations by analysing the arrival times of blade tips as they pass in front of fixed sensors. By comparing the actual arrival time with the expected time for the rigid body rotation, the system can derive vibration data. BTT systems then use advanced identification methods to process this data, allowing the extraction of modal parameters that define the dynamic behaviour of the vibrating blades. While in most conventional configurations the sensors are positioned radially inward, in the direction of the tips of the blades [4], different probe configurations have been explored to measure different displacement components [5]. However, BTT measurements are inherently one-dimensional and are limited to specific points, typically the blade tips.

The advancement of video-based measurement techniques for dynamic analysis has introduced the use of optical cameras to monitor rotor vibrations. Stereo Digital Image Correlation (Stereo-DIC) enables the measurement of multiple points across a structure's surface. This is accomplished by applying a speckle pattern to the surface, which allows DIC to track points on object surface. In this way, the stereo-DIC is able to provide full-field three-dimensional surface displacements. While DIC is a well-established technique for dynamic experimental analysis of non-rotating structures, applying it to rotating systems presents significant challenges. Traditional DIC algorithms struggle to track points accurately across a rapidly changing field of view, leading to inaccurate displacement estimates when dealing with large displacements and rotations. A common approach to overcome this limitation involves using an incremental procedure [6]. This method analyzes images of objects undergoing large displacements by updating the reference image whenever the displacement between frames exceeds the measurable limit. However, when the reference update occurs, an error is introduced into the calculated displacement [7]. Thus, this procedure applied on rotating structure, where high displacement are due to the rigid body rotation, results in frequent updating of the reference image and to non-negligible noise accumulation. Moreover, the incremental procedure requires small displacements between consecutive frames, necessitating the use of high-speed cameras to capture the blade rotation with precise time resolution. High-speed cameras may initially appear to be the most suitable solution for obtaining measurements at higher frequencies, but they result in low resolution and high setup costs. For this reason, methods have been developed to use DIC with cameras at low frame rate to measure the dynamics of structures at frequencies above the Nyquist-Shannon limit. For example, using the down-sampling methods presented in [8, 9], it is possible to correctly measure the frequency content of steady state signals while avoiding aliasing even if the sampling rate does not comply with the Nyquist-Shannon sampling theorem.

This paper explores the capability of the Digital Image Correlation (DIC) technique as a measurement method for analyzing full-field, high-frequency synchronous vibrations of a bladed disk using a cost-effective, low-speed stereo-camera system. The study employs a down-sampling technique to prevent aliasing errors when acquiring samples at frequencies below the Nyquist limit. The presented methodology was tested on a dummy bladed disk mounted on a spinning test rig. The bladed disk has a simplified flat geometry and the vibration modes are mainly out-of-plane bending.

Measurement Methodology

The disk was tested in a spinning rig, where a set of permanent magnets, evenly spaced around the casing, induced synchronous vibrations in relation to the rotor speed. Synchronous vibration is a common phenomenon in rotating machinery blades, where the blade vibration frequencies are integer multiples of the rotational frequency.

During each revolution, two cameras captured a single image of the rotating disk at a specified angular reference position, triggered by an optical probe. These images were then processed using the Stereo-Digital Image Correlation (Stereo-DIC) technique, enabling the calculation of the three-dimensional full-field displacement of the disk's surface as it passed through the reference position. However, since the sampling frequency matched the rotational frequency, the displacement data were downsampled, according to the sampling theorem. To extract meaningful information from the original downsampled data, appropriate identification and reconstruction algorithms were required. Among the various algorithms available for this task [10–13], the post-processing was performed by using a fitting method called Single Parameter Technique (SPT) [14].

Stereo-Digital Image Correlation

The stereo-Digital Image Correlation (stereo-DIC) algorithm operates by processing images taken from different perspectives using DIC techniques. This process tracks surface displacement and deformation by analyzing how the position and shape of specific patterns change over time. In this study, the open-source Matlab program Ncorr [15] was used to implement the DIC algorithm. Another essential aspect of a stereo-DIC setup is the calibration of the stereo-camera system. The calibration process for each camera in a stereo pair establishes the transformation needed to map 2D image points from the camera sensors to their corresponding 3D coordinates in the global system, based on the pinhole optical model [16]. For this work, the calibration was carried out using the MATLAB Stereo Camera Calibrator App [17]. Once the positions of image subsets are identified in both camera frames and the stereo-camera model is defined, triangulation is used to compute the 3D positions of each point over time. This allows the calculation of spatial displacement and deformation fields.

Single Parameter Technique (SPT)

The Single Parameter Technique (SPT) assumes that each point on the disk behaves like a single-degree-of-freedom (SDOF) vibrational system, with damping and subject to a constant-amplitude harmonic excitation force. The system is described by the following differential equation:

$$\ddot{x} + 2\zeta\omega_n\dot{x} + \omega_n^2 x = \frac{F_0}{m}\cos\left(\omega t + \varphi_0\right) \tag{1}$$

where x is the point displacement; ζ is the damping ratio; ω_n is the natural angular frequency of blade; F_0 is the amplitude of excitation force; m is the mass; ω is the angular frequency of excitation force; φ_0 is the initial phase. The excitation frequency ω is directly related to the rotational speed of the disk Ω:

$$\omega = \Omega \cdot EO \tag{2}$$

where EO is the engine order. Thus, the steady state solution for the vibration equation is:

$$x(t) = X(\omega)\cos\left(\Omega t \cdot EO - \phi(\omega) + \varphi_0\right) \tag{3}$$

where,

$$X(\omega) = \frac{\delta}{\sqrt{\left[1 - (\omega/\omega_n)^2\right]^2 + \left(2\zeta \cdot (\omega/\omega_n)\right)^2}} \qquad \tan(\phi(\omega)) = \frac{2\zeta \cdot (\omega/\omega_n)}{1 - (\omega/\omega_n)^2} \tag{4}$$

Here, δ represents the static displacement of the blade tip under the action of the force F_0. Thus, assuming a certain response order, corresponding to the selected EO, for each point downsampled response the function $x(t)$ is fitted to the data in order to compute model parameters.

Test Rig

Figure 1 illustrates the test rig. The tested dummy disk is made of aluminum and features a simple design with 12 blades. Each blade is shaped like a cantilever beam, measuring 150 mm in length and 25 mm in width, with a thickness of 5 mm. The disk has an outer diameter of 400 mm. At the tip of each blade, cylindrical magnets are glued into housings drilled for this purpose. The dummy disk was tested in a laboratory spinning rig. The test rig has a vertical axis and the rotating shaft is positioned under the floor. The disk is fixed between two flanges at the top of the shaft. The excitation system of the rotation setup employs cylindrical permanent magnets that interact with the magnets mounted on the disk during its rotation. This interaction generates impulsive repulsion forces, as the magnets on the rotating disk pass by the fixed magnets, creating periodic excitation. Two coaxial stationary disks were designed to lock these magnets in fixed angular positions around the rotation axis. The number of magnets used in a the test has to match the main engine order (EO) corresponding to the excitation force being simulated. A modal shape associated with a particular nodal diameter number (ND) can be excited by different values of EO. For a disk with N blades, the EO the engine orders capable of exciting a nodal diameter ND can be calculated as:

$$EO = k \cdot N \pm ND, \quad \forall k = 0, 1, 2, 3, \ldots \tag{5}$$

As expressed in Eq. 2, the excitation frequency is related to both the engine order, EO, and the rotational frequency, Ω. Based on Eq. 5, 16 magnets were equally spaced around the test rig to excite the nodal diameter $ND = 4$ of the bladed disk.

Given the limitations of high-speed cameras in terms of cost and sensor resolution, the disk's motion was measured using a low-speed 3D-DIC system, as shown in Fig. 1 This system consists of two Optomotive Spinosaurus cameras, each equipped with a 23 MP Fujinon lens featuring a 12 mm focal length. These cameras are equipped with high-resolution sensors capable of capturing 7.1 MP images at a maximum frame rate of 200 fps and with a minimum shutter time of 2 μs. To prevent motion blur, high-intensity illumination was required to compensate for the reduction in image brightness due to the short shutter time. For this purpose, two 200 W Stratus LED modules were used to illuminate the blade surface effectively. The data acquisition for both cameras was synchronized using a common trigger signal. To achieve this, an optical probe was mounted on the non-rotating part of the rig, pointing towards the upper flange that secures the disc. A reflective patch was attached to this flange, which rotates along with the disk. Each time the reflective patch passed in front of the probe, the optical probe generated a trigger signal, ensuring that both cameras were synchronized with each other and with the shaft's rotation.

Results

First, the dummy disk was modeled in Ansys APDL to determine the rotational velocity range that excites the first modal family. The nodes were constrained to replicate the actual boundary conditions. For this modal family, the frequency range

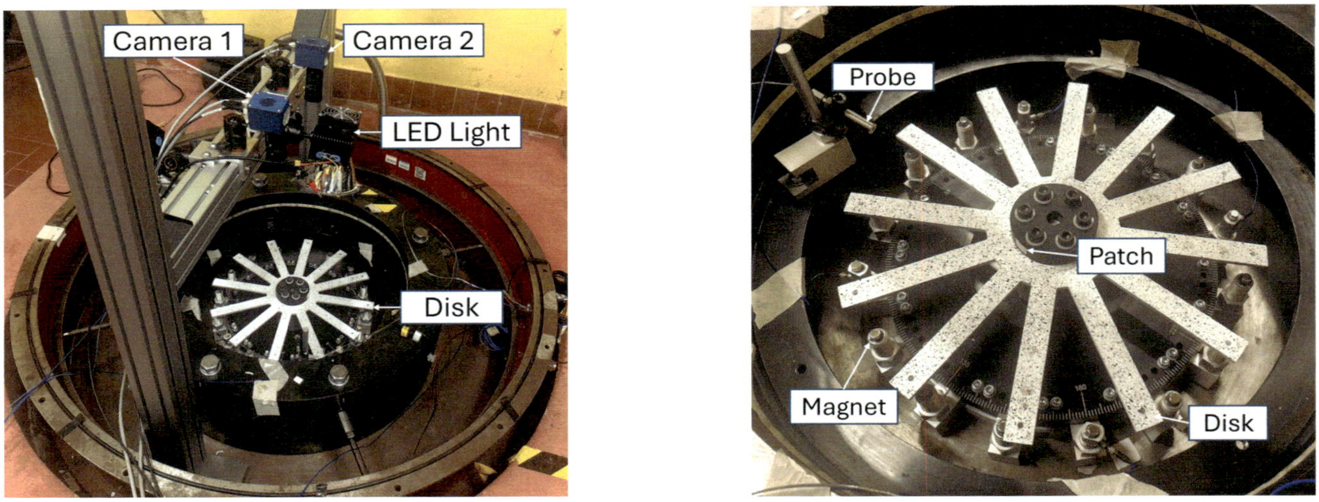

Fig. 1 Test rig setup (left) and close-up view of the disk (right).

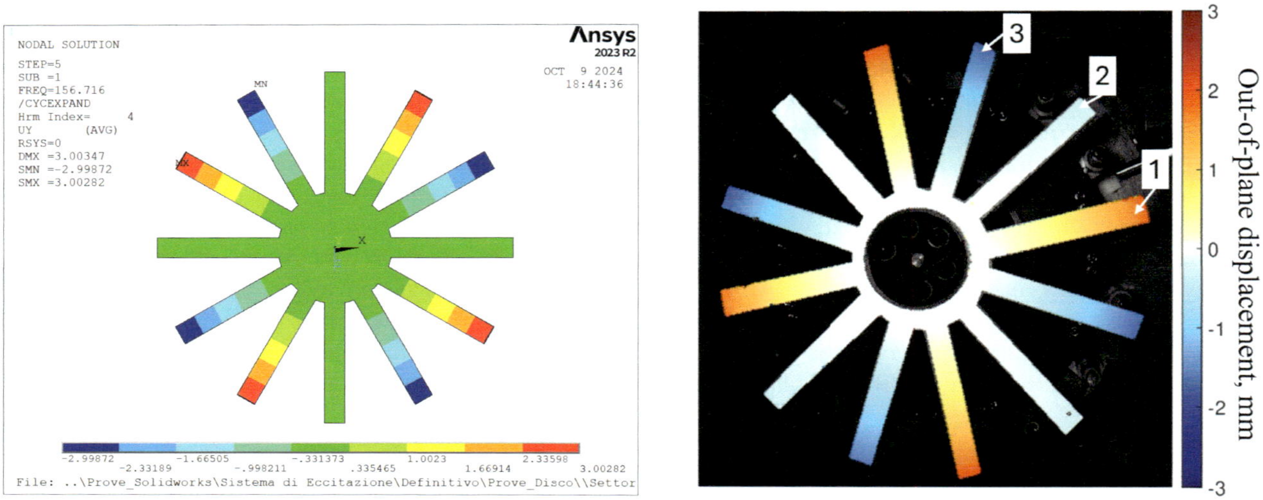

Fig. 2 Out-of-plane bending mode for the dummy disk at ND=4 in the first modal family: numerical simulation (left) and experimental results (right). The numbers on the right figure indicate the points for which the displacement is plotted over time in Fig. 3.

of interest was identified between 148 Hz and 158 Hz. Specifically, the mode with nodal diameter $ND = 4$, shown in the left side of Fig. 2, was found at 154 Hz. Accordingly to Eq. 2 the excitation force was estimated to be around 580 rpm.

Therefore, a linearly increasing speed profile was applied to the disk, ramping up from 0 rpm to 600 rpm over a duration of 11 minutes. Cameras capture the disk motion with a sampling frequency, imposed by the external optical probe, and equal to the rotation frequency. Thus, the maximum sampling frequency is approximately 10 Hz. The color map on the right side of Fig. 2 illustrates the displacement measured by DIC as the reflective patch passed in front of the optical probe at a rotational speed of 567 rpm, highlighting that the mode shape with a nodal diameter of 4 was effectively excited.

The downsampled out-of-plane displacement signal in three points located at the tips of three blades are illustrated in blue Fig. 3. Due to the limitations of having downsampled data, it is not possible to directly compute the resonance frequencies and damping ratio. Consequently, the Single Parameter Technique (SPT) was applied, with the red curves representing the results of the fitting procedure. The parameters obtained for these three points are summarized in Table 1. SPT was performed separately for each of the 27,000 DIC points on the disk surface, yielding a mean resonance frequency of $151.42Hz$ with a standard deviation of $0.07Hz$, and a mean damping ratio of 2% with a standard deviation of 0.06%.

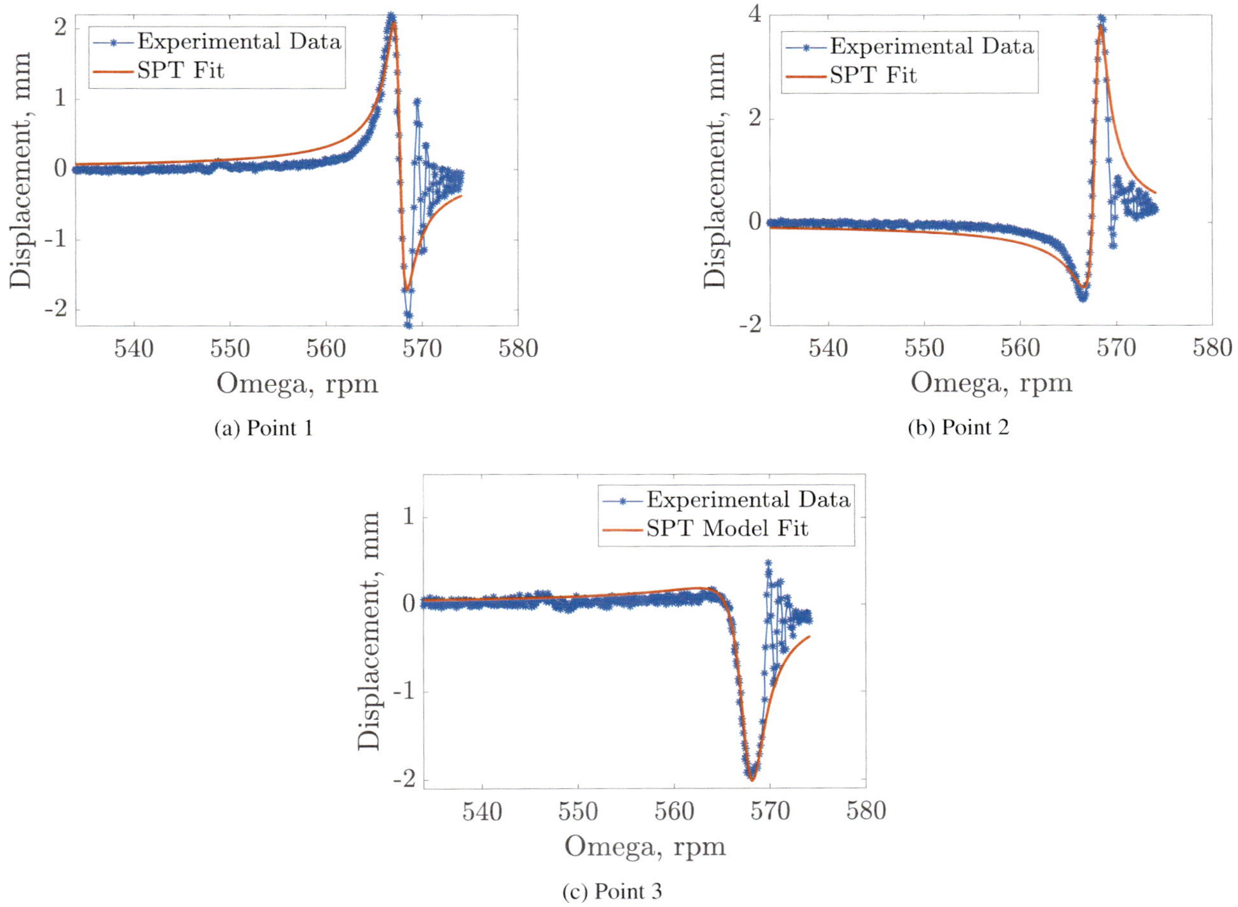

Fig. 3 The blue line depicts the out-of-plane displacement, down-sampled using Digital Image Correlation (DIC), for three points located at the tips of three adjacent blades, as shown in Fig. 2. The red line represents the fitted curve obtained through the Single Parameter Technique (SPT).

Table 1 SPT results for the three points located at the tips of three adjacent blades, as shown in Fig. 2.

	Point 1	Point 2	Point 3
ω_n (rad/s)	951.18	951.58	951.18
ζ	1.1 %	1.4 %	2.7 %
$\delta(\mu m)$	8.7	13.6	11.8
$\varphi_0(rad)$	0.1	2.6	5.3

Conclusion

In this paper, a new methodology for measuring the synchronous vibration of bladed disk was proposed. The methodology was tested on a dummy disk characterized by a simplified geometry. Utilizing a spinning rig equipped with permanent magnets to induce synchronous vibrations, the bending mode with a nodal diameter of 4 was captured by two low-frame rate, high resolution cameras. Thus, the Digital Image Correlation (DIC) technique was employed to obtain detailed full-field displacement data from the disk surface. To avoid aliasing error, the downsampled data computed by DIC were post-processed by the Single Parameter Technique (SPT), allowing for an in-depth analysis of the vibration mode. Overall, the results open the way for further exploration of advanced vibration monitoring techniques that can improve the reliability and safety of industrial applications involving rotating components.

References

1. Polytec. "Surface vibration measurement on rotating components" (2018) Application Note.
2. Boedecker, S., Dräbenstedt, A., Heller, L., Kraft, A., Leonhardt, A., Pape, C., Ristau, S., Reithmeier, E., and Rembe, C. "Optical derotator for scanning vibrometer measurements on rotating objects". In *SPIE Proceedings*. SPIE (2006)
3. Lomenzo, R.A., Barker, A.J., and Wicks, A.L. "Laser vibrometry system for rotating bladed disks". In *Proceedings of the 18th IMAC*, pages 277–282 (1999)
4. Zielinski, M. and Ziller, G. "Noncontact vibration measurements on compressor rotor blades". *Measurement Science and Technology*, 11(7):847–856 (2000)
5. Battiato, G., Firrone, C., and Berruti, T. "Forced response of rotating bladed disks: Blade tip-timing measurements". *Mechanical Systems and Signal Processing*, 85:912–926 (2017)
6. Minervini, D., Mastrodicasa, D., Geluk, T., and Di Lorenzo, E. "Novel methodology for isolating rotational phenomena in tire testing". *INTER-NOISE and NOISE-CON Congress and Conference Proceedings*, 268(7):1845–1856 (2023)
7. Occhipinti, S., Mastrodicasa, D., Manzato, S., and Di Lorenzo, E. "Application of digital image correlation in operational modal analysis of rotating structures". In *ISMA USD Conference Proceedings* (2024)
8. Barone, S., Neri, P., Paoli, A., and Razionale, A.V. "Low-frame-rate single camera system for 3d full-field high-frequency vibration measurements". *Mech Syst Signal Pr*, 123:143–152 (2019)
9. Botto, D., Occhipinti, S., Firrone, C.M., and Neri, P. "High-frequency nonlinear vibration analysis through low-frequency stereo-camera systems". *Mechanical Systems and Signal Processing*, 223:111821 (2025)
10. Heath, S. "A new technique for identifying synchronous resonances using tip-timing". *J. Eng. Gas Turbines Power*, 122(2):219–225 (2000)
11. Dimitriadis, G., Carrington, I.B., Wright, J.R., and Cooper, J.E. "Blade-tip timing measurement of synchronous vibrations of rotating bladed assemblies". *Mechanical Systems and Signal Processing*, 16(4):599–622 (2002)
12. Joung, K.K., Kang, S.C., Paeng, K.S., Park, N.G., Choi, H.J., You, Y.J., and Von Flotow, A. "Analysis of vibration of the turbine blades using non-intrusive stress measurement system". In *Asme Power Conference*, volume 42053, pages 391–397 (2006)
13. Wang, S., Chai, P., Xiao, Z., Tian, J., Chen, Y., and Ouyang, H. "Parameter identifications of synchronous vibration of rotating blades with large amplitude based on blade tip timing". *Journal of the Global Power and Propulsion Society*, 8:39–51 (2024)
14. Zablotskiy, I.Y. and Korostelev, Y.A. "Measurement of resonance vibrations of turbine blades with the elura device". *Energomashinostroneniye,(USSR)*, 2:36–39 (1970)
15. Blaber, J., Adair, B., and Antoniou, A. "Ncorr: Open-source 2d digital image correlation matlab software". *Experimental Mechanics*, 55(6):1105–1122 (2015)
16. Hartley, R. and Zisserman, A. *Multiple View Geometry in Computer Vision*. Cambridge University Press (2004)
17. Inc., T.M. "Computer vision toolbox" (2023)

Chapter 12
Experimental Modal Analysis of a Twin-disc Tribometer using a 3D-scanning Laser Doppler Vibrometer

A. Mario Puhwein, Andreas Biederbeck, Balazs Jakab, Markus Varga, and Markus J. Hochrainer

Abstract Tribometers are employed to analyse the friction and wear properties of material pairings. A twin-disc tribometer enables the dynamic analysis of disc pairs. The dynamic behaviour of the testing machine can significantly influence the tribological results. To quantify these influences and precisely model the systems, the dynamic parameters of the test stands, must be precisely determined. The latest measuring technologies, such as 3D scanning laser Doppler vibrometers, are employed for this purpose. One of the key advantages of this technology is its ability to perform non-contact measurements on rotating components, which enables high-resolution and precise recording of the vibrations. Another advantage of this innovative technology is, that it allows the detection of complex displacements and vibration modes in three dimensions, which leads to significantly improved modal parameters. The primary objective of this study is to identify the modal parameters of a twin-disc tribometer through experimental and numerical modal analysis. These findings provide the foundation for the development of a digital twin test rig. It enables a more precise prediction of the tribometer's behaviour under different operating conditions, thus contributing to optimisation of tribological tests and improvement of material pairings. The utilisation of such sophisticated measurement system and analytical techniques represents a significant advancement in tribology, thereby facilitating the development of more efficient and durable materials.

Keywords 3D-scanning laser Doppler vibrometer · Twin-disc tribometer · Experimental modal analysis · Tribology

Introduction

A twin-disc tribometer is a specialized testing apparatus used in tribological research and material testing to study friction and wear between two rotating surfaces. This device enables realistic simulations of the contact conditions encountered in many engineering applications, such as in the automotive industry, mechanical engineering, and materials science. The basic operation of a twin-disc tribometer involves two discs, which can be made from same or different materials, rolling or roll-sliding contact against each other while various parameters such as pressure, speed, and temperature are controlled. [1]

An important application of the twin-disc tribometer is in researching wheel-rail contact, particularly in the study of railway vehicles. In this context, the slip ratio can be varied to simulate the interaction between the wheel and rail under different operating conditions. These tests are crucial for understanding phenomena like corrugation and general rail surface wear, which directly impact the safety and efficiency of rail transport [2]. Moreover, the twin-disc tribometer is frequently used in the automotive industry to test the friction properties of brake pads and discs. These tests are essential for evaluating the performance and durability of brake components. By analysing friction under various conditions, such as different temperatures and pressure levels, engineers can optimize material composition and treatment to ensure improved braking performance [3].

In the mechanical engineering sector, the twin-disc tribometer plays an important role in examining the fatigue behaviour of transmission components. This test method is used to analyse the dynamic properties and their behaviour under repeated

A. Mario Puhwein · Andreas Biederbeck · Balazs Jakab · Markus Varga
AC2T research GmbH,Viktor Kaplan-Straße 2/C, 2700 Wiener Neustadt, Austria
e-mail: anton.puhwein@ac2t.at; andreas.biederbeck@ac2t.at; balazs.jakab@ac2t.at; markus.varga@ac2t.at

Markus J. Hochrainer
University of Applied Sciences Wiener NeustadtJohannes Gutenberg-Strasse 3, 2700 Wiener Neustadt, Austria
e-mail: markus.hochrainer@fhwn.ac.at

Janko Slavic et al. (eds.), *Computer Vision & Laser Vibrometry, Vol. 6*, Conference Proceedings of the Society for Experimental Mechanics Series, https://doi.org/10.13052/9788743801511_12

loads. These insights are particularly important for the development of durable and reliable drive systems, used in a wide range of applications, from vehicles to industrial machinery [4]. Another example of the twin-disc tribometer applications is the study of wear mechanisms in polymer materials [5]. When developing new plastics, it is important to understand how these materials behave under different contact conditions. The twin-disc tribometer allows for testing specific properties such as wear resistance and temperature stability, thereby facilitating the selection of suitable materials for various applications.

A key advantage of the twin-disc tribometer is its ability to provide precise and reproducible measurements. By continuously monitoring parameters such as torque, speed, and temperature, researchers can gain a detailed understanding of the tribological processes occurring in the contact area. These data are essential for predicting material behaviour in real-world applications and for developing effective strategies to minimize wear.

However, it is important to note that the twin-disc tribometer exhibits eigen-dynamics that must be considered. These dynamics refer to the internal vibrations and movements of the system that can occur during testing. To accurately assess the influence of the device's dynamic behaviour on the tribological results and reduce their impact, the dynamic behaviour of the tribometer must be thoroughly understood and described. A comprehensive analysis of these aspects can help minimize systematic errors and improve the validity of the test results. [6]

In summary, the twin-disc tribometer is an indispensable tool in tribological research. With its wide range of applications across various industries, it contributes to a deeper understanding of friction and wear, supports the development of new materials, and improves the efficiency of technical systems. The ongoing development of these testing methods and technologies will continue to play a central role in materials research and testing, helping to meet the challenges of the future industrial demands.

In this context, the development of a digital twin of such a tribometer is becoming increasingly important. A digital twin enables the virtual simulation of real test conditions, allowing for the analysis of system behaviour under various conditions without the need for extensive physical testing. This not only saves time and costs but also opens up new possibilities for optimizing materials and systems. [7]

However, to develop a digital twin, a highly precise and well-calibrated simulation model is essential. Only such a model can accurately replicate the complex dynamic and tribological processes of the real tribometer and provide reliable predictions. Precise calibration of the model using experimental data is critical to ensuring alignment between simulation and reality, thereby unlocking the full potential of the digital twin. [8]

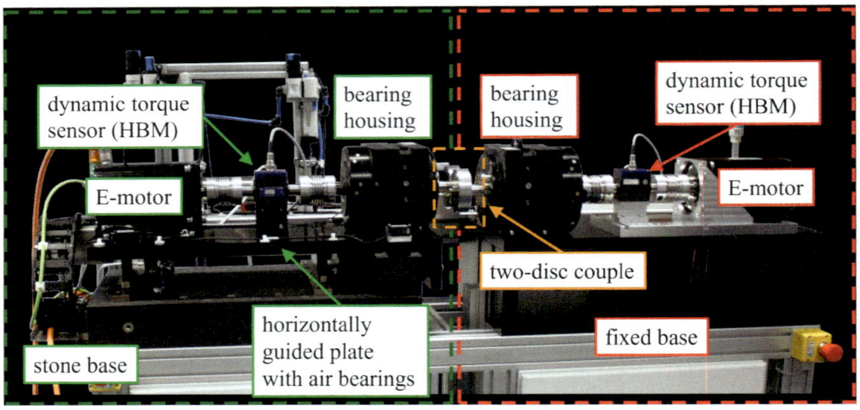

Fig. 1 Twin-disc tribometer and its main components to be analysed

The aim of this work is to investigate the eigen-dynamics of the twin-disc tribometer [3], with particular emphasis on certain components of this test apparatus, as illustrated in Fig. 1. The discs are mounted on two parallel drive trains, each powered by BHM1403P06A2A E-motors (Schneider Electric, Austria) with a nominal torque of $14.4\ Nm$. The speed of the two discs can be controlled continuously and independently of each other, allowing for small slip ratios from $< 1\%$ up to 100%, at rotational speeds from 0 to $1500\ rpm$. To minimize internal friction, the moving side is supported in the horizontal plane with air bearings EZ-0053-080 (Eitzenberger, Germany). The load is applied by a pneumatic cylinder, ensuring that the moving side can be pressed horizontally against the fixed side in the horizontal plane. The acting normal force is measured by a force transducer C9C (Hottinger Brüel & Kjaer, Germany) The coefficient of friction is determined via the torque, which is measured in both drive trains using a torque sensor T21WN (Hottinger Brüel & Kjaer, Germany).

In order to accurately investigate the effect of the eigen-dynamics of the twin-disc tribometer on the tribological results, Experimental Modal Analysis (EMA) [9] is used to analyse the vibration behaviour and identify the modal parameter of the system. This method allows a detailed capture and understanding of the system's natural modes of vibration. Advanced technologies such as laser Doppler vibrometers (LDV) [10] are used in combination with an automated modal hammer. This setup allows for precise measurement of the vibration modes of the tribometer.

The collected data are then analysed using AbraVibe [11] in MATLAB©, which allows the determination of various modal parameters. These parameters include effective mass, damping, stiffness, eigenvalues, and eigenvectors. Such modal parameters are critical for understanding the dynamic properties of the system and for identifying potential weaknesses and unwanted resonances that could affect the quality of tribological test results.

With EMA, engineers and researchers can capture the natural frequencies of the mechanical system and take targeted action to mitigate the effects of its eigen-dynamics. This not only improves the accuracy and reproducibility of measurement results, but also helps to improve the reliability and performance of the entire test setup. Another key benefit of these approaches is their potential to promote environmentally friendly, climate-neutral, and resource-efficient solutions. By increasing the efficiency of materials and processes, both material consumption and environmental impact can be reduced. These analyses are essential to ensure the quality of tribological research and to develop innovative materials and technologies that meet the demands of modern applications.

Methodology

In this work, LDVs are mainly used to record measurement data, offering the advantage that they do not need to be physically connected to the structure due to the non-contact measurement process and therefore do not influence the dynamic behaviour of the system [12].

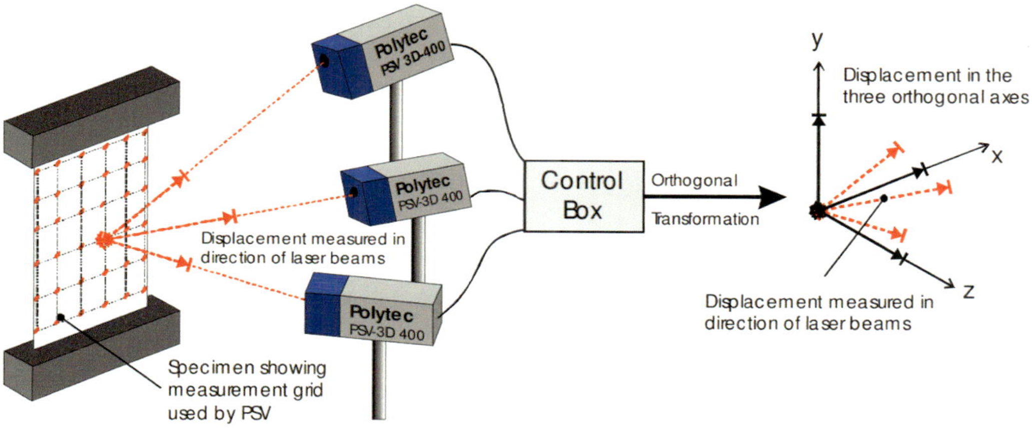

Fig. 2 Schematic of the vibration measurement with the 3D scanning LDV and the determination of the Cartesian components from the 3 laser measurements [13]

LDVs operate on the principle of the Doppler effect, allowing them to directly and accurately measure surface velocity by detecting the frequency shift between the incoming and reflected light received by the measurement system. Similar to a 3-axis accelerometer, multiple LDVs can be used to measure three-dimensional displacement by arranging three scanning LDVs in a specific configuration. When properly aligned, LDVs provide measurement data comparable to that of commonly used accelerometers [9]. Fig. 2 schematically illustrates the operation of a 3D scanning LDV, showing how the vector sum of the three individual laser heads produces a result that can be easily transformed into the three components of a cartesian coordinate system. This capability proves to be a significant advantage when interpreting vibration modes in practice [14]. Thanks to the scanning function of LDVs, it is possible to quickly acquire large amounts of data from extensive structures and analyse large structures with a high density of measurement points [15].

In this study, classical EMA is used to identify the modal parameters of the mechanical system. EMA uses known or specially generated excitation forces (modal hammer, shaker) and the resulting accelerations, velocities or displacements to analyse the dynamic behaviour. To perform a modal analysis, measurement data are first obtained and frequency response functions $H(\omega) = V(\omega)/F(\omega)$ where V, F, ω represent the frequency spectrum of the measured velocity, the excitation force, and the excitation frequency, respectively. Following this, more general system identification methods are applied to determine the modal parameters of the studied model.

From these transfer functions, the modal parameters can be extracted using the following polynomial approach:

$$H(\omega) = \sum_{r=1}^{N} \frac{\phi_r \, \phi_r^T}{\zeta_r \omega_r + j\left(\omega - \omega_{d,r}\right)} + \frac{\phi_r^* \left(\phi_r^*\right)^T}{\zeta_r \omega_r - j\left(\omega - \omega_{d,r}\right)} \tag{1}$$

where ϕ_r and ϕ_r^* represent the eigenvector and its complex conjugate, ζ_r is the damping ratio, ω_r the undamped natural frequency, $\omega_{d,r} = \omega_r \sqrt{1-\zeta_r^2}$ the damped natural frequency of the r-th mode of the system, ω is the excitation frequency, and N represents the number of modes. [9]

As the modal order (N) is increased, there is a systematic expansion of the model framework which is compared with the measured transfer function. This comparative analysis includes an assessment of the stability of the pole pairs with respect to selected parameters such as eigenfrequencies and damping ratios. The results of these stability investigations are presented in stability diagrams, from which the stably identified modes and their modal parameters can be derived. [11]

Another essential instrument in the EMA is the Modal Assurance Criterion (MAC) matrix, which is used to assess the similarity between mode shapes obtained from the same or different measurements or to compare measurement results with numerical simulations. The MAC between two mode shapes ϕ_r and ϕ_s is defined as:

$$\mathrm{MAC}_{r,s} = \frac{\left|\phi_r \, \phi_s\right|^2}{\left(\phi_r^H \phi_r\right)\left(\phi_s^H \phi_s\right)} \tag{2}$$

where ϕ_r and ϕ_s and ϕ_r^H and ϕ_s^H are the r-th and s-th modes and their Hermitian transposes.

The MAC values range is between 0 and 1, with a MAC value of 1 indicating that the mode shapes are identical, while a value of 0 means that they are orthogonal. Values between 0 and 1 indicate partial similarity. To determine the MAC for a system with several mode shapes, a MAC matrix is created in which the MAC values for all possible pairs of mode shapes are mapped. Dependencies between modes in one measurement or different measurements can be determined. Furthermore, the matrix is crucial for the validation of experiments and numerical models, as high MAC values indicate successful modelling of the dynamic properties of a system. Thus, the MAC is an important tool in dynamic analysis that contributes to the quality assurance of models and is essential for the development of digital twins. [9]

Measurement Setup

In order to facilitate the detection of the various vibration modes, the test stand is subjected to measurements from two distinct directions: from the "front" and from the "top." The front measurement direction is chosen to capture movements mainly in the x- and y-axes of the components, while the top measurement direction allows movements in the z-x plane to be recorded, focusing on the torsional modes of the horizontally mounted plate on the flexible side of the test rig. The measurement grids used are shown in Fig. 3. Excitation by a modal hammer was mainly applied to the two discs to stimulate the modes that would also be excited by frictional contact during the operation of the test rig.

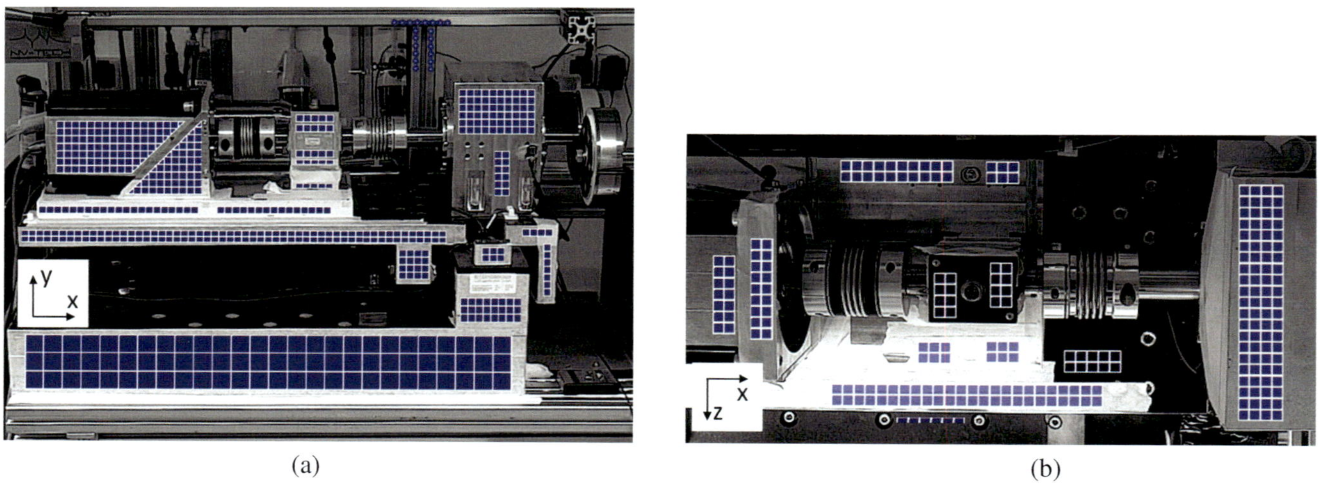

Fig. 3 Measurement grids on the twin-disc tribometer a) "front" direction b) "top" direction

For the measurements, a steel disc with a diameter of $145\ mm$ and a width of $40\ mm$ was mounted on the rigid side, while a steel disc with a diameter of $140\ mm$, a width of $40\ mm$ and a composite layer of approximately $5\ mm$ thickness and $30\ mm$ width was used on the flexible side. These discs were pressed together with a normal force of $800\ N$.

Advanced measuring instruments were used to perform the modal experiments and collect data, including a SAM1, a scalable modal hammer (NV-Tech, Germany), high-precision HBK 4524 and HBK 4507-B accelerometers (Hottinger Brüel & Kjaer, Germany), and state-of-the-art 1D and 3D laser Doppler vibrometers (LDV), including the VibroGo and the 3D scanning laser Doppler vibrometer PSV-3D QTec-600 (Polytec, Germany) [16].

The measurements were conducted for all scanning points with a bandwidth of $12.8\ kHz$, a sampling rate of $32\ kHz$, a measurement time of $4\ s$, and a frequency resolution of $0.25\ Hz$. A complex averaging over three repetitions was applied, and the calibration factor was set to $25\ mm/s$. Velocities and accelerations were measured using an exponential window, while force was measured with a force window. Speckle tracking was set to "very good," and the tracking filter was configured to "slow."

Results

This chapter presents the results of the experimental modal analysis of the two-disc tribometer. The measurement data described in the previous chapter are analysed to identify the different vibration modes and their associated modal parameters. The data recorded with the PSV-3D QTec-600 are analysed using the AbraVibe toolbox in MATLAB [11], as mentioned before, to investigate the relationships between the excitation frequencies and the resulting natural frequencies of the system and to describe the dynamic behaviour of the tribometer. It is important to note that the presented results represent only a very small excerpt of the entire measurement campaign, which was extensive in scope.

As discussed in the Methodology chapter, the results of the parameter identification are visualised in stability diagrams (see Fig. 4 to provide a clear understanding of the identified modal properties. In these diagrams, the frequency is plotted on the x-axis and the number of poles on the y-axis (see equation 1). The red 'o' indicates unstable pole pairs and the green '+' stable pole pairs. For better visualisation, the transfer function used for identification (blue line) is also shown. Fig. 4a) shows four identified modes in the frequency range from $240\ Hz$ to $290\ Hz$, which are labelled with green '+' at the upper end of the y-axis. These were detected by the measurements from the "front" direction. Furthermore, the stability diagram in Fig. 4b) shows a stably identified vibration torsion mode at $140\ Hz$, which was recorded from the "top" direction.

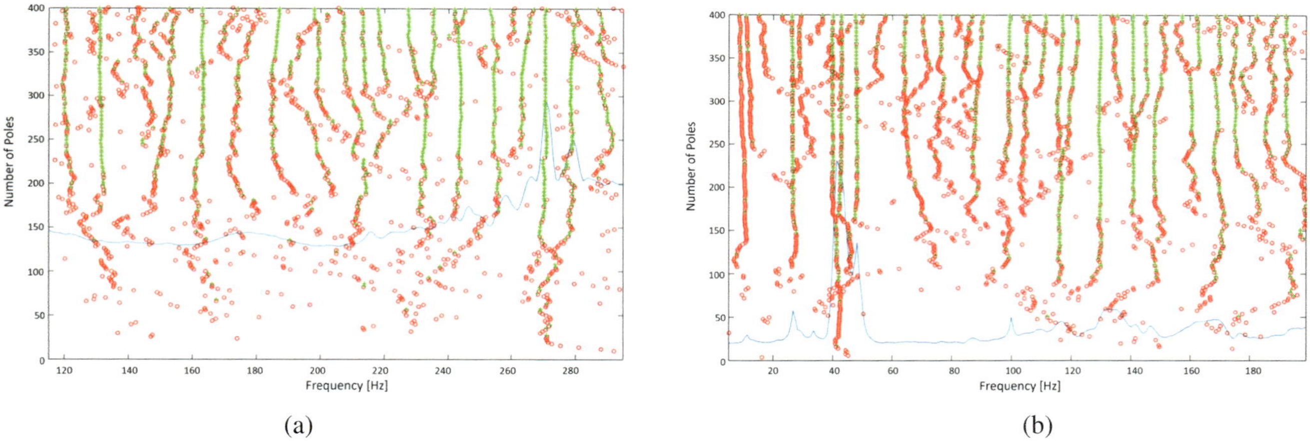

Fig. 4 Stability diagrams of the identification process for one measurement result in the a) 'front' direction b) 'top' direction

In the 3D representation of the vibration modes, the various vibration modes of the tribometer are visualised to illustrate their spatial distribution and intensity. These visualisations provide a clear insight into the dynamic properties of the system and show how the individual components behave in different frequency ranges. The 3D animations help to recognise relevant vibration modes such as torsional modes, bending modes and others that can occur during operation. This makes it possible to better understand and analyse the behaviour of the tribometer under realistic operating conditions.

Fig. 5 a) shows a mode shape at $270,4\ Hz$ with a damping ratio $0,56\%$. It can be observed that the stone slab foundation moves up and down on the right-hand side, while the flexible slab mounted on it with air bearings shows a bending mode. In addition, the movements of the attachments on this plate indicate a bending mode, as the motor moves up and down with its

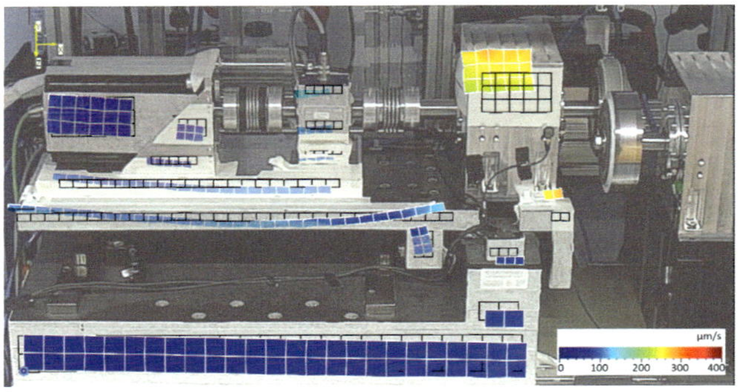

(a)

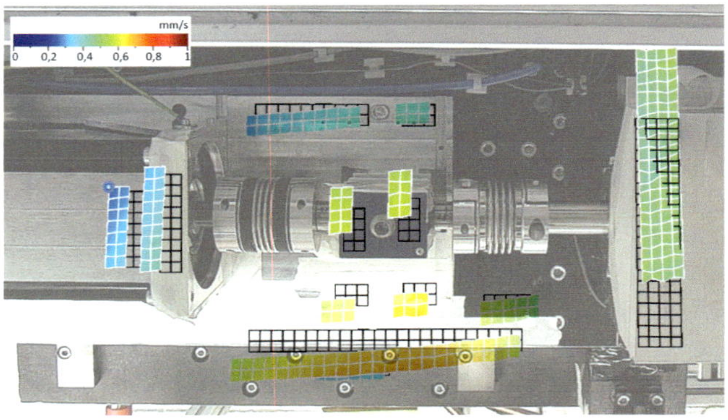

(b)

Fig. 5 Visual representations of mode shapes at a) $270,4\ Hz$ form "front" direction b) $140,7\ Hz$ from "top" direction

free end. The bearing housing on the right-hand side also shows a rotational movement around its mounting. Fig. 5b) shows a mode shape at $140,7\ Hz$ with a damping ratio $0,92\%$. The 3D animation clearly shows that the flexible plate exhibits a torsional mode. This is clarified by the movement of the attachment parts, in particular by the counter-movements of the bearing housing and the motor. In both the "front" and "top" measurements, the discs with the composite cover are excited by an impact from the automatic modal hammer onto the support steel disc.

Fig. 6 presents the MAC matrices (see equation 2) determined for the measurements shown in Fig. 4 and 5, comparing the modes of the respective measurements. The matrix in Fig. 6a) shows the results of the "front" direction measurement, where it can be observed that the modes at $41.7\ Hz$ and $270.4\ Hz$ exhibit a similarity of approximately 40%. This can be attributed to the tilting motion of the bearing housing around the plate in both modes. In the "top" direction measurement, mode similarities are also present, which were confirmed through detailed analyses of the mode animations. Notably, the modes at $42.7\ Hz$ and $140.7\ Hz$, as well as at $89.7\ Hz$ and $229.0\ Hz$, show a similarity of 55% and 45%. These similarities were again traced back to the pronounced movements of the bearing housing.

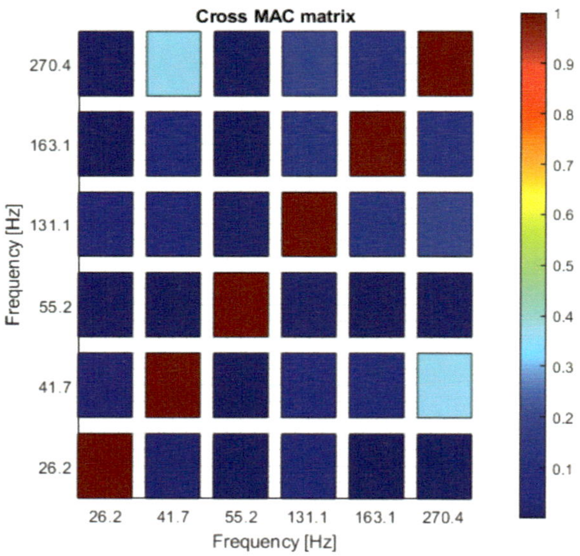

(a)

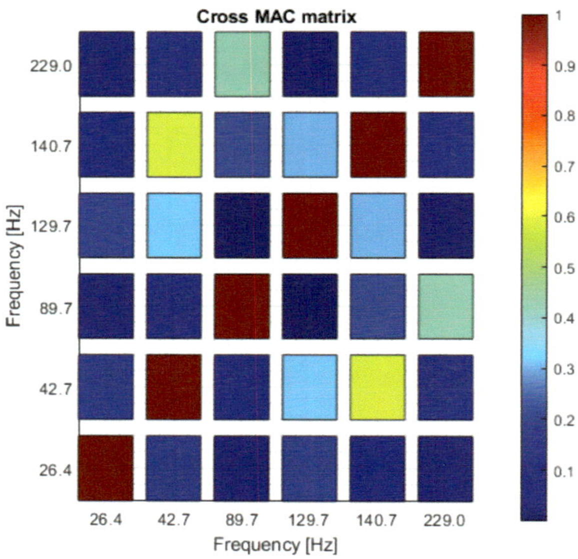

(b)

Fig. 6 MAC matrix for the respective measurement to compare the modes with one another for a) 'front' direction b) 'top' direction

Conclusion

In this thesis, experimental modal analysis was successfully performed on a two-disc tribometer using a 3D scanning laser Doppler vibrometer. The modal parameters of the system were precisely determined in order to better understand the natural dynamic behaviour of the test bench. The non-contact measurement method of the LDV proved to be advantageous, as it enabled the precise recording of the vibration modes without influencing the system itself.

The results obtained provide important information about the natural frequencies, damping rates and mode shapes of the tribometer. This represents an essential step in the development of a realistic simulation model (virtual prototype), which forms the basis for the creation of a successful digital twin [17]. Such a digital twin allows the accurate simulation of the real behaviour of the test rig under different operating conditions and thus contributes to the continuous monitoring and prediction as well as the optimisation of tribological tests.

The experimental identification of the dynamic properties allows the simulation model to be precisely calibrated [18], which is a significant aspect for the further development of a digital twin. Understanding the natural dynamics of the test bench is also an essential step in being able to correctly interpret the frequency-dependent tribological results. These results are an important contribution to improving the predictive accuracy and reproducibility of tribological tests and support the further development of tribological measuring devices, material pairings and lubricants.

Acknowledgments This work was carried out as part of the COMET Centre InTribology (FFG no. 906860), a project of the "Excellence Centre for Tribology" (AC2T research GmbH). InTribology is funded within the COMET – Competence Centres for Excellent Technologies Programme by the federal ministries BMK and BMAW as well as the federal states of Niederösterreich and Vorarlberg based on financial support from the project partners involved. COMET is managed by The Austrian Research Promotion Agency (FFG).

References

1. Y. Hu, W. Wang, M. Watson, K. Six, H. Al-Maliki, A. Meierhofer, and R. Lewis, "Wear of driving versus driven discs in a twin disc rolling-sliding test," *Wear*, vol. 512–513, p. 204528, 2023, doi: 10.1016/j.wear.2022.204528.
2. M. Freisinger, B. Jakab, K. Pichelbauer, G. Trummer, K. Six, and P. H. Mayrhofer, "Fatigue crack initiation in the presence of stratified surface layers on rail wheels," *International Journal of Fatigue*, vol. 177, no. 107958, Dec. 2023, doi: 10.1016/j.ijfatigue.2023.107958.
3. B. Jakab and T. Russ, "Novel twin-disc tribometer setup with small slip conditions for composite materials," in *Proc. 4th Int. Conf. Polym. Tribol.*, Stockholm, Sweden, 2022.
4. G. Meneghetti, A. Terrin, and S. Giacometti, "A twin disc test rig for contact fatigue characterization of gear materials," *Procedia Structural Integrity*, vol. 2, pp. 3185–3193, 2016, doi: 10.1016/j.prostr.2016.06.397.
5. S. Matkovič, A. Pogačnik, and M. Kalin, "Wear-coefficient analyses for polymer-gear life-time predictions: A critical appraisal of methodologies," *Wear*, vol. 480–481, p. 203944, 2021, doi: 10.1016/j.wear.2021.203944.
6. A. M. Puhwein, B. Jakab, C. Haslehner, F. Widder, V. Thalhammer, M. J. Hochrainer, G. Vorlaufer, and M. Varger, "Investigation of the dynamic influences of a two-disc tribometer on wear via high-end cameras and vibrometers," in *Proc. 42nd IMAC, A Conference and Exposition on Structural Dynamics 2024, Vol. 6: Computer Vision & Laser Vibrometry*, J. Baqersad, D. Di Maio, and D. Rohe, Eds. Cham, Switzerland: Springer, 2024.
7. Z. Zhang, Y. Ma, K. He, R. Hu, and M. Hu, "Digital twin enhanced tribo-test service," in *Digital Twin Driven Service*, 1st ed., Amsterdam, Netherlands: Elsevier, 2022, pp. 279–302, doi: 10.1016/B978-0-323-91300-3.00009-7.
8. D. J. Wagg, K. Worden, R. J. Barthorpe, and P. Gardner, "Digital twins: State-of-the-art and future directions for modeling and simulation in engineering dynamics applications," *ASME J. Risk Uncertainty Part B*, vol. 6, no. 3, pp. 030901, Sep. 2020, doi: 10.1115/1.4046739.
9. D. J. Ewins, *Modal Testing: Theory, Practice And Application*, 2nd ed., Research Studies Press LTD., 2000.
10. P. Chiariotti, C. Rembe, P. Castellini, and M. Allen, "Laser doppler vibrometry measurements in structural dynamics," in *Handbook of Experimental Structural Dynamics*, 2021, pp. 1–45. [Online]. Available: https://doi.org/10.1007/978-3-030-65659-3_8.
11. A. Brandt, *Noise and Vibration Analysis: Signal Analysis and Experimental Procedures*, 2nd ed., Wiley, 2023.
12. J. Dank, "Ohne massebeladung wirkung und kompensation in der schwingungsanalyse - applikationsnote," *Polytec Magazine*, Oct. 2016.
13. B. Cazzolato, S. Wildy, J. Codrington, A. Kotousov, and M. Schuessler, "Scanning laser vibrometer for non-contact three-dimensional displacement and strain measurements," in *Proc. Acoustics 2008*, Nov. 2008.
14. S. J. Rothberg, M. S. Allen, P. Castellini, D. Di Maio, J. J. J. Dirckx, D. J. Ewins, B. J. Halkon, P. Muyshondt, N. Paone, T. Ryan, H. Steger, E. P. Tomasini, S. Vanlanduit, und J. F. Vignola, "An international review of laser Doppler vibrometry: making light work of vibration measurement," *Optics and Lasers in Engineering*, vol. 99, pp. 11–22, 2017, doi: 10.1016/j.optlaseng.2017.07.019.
15. J. Dank, "Hohe Punktdichte Optisch das wahre Schwingverhalten messen," *Polytec Magazine*, Aug. 2018.
16. Polytec, "PSV-QTec 3D scanning vibrometer," [Online]. Available: https://www.polytec.com/de/vibrometrie/produkte/full-field-vibrometer/psv-qtec-3d-scanning-vibrometer. Accessed: Sep. 13, 2023.

17. M. Haywood-Alexander, R. S. Mills, M. D. Champneys, M. R. Jones, M. S. Bonney, D. Wagg, und T. J. Rogers, "Full-scale modal testing of a Hawk T1A aircraft for benchmarking vibration-based methods," *Journal of Sound and Vibration*, vol. 576, Art. 118295, Apr. 2024, doi: 10.1016/j.jsv.2024.118295.
18. B. Weekes und D. Ewins, "Effective Use of Scanning Laser Doppler Vibrometers for Modal Testing," in *Topics in Modal Analysis II*, R. Allemang, Ed. Cham: Springer, 2014, vol. 8, pp. 63–72, doi: https://doi.org/10.1007/978-3-319-04774-4_5

Chapter 13
A Neuromorphic, Event-Based, Two-Color Pyrometry for In-process Monitoring of Temperature Change in Welding and Additive Manufacturing

David Mascareñas, Andre Green, Mahtab Heydari, and Allison Davis

Abstract There is currently an increased need to make in-situ measurements of welding and metal additive manufacturing processes to ensure the quality of the resulting fabrications and to predict the resulting mechanical properties. The microstructure of metal parts resulting from these processes plays a large role in determining the resulting mechanical properties. In order to predict the microstructure of these components, it is helpful to measure the solidification rate/time-rate-of-change of temperature in these components. Two-color pyrometry is often used to measure temperature for these types of industrial applications. However, the resulting measurement is often a spatial point measurement or an average over the field of view. It would be preferable to measure temperature changes with high spatial resolution over the entirety of the melt pool and surrounding material in a full-field manner. In prior work researchers have tried to make two- color pyrometry imagers that measure the full-field melt pool, however, these efforts were hampered by the limited dynamic range of conventional 8 bit imagers. Furthermore, the timescales associated with melt pools tend to be on the order of 10^{-3} to 10^{-5} seconds which means that an imager must have a very high frame rate to capture the phenome of interest. Furthermore, the imager would generate large amounts of data that subsequently need to be stored and processed. This large data generation can be impractical on the shop floor. In this work we present a path towards the use of high dynamic range, high-speed, efficient sampling neuromorphic change detection event imagers for the purpose of estimating the time rate of change of melt pool temperatures at high spatial resolution. We anticipate that the development of neuromorphic two-color pyrometer could enable practical in-process monitoring of solidification of melt pools.

Keywords Pyrometer · Dynamic vision sensor · Neuromorphic imager · Event based imager

Introduction

Two-color pyrometry (a.k.a. ratio pyrometry) was originally introduced for the purpose of measuring temperature in a stand-off fashion for the case of unknown emissivity values [1], [2]. Early two-color pyrometers made use of a pair of photodetectors/bandpass filters to capture an average of the "radiative emissions" over the field of view associated with the optics used [3], [4]. These two-color pyrometers were used for measuring the emissions of flames. In 2023 Meyers et al. [5] demonstrated the use of two-color pyrometry with a complimentary metal-oxide semiconductor (CMOS) imager featuring a Bayer filter to sense visible colors. In this work temperature imaging was achieved by examining the ratio of the response for pixels with different Bayer filter color values. The imager was testing in the context of Laser Powder Bed Fusion (LPBF) additive manufacturing. This imager had a number of issues and tradeoffs. First, the pixels were prone to oversaturation, particularly in the middle of the melt pool, making it impossible to recover these temperature measurements. The researchers then tried to reduce saturation by varying the exposure time. However, the researchers report that reducing exposure would reduce the ability to measure the full length of the melt pool. Furthermore, the researchers reported motion blur phenomena in

David Mascareñas · Andre Green · Mahtab Heydari · Allison Davis
Engineering Division, Los Alamos National Laboratory
e-mail: dmascarenas@lanl.gov; andre_green@lanl.gov; mahtabh@lanl.gov; allisonmdavis@lanl.gov

Mahtab Heydari
Department of Mechanical Engineering, Texas A&M
e-mail: mahtabh@lanl.gov

Janko Slavic et al. (eds.), *Computer Vision & Laser Vibrometry, Vol. 6*, Conference Proceedings of the Society for Experimental Mechanics Series, https://doi.org/10.13052/9788743801511_13

the measurements when using a longer exposure time. Ultimately the researchers simply ran a single experiment, multiple times over, at different exposure times, in order to try to work around the tradeoffs between pixel saturation, quality of melt pool geometry measurement and motion blur. These mitigation strategies are not suitable for in-process monitoring of melt pool. However, using asynchronous, neuromorphic event-based imagers should greatly mitigate all the current issues with achieving in-process temperature measurements of melt pools.

Background

Asynchronous neuromorphic imagers are fundamentally different than conventional imagers at the hardware level. An event-based imager does not capture individual photographs or frames. Instead, an event based imager only detects information associated with a change in log light intensity value at an individual pixel. Furthermore, this event detection is done directly in hardware at the pixel level. It must be stressed that the event detection is not the result of an image processing algorithm performed on a conventional image. Lichtsteiner [6] provides a representative schematic of the circuit that performs event detection at each pixel that emits events when a change in log light intensity is detected. The data for each event consists of an event timestamp that typically has ∼10s of microseconds accuracy [7] as well as the x-y position of the pixel that detected the event and a polarity value which indicates if the pixel detected an increase or decrease in light intensity. Data is only reported from the imager if the light intensity level at a given pixel changes. Furthermore, only the pixels whose light intensity value has changed report data. If the pixel event data is arranged into the form of a static image, the resulting data looks like a conventional image that has been processed using an edge detection algorithm. Figure 1 shows an example of an office scene consisting of multiple monitors being observed using an event-based imager. The grey background indicates pixels where no data is reported. In contrast, the white and black pixels indicate areas where the light intensity value has either increased or decreased beyond the threshold. Generally the black and white pixel events occur at the edges of objects because this is where the strongest light intensity changes tend to occur. The result of this detection-at-the-pixel approach, is that event-based sensors consume less power, less memory/bandwidth, and they operate across a wide range of timescales and dynamic ranges [6].

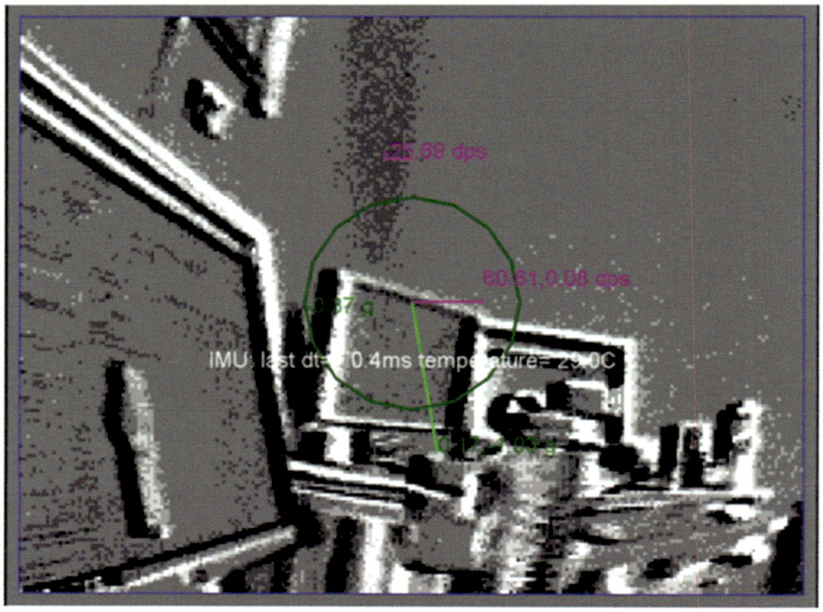

Fig. 1 Example of event data captured from an office scene and formed into a frame [10].

Event imagers have four properties that make them particularly attractive for melt-pool two-color pyrometry. First, the event-detection pixel circuit detects changes in log light intensity which means that imager naturally has a high dynamic range of ∼120 dB, [6], [7]. In prior work the team has found that this high dynamic range is sufficient for observing melt pools [8]. Second event imagers are able to capture phenomena on the sub-millisecond timescale. Conventional high dynamic range weld cameras [9] have framerates on the order of tens of Hertz which is not sufficient for observing melt pools with sub-millisecond dynamics. Third event imagers are very efficient at sampling. In the case of analyzing data from tests associated with observing laser welding melt pools, these imagers have been observed to require approximately 35 times less memory to store data comparable to that of a conventional imager . These memory savings are important because they can potentially help lead to online process control of AM as well as for making it possible to store more data associated with the process parameters during a build. Currently building components requires storing data files on the terabyte scale of memory which is not amenable to storing, processing or communicating over a network. Fourth, event imagers provide a timestamp for each event recorded with 65 - 200 μs temporal accuracy [7]. The high resolution timestamps opens up interesting possibilities for changing the effective exposure time associated with forming frames from the events to better estimate melt pool geometry and for motion blur control. The concept of the digital coded exposure was introduced by Gothard [8], which allows for arbitrary modification of exposure time using event data in post-processing.

Analysis

Experimental Setup

The prototype neuromorphic two-color pyrometer was is shown in Figure 2. Two DVXplorer silicon retina imagers were attached to a beam-splitter sharing a 100 mm lens (Thorlabs MVL100M23) and aimed at a halogen lightbulb. Note that the 100 mm lens/event imagers are being used in a non-standard configuration such that the image comes into focus when the object is located ~21 centimeters from the front of the 100 mm lens. The field-of-view of this optical setup is slightly less than 10 mm on a side. A green 532 nm band-pass filter (ThorLabs FLH532-10) was placed in the optical path of one imager and a red 650 nm band-pass filter (ThorLabs FBH650-10) was placed in the optical path of the other imager; both filters had a full-width-half-maximum (FWHM) of 10 nm. The filament of a halogen lightbulb served as an approximate blackbody radiator and was exposed to voltages across its terminals at integer voltages in the 5V to 10V range (inclusive).

Figure 3 shows a plot of the response ratio vs temperature for the optical setup associated with our neuromorphic two-color pyrometer. Although the mapping between temperature and the response ratio between the spectrum intensity at 532 nm and 650 nm is non-linear, it is one-to-one and so can be inverted to recover the temperature from the spectrum intensity ratio.

The spectrum emitted by the halogen bulb was approximately black-body, but did not conform perfectly to the theoretical predictions (Figure 4); this may be an effect of our spectrometer's diffraction grating or non-idealities associated with the halogen bulb. Consequently the decision was made to track the peak-intensity wavelength rather than use the temperature of the most-fit the blackbody curve to track changes in the filament temperature.

Fig. 2 The beam-splitter (center) has two DVXplorer silicon retina imagers attached (right and bottom); the imager on the right is equipped with a 532 nm filter; the imager at the bottom is equipped with a 650 nm filter. Both imagers see through the beam-splitter to share a 100 mm lens.

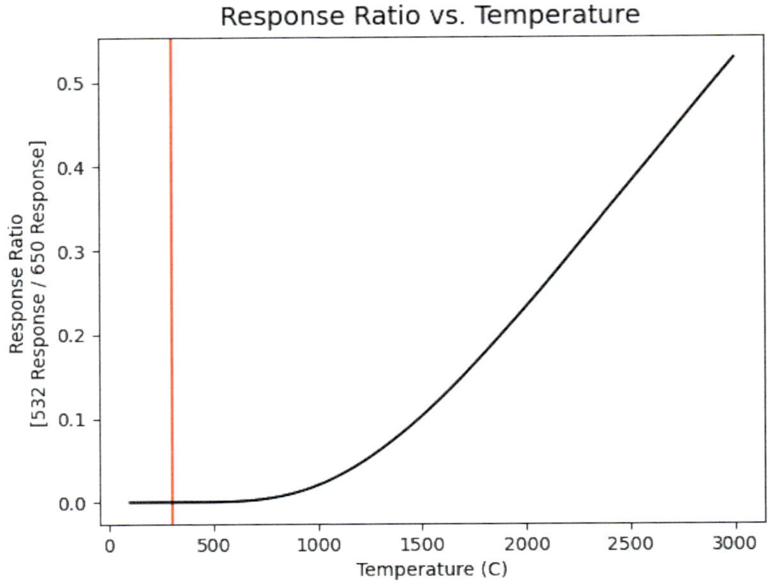

Fig. 3 The mapping between temperature and the ratio of irradiance at 650 nm and 532 nm is invertible.

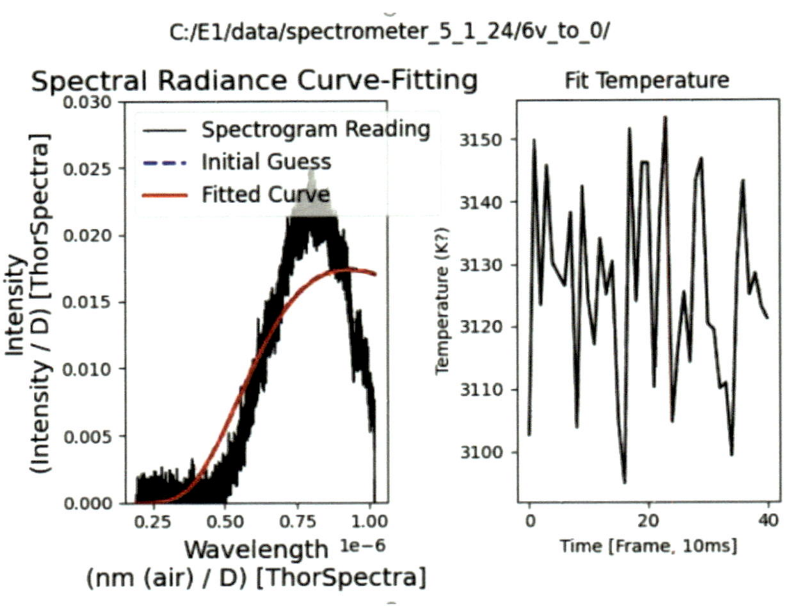

Fig. 4 The mapping between temperature and the ratio of irradiance at 650 nm and 532 nm is invertible.

Experimental Results

Next the wavelength of highest intensity detected by the two-color pyrometer changed in response to the tungsten filament being excited with different voltages is considered. Note, the voltage scale for Figure 5 through Figure 7 is shown in Figure 6. Figure 5 shows the response of the spectrometer to the different initial voltage levels. In these tests the tungsten filament was excited using the specified voltage levels and brought to steady state. Then the voltage was abruptly set to zero in order to induce a temperature change, and thus a light intensity change at the pixels that could be detected using the event-based imagers. Figure 6 shows the event ratio time series associated with the different voltage levels. These ratios are calculated by taking all of the events that occur in the pixels that are associated with the image of the halogen blub. For this initial proof-of-concept the ratio of events at a pixel was used to try to indicate a change in temperature at a pixel or over a set of pixels. Ideally, temperature estimates for each pixel based on the events at that pixel/spatial location should be obtained. However,

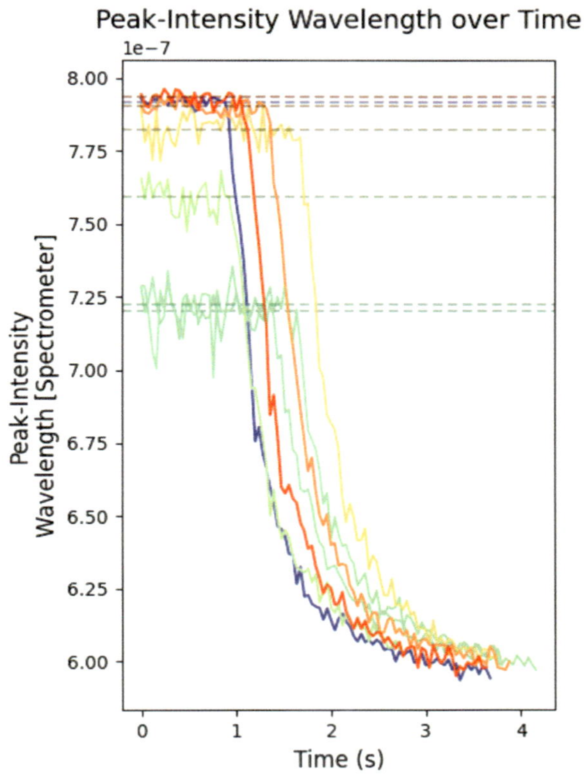

Fig. 5 The spectrogram records different peak intensities for each voltage, with higher peaks for higher voltages as expected. However, the filament appears to reach its peak temperature at 8V.

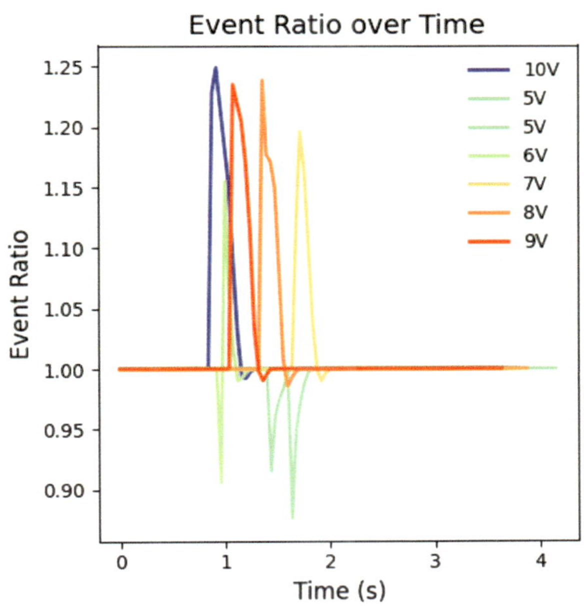

Fig. 6 The event-ratio time-series from each test are overlaid in time. The offsets between the peaks are merely from experimental variation and not of particular relevance.

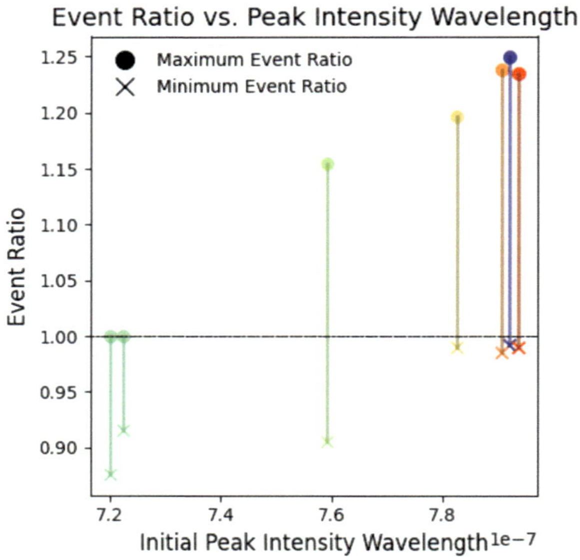

Fig. 7 The minimum and maximum event ratios from each experiment's time-series. The maximum peak intensities for tests above 8V appear saturated as is also seen with the spectrometer.

in initial tests, wide variance in event ratios observed at pixels/spatial locations that should presumably exhibit similar temperature changes, were observed. This observation may be caused by variation in the contrast threshold as mentioned in prior studies of the uniformity of response of pixels in event imagers [6]. To enable per-pixel temperature change estimates it will most likely be necessary to calibrate the imager in order to remove this variance in pixel behavior. The variation in pixel behavior has some similarity to fixed pattern noise in conventional imagers [6].

The wavelength with the greatest intensity as detected by the spectrometer, increases with voltage, although above 9V the response appears to saturate; this may be a limitation of the bulb used. Similarly, the ratio of events between the camera with the 650 nm filter and the 532 nm filter increases as the halogen bulb is supplied a higher voltage (and hence is at a hotter temperature), meaning that the halogen bulb at different temperatures can be differentiated by this method.

At the lowest voltage (5V), events only triggered on a single camera; it is believed that this is because the brightness at 5V was sufficient to trigger events from the 532 nm filtered camera, but insufficient to trigger events from the 650 nm filtered camera. It is also possible this was due to manufacturing variations in the event cameras.

Simulation

In addition to the experimental investigation, a 3D heat-diffusion model was made to test the silicon retina response in simulation. Each pixel of the simulation has an associated temperature; each temperature is mapped to black-body curve; the dot-product of this curve with the frequency-response curve for each simulated 532 nm and 650 nm is the corresponding intensity for the simulated input to the virtual silicon retinas. Then a simulation of the silicon retina response to this irradiance excitation in order to generate events using an approach primarily based on threshold crossings is performed; finally, the ratio of these events is used to estimate the changes in temperature which are combined to generate the final temperature field reconstruction.

In this work the case where the threshold for triggering an event was set at both a low (Figure 10) and high (Figure 11) value is considered.

In simulation, the temperature reconstructions approximate the temperature within the area of interest (within the heated region) well. In both simulated reconstructions, the room-temperature outside border of the region outside the heated area that receives heat is not accurately reconstructed because there is insufficient light being generated by the blackbody radiator to generate events (Figure 12); in future work, this may be rectified with the use of a short-wave infrared (SWIR) event-driven imager.

In the event reconstruction which uses a low event threshold, there is also an issue on the corners within the sample: here the temperature is changing too rapidly for the simulated imager to generate events proportional to the change. A higher threshold can be used to eliminate this issue at the cost of reduced temperature accuracy for slower changes in the center

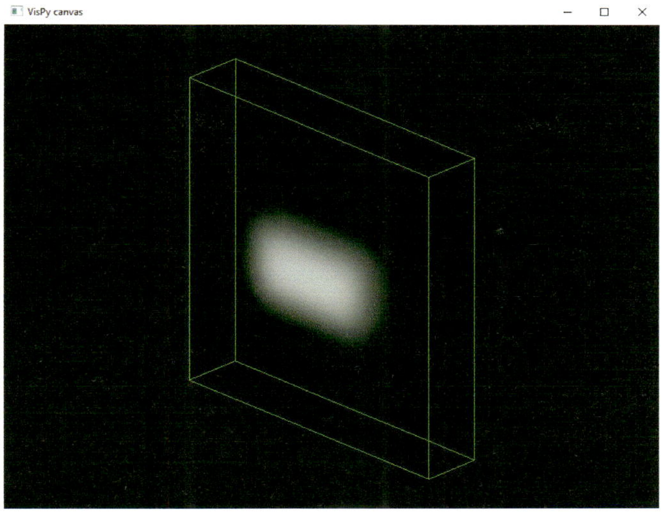

Fig. 8 The 3D heat-diffusion simulation; a block of steel with no cross-medium (air) interaction was used.

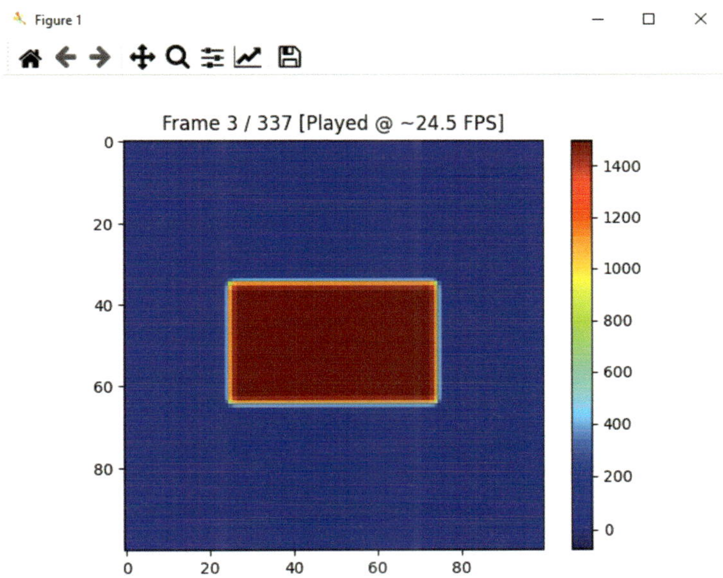

Fig. 9 A 2D view of the top of the simulated sample; this is what would be visible to a thermal imager.

of the heated sample: in short, with a higher threshold the real temperature is changing, but not quickly enough to generate an event such that the reconstruction aligns with the simulation at all times. However, please note, event imagers have a temporal resolution on the order of 65-200 μs [7]. Some of the issues mentioned here, with respect to the simulated imager not being fast enough to generate events, are most likely an artifact of the approach used to generate this simulation and are not necessarily present in observing actual physical systems. Prior work has been done on generating simulations of events, but these techniques typically rely on working from a uniform-in-time simulation framework which is not necessarily well-suited to event-based sensing, and can struggle with phenomena such as noise and the behavior of the arbiter [11], [12], [13]. The problem of simulating event-based imagers is still an open research problem worth of further investigation.

Future Experimental Setup

Simulating the process of event generation is challenging. To alleviate this challenge the researchers developed an alternative sensor-in-the-loop approach wherein the DVXplorer event-imager is coupled with a high-speed monitor (Alienware

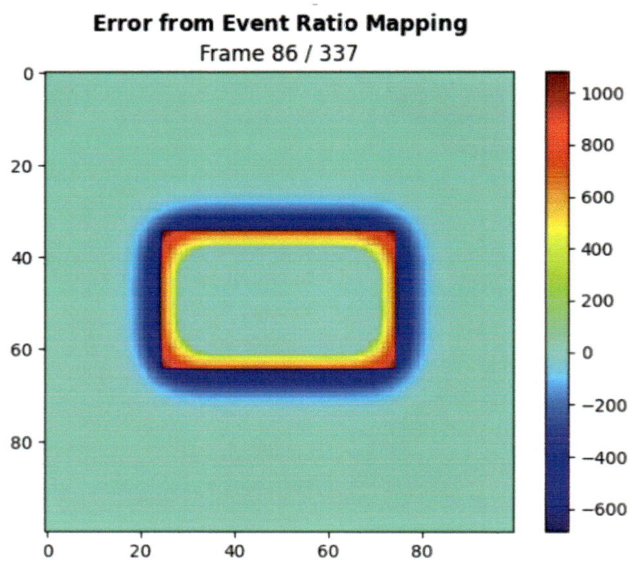

Fig. 10 The error of the reconstruction (using a low threshold for event generation) against the simulated temperature. The inner borders of the heated region cool quickly, over-saturating a low-threshold imager and lead to an over-estimation of temperature.

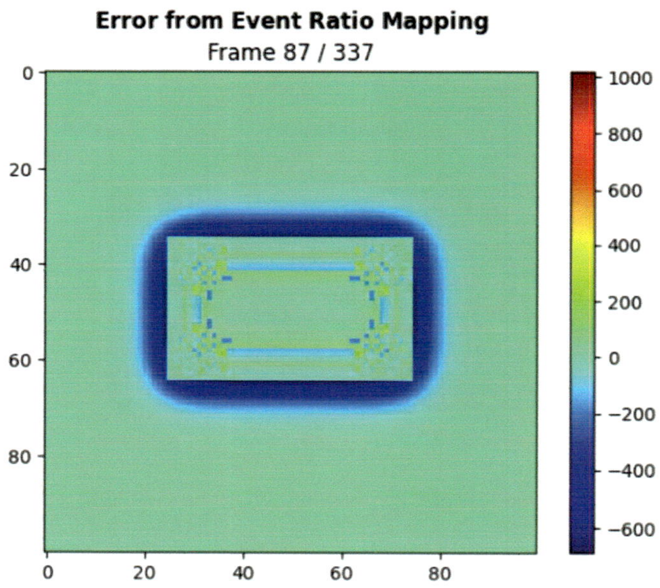

Fig. 11 The error of the reconstruction (using a high threshold for event generation) against the simulated temperature. The issue with the inner borders has been resolved, at the cost of minor error within the regions of the sample whose temperatures change more slowly.

AW2523F, 360 Hz refresh rate) in an optically isolated environment (Figure 13), where a simulation of the desired dynamic experiment (e.g. heat-diffusion) is displayed on-screen. The event-based imager was attached to a beam splitter and mounted to a frame with the front of the beam splitter placed at a distance of 33.5 inches away from the face of the monitor. A 12mm lens was used to observe the monitor using the event imager. The frame used to hold the monitor and imager is of dimensions 23 5/8" x 15 3/4" x 39 1/8", and the bar holding the imager was set to a height of 5 9/16" above the bottom of the frame. Figure 14 shows examples of a diffusion simulation along with examples of events formed into frames captured by observing high-speed monitor displaying the simulation.

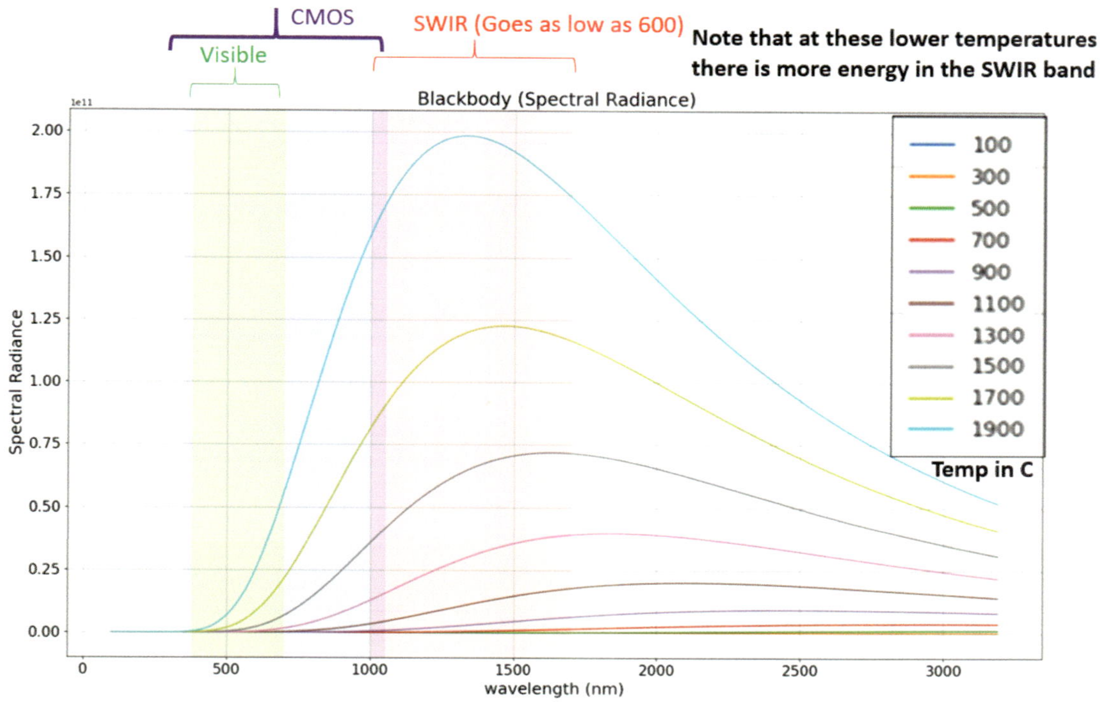

Fig. 12 At low C, the intensity of light at 532 nm and 650 nm is low enough to not be registered by our event-driven imagers; future development of event-based SWIR imagers may rectify this issue.

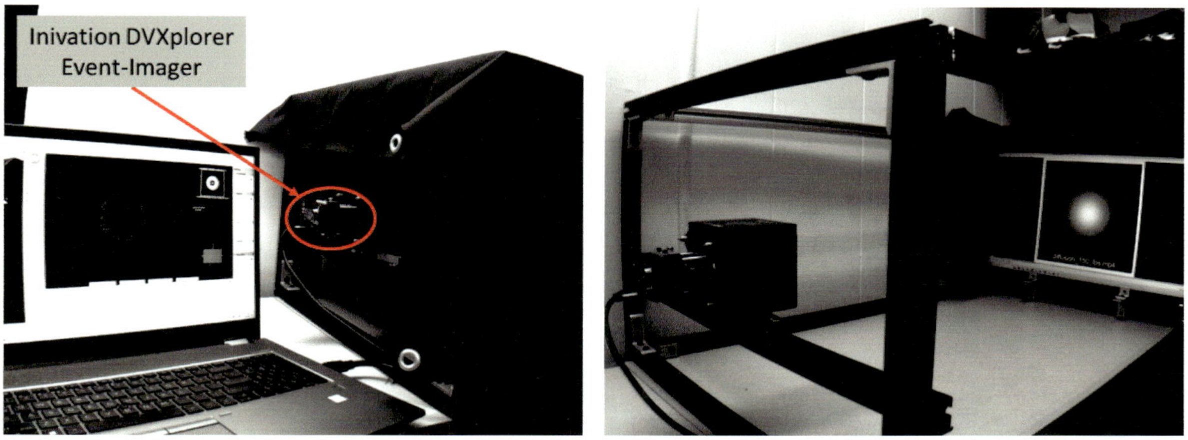

Fig. 13 The future experimental setup; a DVS imager (to be replaced by the beam-splitter assembly shown previously) is mounted on a cage and aimed at a high-speed monitor. The assembly is covered to isolate outside light.

This hybrid sensor-in-the-loop approach can be employed to develop more sophisticated event/image processing algorithms that can better estimate the change in temperature. In addition, this approach should help facilitate providing controlled stimulus to the event imager that can be used to help account for variations in the response of individual pixels. The advantage of this hybrid setup is that the user has full control over the simulation and has access to the full-field ground truth information that led to the events generated by the event imager. The disadvantage of this hybrid setup is that the dynamics that can be emulated is limited by the framerate of the monitor and the dynamic range of the light that can be generated is limited by the light output of the monitor. As such in its current form the hybrid sensor-in-the-loop simulator cannot emulate high light intensity environments.

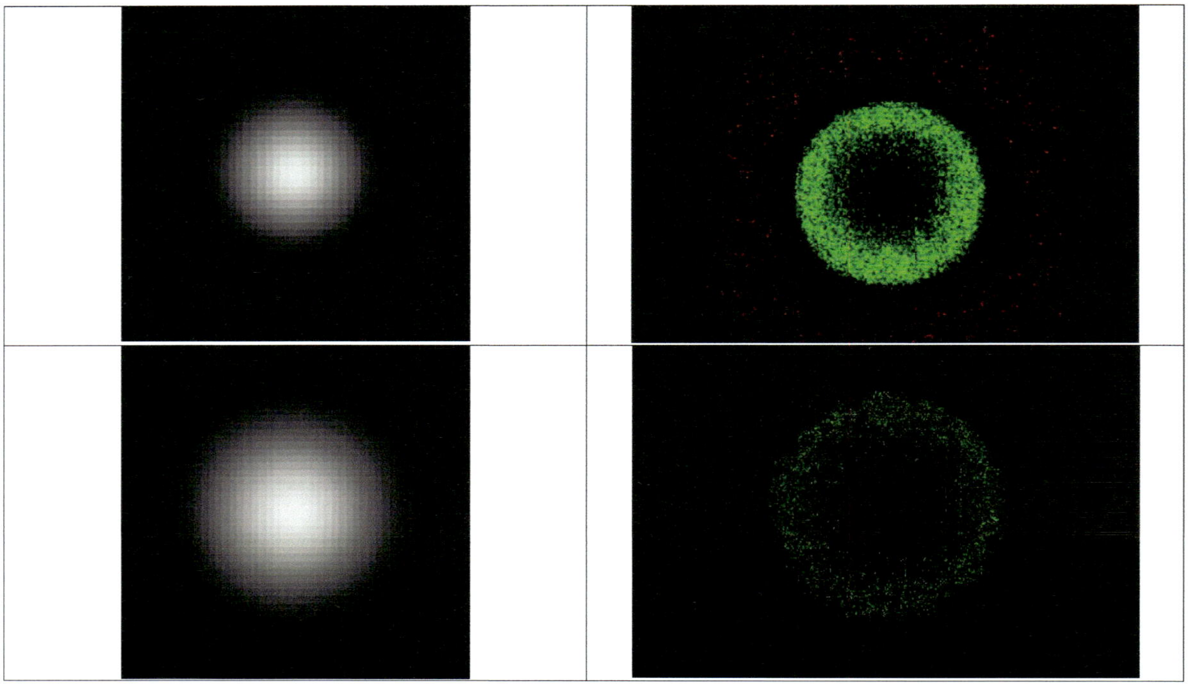

Fig. 14 The monitor-displayed image (photographed on left) is displayed next to examples of the DVXplorer silicon retina imager events, which produces the events shown on the right. Note that the images on the left are not strictly synchronized with the event data on the right.

Conclusion

This work demonstrates a proof-of-concept for the use of event-based imagers for enabling two-color pyrometer imagery in circumstances that would involve a high light intensity environment that cannot be directly observed using conventional 8 bit imagers. Such high intensity light environments are commonly encountered for in-process monitoring of welding as well as metal additive manufacturing. A proof-of-concept of neuromorphic two-color pyrometry has been demonstrated both in simulation as well as with an experimental prototype. The experimental implementation of the neuromorphic two-color pyrometer currently consolidates the response across many pixels as opposed to capturing an image. In order to capture an image associated with temperature change it appears it will be necessary to calibrate out variations in the response of individual pixels. However, the simulation of the event-imaging two-color pyrometer does provide evidence the concept for an event-driven neuromorphic imager is feasible. Future work will consider how variations in response across the pixels can be accounted for to get better estimates of temperature change at high spatial resolution across the full field of view. Variants of the proposed neuromorphic two-color pyrometer could also find application in use cases such as dynamic in-process monitoring of flame temperature during biomass combustion for the generation of clean energy [14], [15].

Acknowledgments Funds for this work were provided by the Los Alamos National Laboratory Technology Evaluation & Demonstration program. Los Alamos National Laboratory is operated by Triad National Security, LLC, for the National Nuclear Security Administration of US Department of Energy (Contract No. 89233218CNA000001). The DVS/DAVIS technology was developed by the Sensors group of the Institute of Neuroinformatics (University of Zurich and ETH Zurich), which was funded by the EU FP7 SeeBetter project (grant 270324).

References

1. A. S. Morris and R. Langari, "Two-color pyrometer (ratio pyrometer)," in Measurement and Instrumentation: Theory and Application, Elsevier Inc., 2012, pp. 347–396.W409W10080
2. J. Dixon, "Industrial radiation thermometry," Journal of Measurement and Control, vol. 20, no. 6, pp. 11–16, 1987. W409W10080
3. S. A. Reggeti, A. K. Agrawal and J. A. Bittle, "Two-color pyrometry system to eliminate optical errors for spatially resolved measurements in flames," Applied Optics, vol. 58, no. 32, pp. 8905–8913, 2019. W409W10080

4. H. Hottel and F. P. Broughton, "Total Radiation from Luminous Gas Flames," Industrial & Engineering Chemistry Analytical Edition, vol. 4, no. 2, pp. 166–175, 1932. W409W10080
5. A. J. Myers, G. Quirarte, F. Ogoke, B. M. Lane, S. Z. Uddin, A. B. Farimani, J. L. Beuth and J. A. Malen, "High-resolution melt pool thermal imaging for metals additive manufacturing using the two-color method with a color camera," Additive Manufacturing, vol. 73, no. 103663, 2023. W409W10080
6. P. Lichtsteiner, P. C. and D. T., "A 128 x 128 120 dB 15 μs Latency Asynchronous Temporal Contrast Vision Sensor," IEEE J. Solid-State Circuits, vol. 43, no. doi:10.1109/JSSC.2007.914337, pp. 566–576, 2008. W409W10080
7. Inivation, "Specifications - Current Models," 2022. [Online]. Available: https://inivation.com/wp-content/uploads/2023/02/2022-09-iniVation-devices-Specifications.pdf. [Accessed 12 May 2023].W409W10080
8. A. Gothard, D. Jones, A. Green, M. Torrez, A. Cattaneo and D. Mascareñas, "Digital coded exposure formation of frames from event-based imagery," Neuromorphic Computing and Engineering, vol. 2, no. 014005, 2022. W409W10080
9. Xiris, "Solutions for Weld Inspection and Monitoring," [Online]. Available: https://www.xiris.com/. [Accessed 20 September 2023].W409W10080
10. C. Schelle, Artist, Event-Based Image of Office Scene. [Art]. Los Alamos National Laboratory, 2018. W409W10080
11. D. Joubert, A. Marcireau, N. Ralph, A. Jolley, A. v. Schaik and G. Cohen, "Event Camera Simulator Improvements via Characterized Parameters," Front Neurosci., vol. 15, no. 702765, 2021. W409W10080
12. H. Rebecq, D. Gehrig and D. Scaramuzza, "ESIM: an Open Event Camera Simulator," in Conf. on Robotics Learning (CoRL), 2018. W409W10080
13. Y. Hu, S.-C. Liu and T. Delbruck, "v2e: From Video Frames to Realistic DVS Events," in IEEE/CVF Conference on Computer Vision and Pattern Recognition Workshops (CVPRW), 2021. W409W10080
14. W. Yan, K. Li, X. Huang, L. Yu, C. Lou and Y. Chen, "Online Measurement of the Flame Temperature and Emissivity during Biomass Volatile Combustion Using Spectral Thermometry and Image Thermometry," Energy and Fuels, vol. 34, no. 1, pp. 907–919, 2020. W409W10080
15. J. P. Silva, S. Teixeira, É. Grilo, B. Peters and J. C. Teixeira, "Analysis and monitoring of the combustion performance in a biomass power plant," Cleaner Engineering and Technology, vol. 5, no. 100334, 2021. W409W10080

Chapter 14
Operational Modal Analysis of a Rotating Structure Using a Novel Image-Based Long-Range Continuously Scanning Laser Doppler Vibrometer

L. F. Lyu and W. D. Zhu

Abstract A novel image-based long-range continuously scanning laser Doppler vibrometer (CSLDV) is developed to measure vibration of a rotating structure with a complex background. The image-based long-range tracking CSLDV can track and scan the rotating structure by processing its images captured by a camera in the image-based long-range tracking CSLDV. A novel image processing method with a background averaging technique is developed to extract the rotating structure from the background in captured images and determine its location. A one-dimensional scan scheme is used to scan along a straight line along the rotating structure. An improved demodulation method is used to process measurements of the image-based long-range tracking CSLDV to estimate modal parameters of the rotating structure, including damped natural frequencies and undamped mode shapes. Experiments of tracking and scanning a rotating blade are conducted to validate vibration measurements of a rotating blade using the image-based long-range tracking CSLDV. Measured responses of the rotating blade are processed to estimate its speeds and modal parameters. The first three one-dimensional damped natural frequencies of the rotating blade and undamped mode shapes along its middle line are successfully estimated and compared with results from the finite element model of the corresponding stationary blade.

Keywords Long-range laser Doppler vibrometer · Continuously scanning laser Doppler vibrometer · Image processing · Rotation tracking · Operational modal analysis

Introduction

A laser Doppler vibrometer (LDV) can measure the velocity of a point on a structure from the Doppler shift of the reflected laser beam of the LDV, which is an efficient way for noncontact vibration measurement of the structure [1]. A single-point LDV can only measure the vibration at one point on the structure each time, and it has to be moved to a new position to measure the vibration at another point. A scanning laser Doppler vibrometer (SLDV) was developed to increase the efficiency of vibration measurement [1–3]. A scanner with a set of orthogonal mirrors is used to reflect the laser beam of a single-point LDV in the SLDV. Mirrors of the scanner can rotate so that the laser spot of the SLDV can be moved to different positions while the SLDV is stationary. The SLDV measures the vibration of a point for some time and moves its laser spot to another point, which can provide vibration measurements on a grid on a structure. Measured responses on the grid can be used to estimate natural frequencies, mode shapes, and operational deflection shapes (ODSs) of the structure. To increase the spatial resolution of vibration measurement of a SLDV, a continuously scanning laser Doppler vibrometer (CSLDV) was developed [4–7]. Mirrors of the scanner in the CSLDV can continuously rotate so that the laser spot of the CSLDV can be continuously swept on the surface of a structure. One-dimensional (1D) and two-dimensional (2D) scan schemes were developed for the CSLDV to scan along different trajectories [7,8]. The CSLDV can be used to obtain full-field mode shapes and ODSs of stationary structures [8–10]. To measure vibration of a rotating structure, an image-based tracking CSLDV was developed [11–15], where a camera is integrated into the CSLDV to capture images of the rotating structure and determine its location. The image-based tracking CSLDV can only track and scan a rotating structure with a short distance from the CSLDV and reflective tapes are required to increase signal-to-noise ratios of measured responses. Hence,

L. F. Lyu · W. D. Zhu
Department of Mechanical Engineering, University of Maryland, Baltimore County, 1000 Hilltop Circle, Baltimore, MD 21250
email: linfenl1@umbc.edu; wzhu@umbc.edu

Janko Slavic et al. (eds.), *Computer Vision & Laser Vibrometry, Vol. 6*, Conference Proceedings of the Society for Experimental Mechanics Series, https://doi.org/10.13052/9788743801511_14

it is difficult to use the image-based tracking CSLDV for a large outdoor rotating structure, such as a rotating wind turbine blade.

Different operational modal analysis methods were developed to process measurement of a CSLDV [16–21]. A demodulation method was developed to estimate ODSs of a structure under sinusoidal excitation [16,17]. Measured responses of the structure are multiplied by sinusoidal signals whose frequencies are excitation frequencies, and processed responses are filtered by a low-pass filter to obtain ODSs of the structure. The demodulation method was improved to process measured responses of a structure under random excitation [12]. Measured responses of the structure are filtered by a band-pass filter that only contains a selected natural frequency, and filtered responses are multiplied by sinusoidal signals whose frequencies are the selected natural frequency. Multiplied responses are filtered by a low-pass filter to obtain undamped mode shapes of the structure. A lifting method was developed to estimate modal parameters of a structure under random excitation, such as damped natural frequencies, undamped mode shapes, and modal damping ratios [20,21]. The lifting method reconstructs measured responses of a CSLDV to obtain multiple measured responses at virtual measurement points as if there were vibration sensors attached to these points, which requires a high scan frequency of the CSLDV to estimate modes of a structure with high natural frequencies [11].

In this work, a novel image-based long-range tracking CSLDV is developed for tracking and scanning a relatively distant rotating structure with a complex background, which has application to a rotating wind turbine blade. A new image processing method with a background averaging technique is developed to remove backgrounds in images of the rotating structure captured by the camera in the image-based long-range tracking CSLDV. A 1D scan scheme is used for sweeping the laser spot of the image-based long-range tracking CSLDV, and the improved demodulation method is used to process its measurements. A rotating blade that is excited by air flow from an excitation fan is set up as a rotating structure that is tracked and scanned by the image-based long-range tracking CSLDV. Damped natural frequencies and undamped modes shapes of the rotating blade with different speeds are estimated.

Methodologies

To extract a rotating structure from a complex background, a novel background averaging method is developed here. A rotating fan is used as an example of a rotating structure for describing the background averaging method. A captured image of the rotating fan with a complex background is shown in Figure 1(a). Let the total number of captured images be N, and p_i is the i-th captured image, where $i = 1, 2, 3, \ldots, N$; the averaged image is

$$p_a = \frac{1}{N}\sum_{i=1}^{N} p_i \tag{1}$$

which is shown in Figure 1(b), where the hub and background remain in the image while blades of the fan are averaged out and would disappear with a sufficiently large N. Note that pixel values in the averaged image p_a should be integers, and one may need to apply a rounding function to the right-hand side of Equation (1). To extract blades of the fan from the background, one needs to obtain the difference between p_i and p_a:

$$p_i^{'} = |p_i - p_a| \tag{2}$$

The result of $|p_i - p_a|$ in Equation (2) is shown in Figure 1(c), where locations of blades can be easily determined by a common edge detection method [22]. Note that the robust edge detection method in [15] can also determine the location of a rotating structure with a complex background; however, this method can only be applied to a rotating structure with a non-zero speed, while the averaged image p_a in the present work can be applied to a stationary structure. Once edges of blades are located, a distance condition is used to track the location of an edge point on one of the blades [14].

When the distance between an image-based tracking CSLDV and the rotating blade that is scanned is long enough, rotation angles of mirrors of the scanner in the image-based tracking CSLDV are small and the location of the laser spot on the blade is generally linearly related to rotation angles of mirrors of the scanner [11]. Pixel coordinates of the laser spot in captured images of the camera in the image-based tracking CSLDV are linearly related to the location of the laser spot on the rotating blade when the camera is close to the scanner; thus pixel coordinates of the laser spot at time instant t in captured images are generally linearly related to rotation angles of mirrors of the scanner at time instant t. Once the location of the rotating blade that needs to be tracked is determined in captured images, a 1D scan path can be generated on the blade and rotation angles of mirrors of the scanner can be controlled so that the laser spot of the image-based tracking CSLDV can be swept along the path. Let the location of the edge point in the processed image $p_i^{'}$ determined by the edge detection method and distance condition be $(x_e,\ y_e)$, where the x axis and y axis are along the horizontal and vertical directions of $p_i^{'}$,

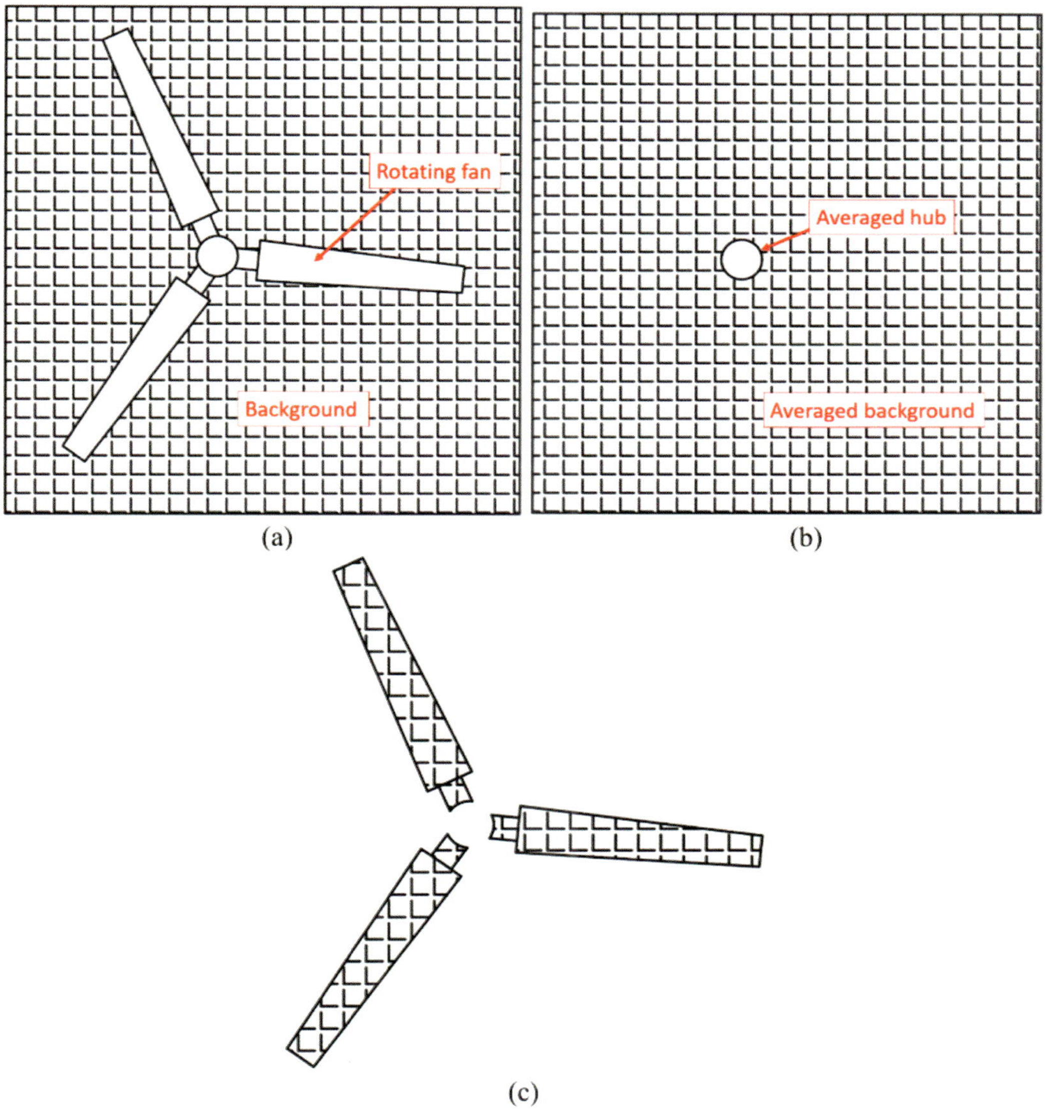

Fig. 1 (a) Schematic of a captured image of the rotating fan, (b) the averaged image of captured images of the rotating fan, and (c) the resulting image by subtracting the image in (b) from that in (a)

respectively, and the location of one end point of a straight scan path along the middle line of the rotating blade is

$$\begin{aligned} x_1 &= x_c + c[(x_e - x_c)\cos\theta + (y_e - y_c)\sin\theta] \\ y_1 &= y_c + c[(y_e - y_c)\cos\theta - (x_e - x_c)\sin\theta] \end{aligned} \quad (3)$$

where (x_c, y_c) is the location of the rotation center of the blade, c is a constant coefficient, and θ is the angle between the line passing through $(x_e,\ y_e)$ and (x_c, y_c) and the middle line of the blade. When locations of two end points of a straight scan path are determined, the laser spot can be swept along the straight path between them (Figure 2).

Once vibration measurement of the rotating blade is completed, measured responses of the rotating blade under random excitation can be processed by the improved demodulation method [12]. Fast Fourier transforms (FFTs) of measured responses are obtained to estimate damped natural frequencies of the rotating blade. Measured responses are filtered by a bandpass filter that only contains a selected damped natural frequency, and filtered responses are multiplied by sinusoidal signals whose frequencies are the selected damped natural frequency. A low-pass filter is applied to resulting responses to obtain the undamped mode shape associated with the selected damped natural frequency.

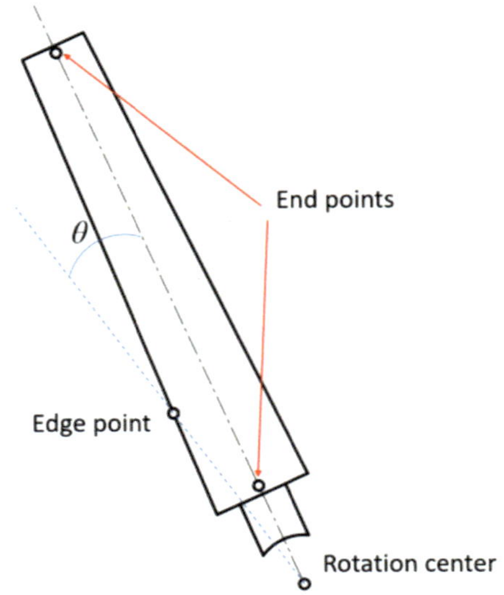

Fig. 2 Schematic of generating two end points of the straight scan path based on the location of the edge point.

Experiment and Results

The image-based long-range tracking CSLDV consists of a Polytec RSV-150 remote sensing vibrometer, a Cambridge 6240H scanner with a pair of orthogonal mirrors, a Basler acA1300-200um camera, and an NI controller (Figure 3(a)). An aluminum base was machined to hold the laser vibrometer, the camera, and the scanner (Figure 3(b)). The image-based long-range tracking CSLDV was placed in front of the rotating blade with a distance of 5.37 m (Figure 3(c)). Three aluminum blades were attached to a fan hub and one of them was scanned by the image-based long-range tracking CSLDV, and the fan hub was mounted on an aluminum frame (Figure 3(d)). A speed controller was connected to the hub so that it could rotate with different constant speeds. Lengths of middle lines, widths, and thicknesses of blades were 0.8 m, 0.1 m, and 3 mm, respectively. A finite element model of a stationary blade was used to estimate its undamped natural frequencies and undamped mode shapes via finite element method (FEM) simulations in ABAQUS (Figure 3(e)). The model was a clamped-free plate and shell elements were used in the model. The Young's modulus of the model was 70×10^9 Pa, the density was 2700 kg/m^3, and the Poisson's ratio was 0.35. A small fan was used to excite the rotating blades by its air flow, which could be considered as random excitation. The frame rate of the camera was set to 50 frames per second and the scan frequency of the scanner, which is the number of times the laser spot is swept back and forth along generated scan paths on the rotating blade in one second, was set to 0.625 Hz.

The position of the image-based long-range tracking CSLDV was adjusted so that the initial position of its laser spot was at the center of the hub. The hub was turned on and rotated with a constant speed, and the camera captured a sufficient number of images of the rotating blade. Captured images were used to obtain an averaged image, which was used as a background. Once the averaged image was obtained, it was used to remove backgrounds in images of the rotating blade when the image-based long-range tracking CSLDV scanned the blade.

Three of 1000 captured images of the three rotating blades are shown in Figures 4(a)-(c). The 1000 images were used to obtain an averaged image (Figure 4(d)). Three blades were eliminated in the averaged image while the hub was still in the image. The averaged image was subtracted from captured images to obtain images that only contain three blades (Figures 5(a)-(c)). The Sobel operator was applied to resulting images by subtracting the averaged image from captured images and their edge maps were obtained (Figures 5(d)-(f)) [22]. Note that there were some detected edges within boundaries of blades (Figures 5(d)-(f)), which did not affect tracking of one of the three blades since they did not appear outside boundaries of blades. There were detected edges at the location of the fan hub (Figures 5(d)-(f)), which was caused by the difference between the averaged hub image and the hub image at a certain instant during the hub's rotation.

A square region whose center was the hub center was used to extract a local region of interest (ROI) of the rotating blade that was tracked from an original captured image (Figure 6(a)). The pixel size of the ROI was 601×601 and the pixel location of the hub center in the ROI was (301, 301). Note that locations of boundary points of the ROI were determined by the pixel location of the hub center in the original image in Figure 6(a), and the pixel location of the hub center in the original image was determined by finding the center of the circular boundary of the hub (Figure 6(b)). Edges of rotating blades in

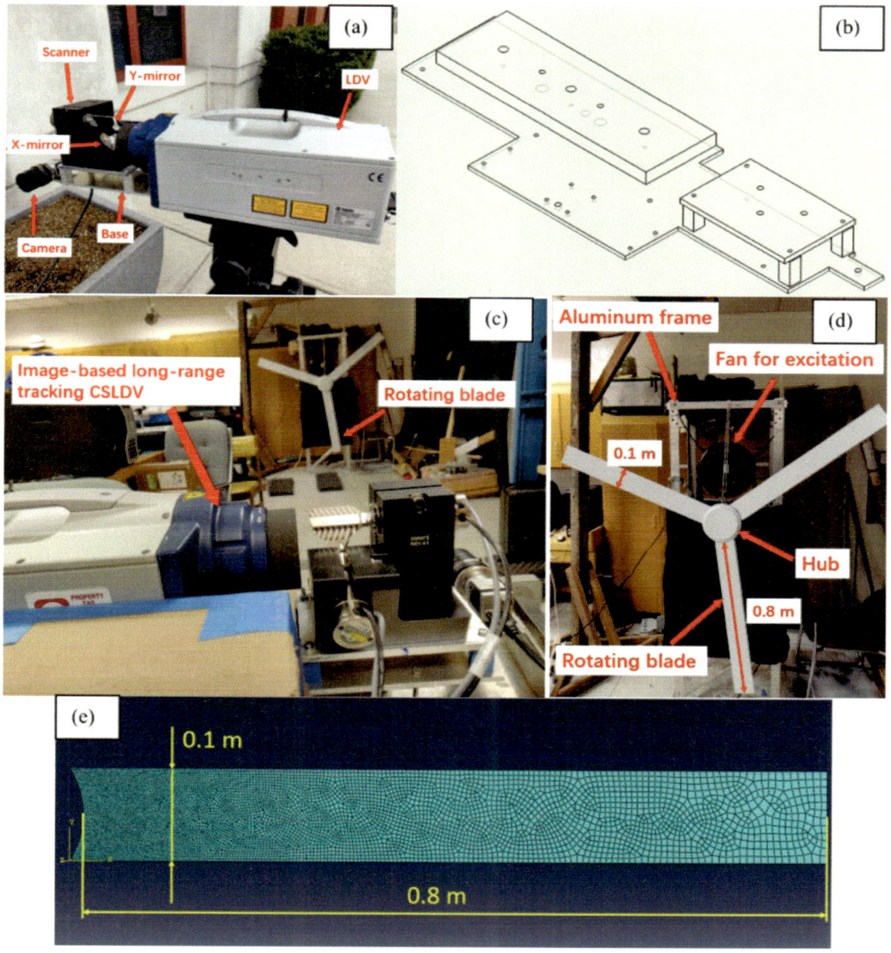

Fig. 3 (a) Picture of the image-based long-range tracking CSLDV, (b) the schematic of the aluminum base in (a), (c) the experimental setup for tracking and scanning a rotating blade using the image-based long-range tracking CSLDV, (d) the setup of the rotating blade, and (e) the finite element model of the rotating blade.

the ROI are shown in Figure 6(c), where an edge detection point whose position was (155, 551) in the ROI was used for the distance condition [14]. A set of inner and outer radial bounds whose centers are the hub center were used to determine edges within the two bounds, and radii of inner and outer radial bounds were 275 pixels and 300 pixels, respectively. Distances between the edge tracking point and edges within the two bounds were calculated and edges with distances smaller than a certain value (50 pixels in this work) were selected. The location of the geometric center of selected edges was calculated and used as an updated edge tracking point for the next captured image after the image shown in Figure 6(a). The updated edge tracking point was used as the edge position of the rotating blade that was tracked.

Estimated edge positions were used to generate scan paths on the rotating blade, and the laser spot of the image-based long-range tracking CSLDV was swept along them by rotating mirrors of the scanner. X- and Y-mirror signals of the scanner when scanning the rotating blade with different speeds, which were normalized by dividing them by their maximum values, are shown in Figure 7. X-mirror signals corresponded to horizontal movements of the laser spot and Y-mirror signals corresponded to vertical movements of the laser spot. Note that the stationary blade that was scanned was essentially at a vertical position so that the Y-mirror signal was much larger than the X-mirror signal.

Mirror signals were processed by calculating root sum squares of X- and Y-mirror signals [11]. Processed mirror signals were used for the improved demodulation method to determine end-to-end undamped mode shapes of the rotating blade [12]. Note that mirror signals when scanning the stationary blade were not processed since their shapes were the same as those obtained by calculating root sum squares of X- and Y-mirror signals. One can see that processed mirror signals with different rotation speeds were nearly the same since scan frequencies were the same when scanning the rotating blade with different rotation speeds (Figures 8(a) and (b)). Mirror signals in Figure 7 were also used to obtain rotation speeds of the

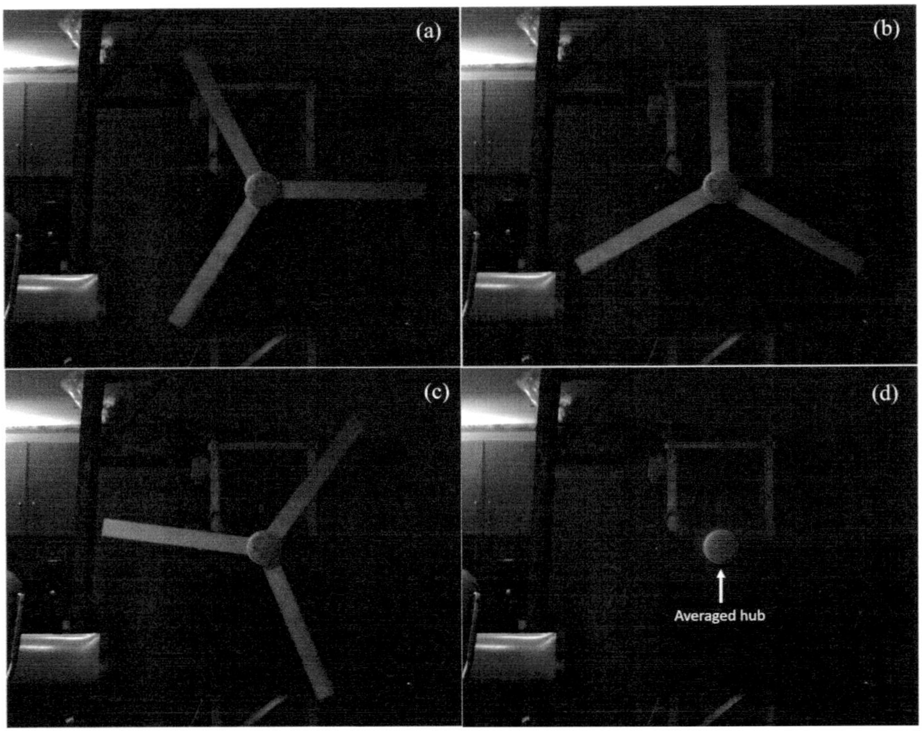

Fig. 4 (a) 50th, (b) 90th, and (c) 150th captured images of the camera, and (d) the averaged image.

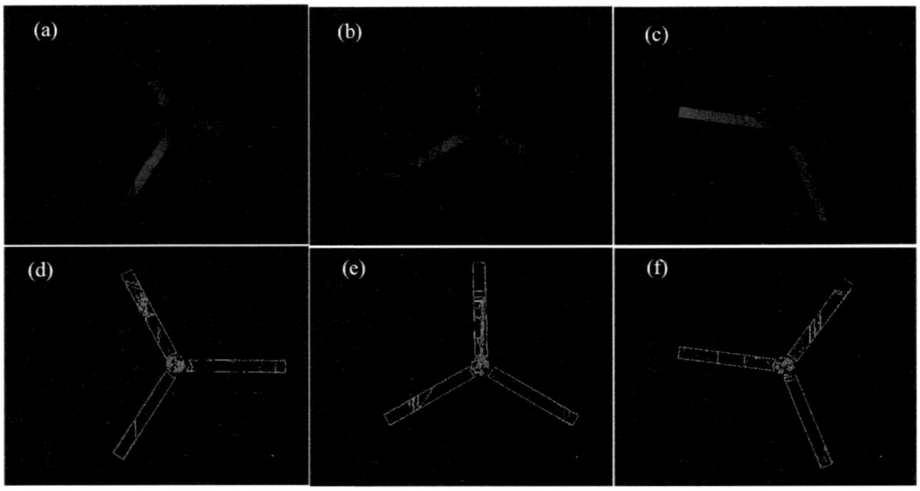

Fig. 5 Resulting images by subtracting the averaged image from (a) the 50th, (b) 90th, and (c) 150th captured images of the camera, and edge maps obtained from (d) the resulting image in (a), (e) the resulting image in (b), and (f) the resulting image in (c).

blade (Figure 8(c)). X- and Y- mirror signals were linearly related to horizontal and vertical positions of the laser spot, and they were converted to angular positions and velocities of the laser spot. Since the laser spot was swept along the blade, its angular velocities could be considered as rotation speeds of the blade. Estimated average rotation speeds of the blade with low and high speeds were 15.38 revolutions per minute (rpm) and 32.22 rpm, respectively.

Measured responses of the rotating blade with different speeds are shown in Figure 9. Note that there were outliers in measured responses that were caused by measurement noise of the long-range LDV, and they could be eliminated by some signal processing methods, such as an interpolation method. Damped natural frequencies of the rotating blade were estimated from FFTs of measured responses (Table 1). Note that damped natural frequencies of the stationary blade were somewhat

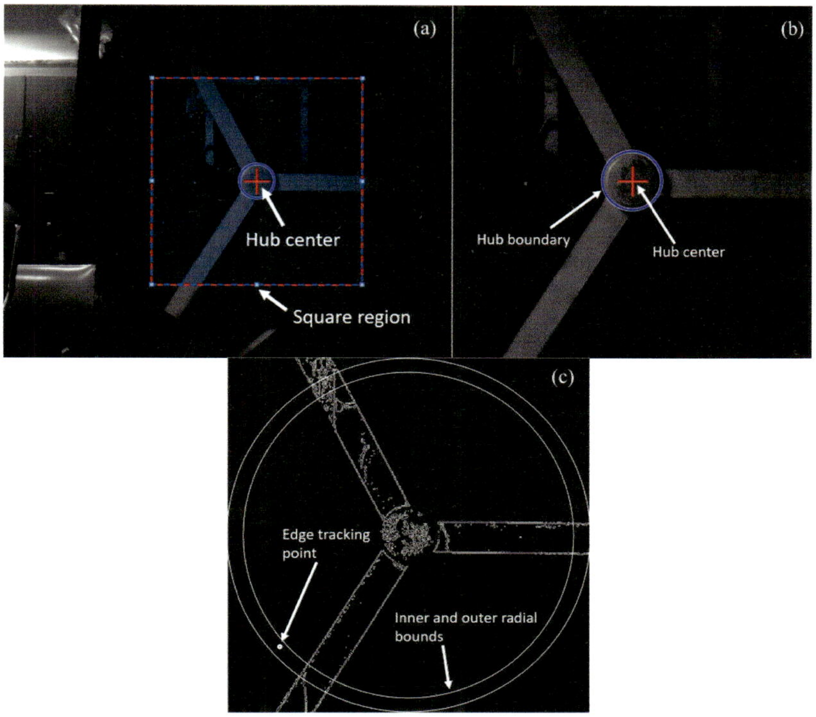

Fig. 6 (a) Original captured image with a square ROI, (b) the ROI, and (c) detected edges in the ROI.

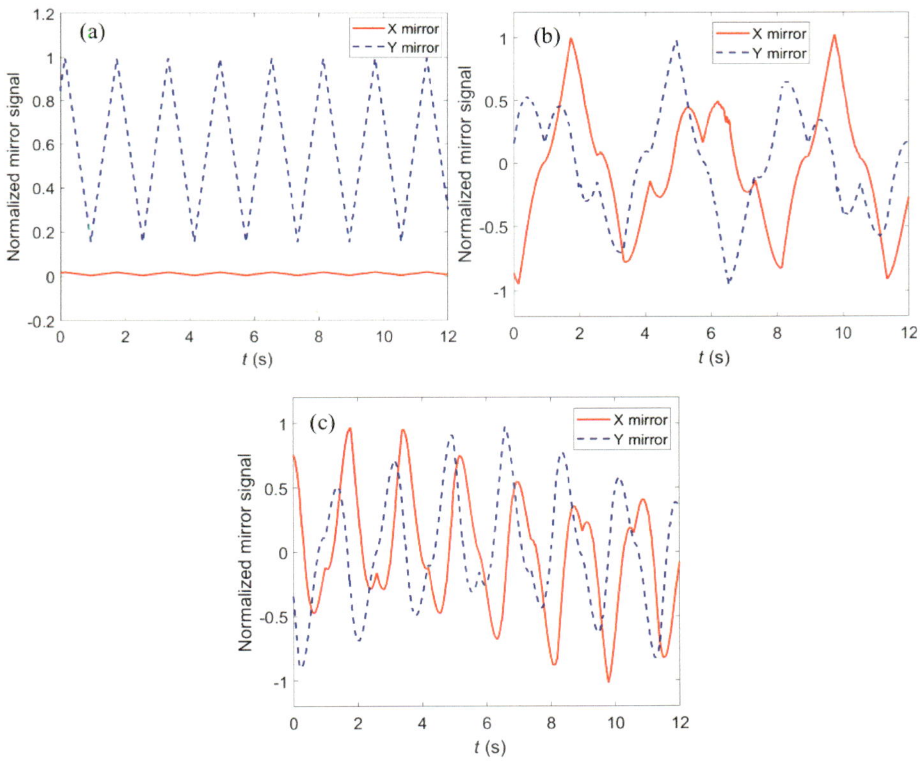

Fig. 7 Normalized mirror signals when scanning (a) the stationary blade, (b) the rotating blade with a low speed, and (c) the rotating blade with a high speed.

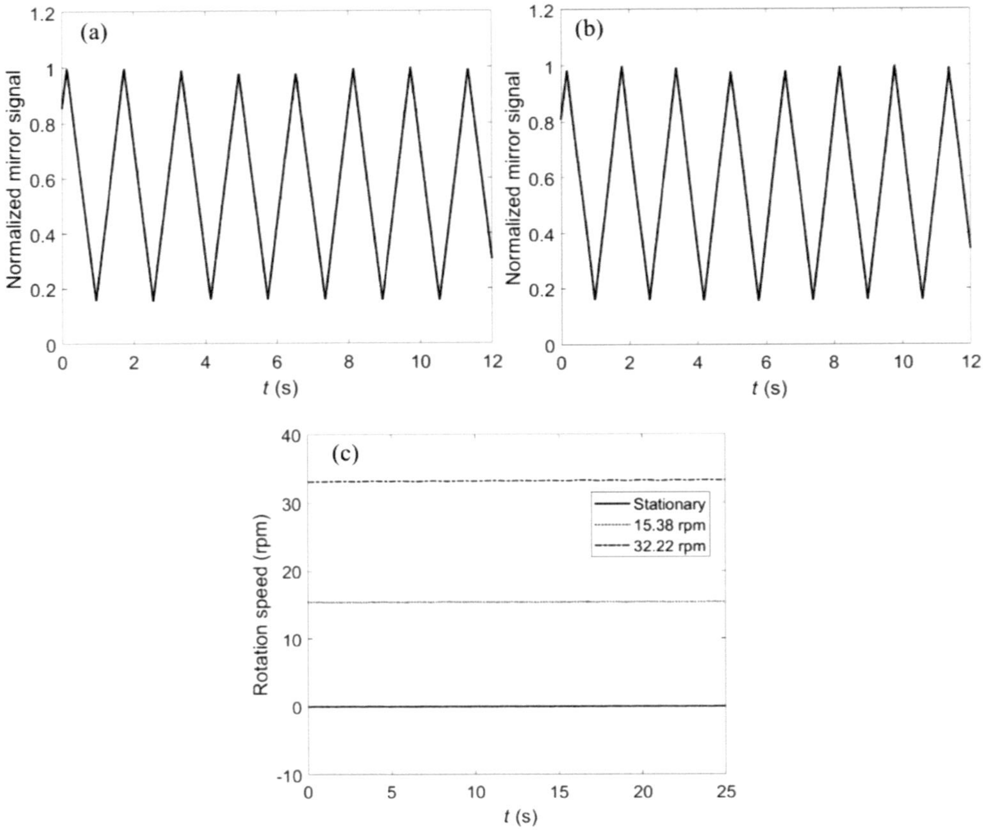

Fig. 8 Normalized processed mirror signals when scanning (a) the rotating blade with a low speed and (b) the rotating blade with a high speed, and (c) estimated rotation speeds of the blade.

Table 1 Estimated damped natural frequencies of stationary and rotating blades and undamped natural frequencies of the stationary blade from the finite element model.

	First damped natural frequency	Second damped natural frequency	Third damped natural frequency
FEM	1.63 Hz	10.19 Hz	28.55 Hz
Stationary	1.31 Hz	8.97 Hz	27.36 Hz
15.38 rpm	1.547 Hz	9.39 Hz	27.96 Hz
32.22 rpm	1.656 Hz	9.55 Hz	28.61 Hz

lower than those from FEM simulations (Table 2), which was due to the facts that the blade in the experiment was not perfectly clamped and there was some damping in the experiment while there was no damping in the finite element model. Estimated damped natural frequencies of the blade increased with its rotation speed, which was caused by the stiffening effect due to rotation [11–15].

Table 2 Differences between estimated damped natural frequencies of stationary and rotating blades and undamped natural frequencies of the stationary blade from the finite element model.

	First damped natural frequency	Second damped natural frequency	Third damped natural frequency
Stationary & FEM	19.63%	11.97%	4.17%
15.38 rpm & FEM	5.09%	7.85%	2.07%
32.22 rpm & FEM	-1.60%	6.28%	-0.21%

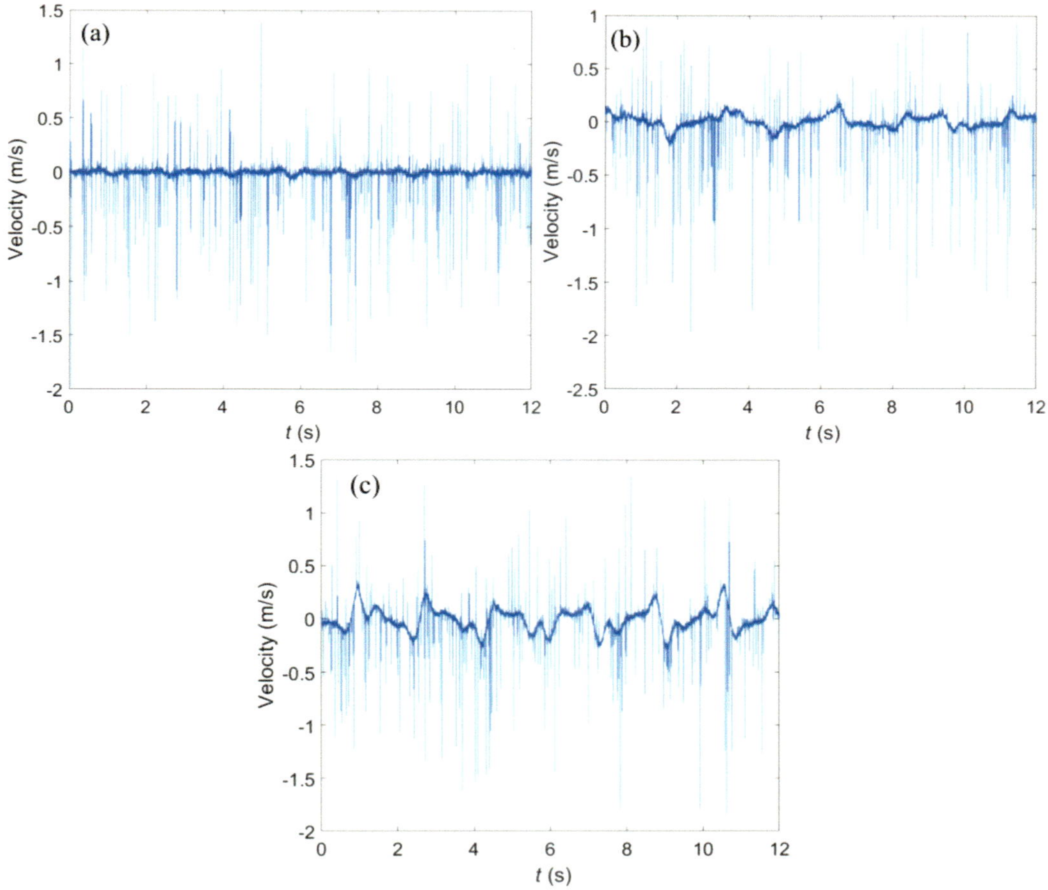

Fig. 9 Normalized processed mirror signals when scanning (a) the stationary blade, (b) the rotating blade with a speed of 15.38 rpm, and (c) the rotating blade with a speed of 32.22 rpm.

The improved demodulation method was applied to measured responses and processed mirror signals to estimate undamped mode shapes of the rotating blade. The estimated first three undamped bending mode shapes of the rotating blade with different speeds and first three undamped bending mode shapes of the stationary blade from the finite element model are shown in Figure 10. Estimated undamped mode shapes of the rotating blade are similar to bending mode shapes of a cantilever beam. Modal assurance criteria (MAC) values between estimated undamped mode shapes of the rotating blade and undamped mode shapes of the stationary blade from the finite element model were calculated (Table 3). MAC values are relatively high (over 95%) except for the value between the estimated third undamped mode shape of the rotating blade with a speed of 32.22 rpm and the undamped third mode shape of the stationary blade from the finite element model (93.85%), which could be caused by the stiffening effect due to rotation, differences in boundary conditions, and some inaccuracy of scanning the same path on the rotating blade with a high speed. Note that only 1D bending mode shapes of the rotating blade were estimated. To estimate torsional mode shapes of the rotating blade, a 2D scan scheme needs to be developed to scan its whole surface [15]. To estimate in-plane modes of the rotating blade, a 3D CSLDV system needs to be developed to measure 3D vibrations of the rotating blade [23–26].

Table 3 MAC values between estimated undamped mode shapes of stationary and rotating blades and undamped mode shapes of the stationary blade from its finite element model.

	First mode	Second mode	Third mode
Stationary & FEM	99.98%	99.83%	99.20%
15.38 rpm & FEM	99.93%	98.95%	96.64%
32.22 rpm & FEM	99.97%	98.37%	93.85%

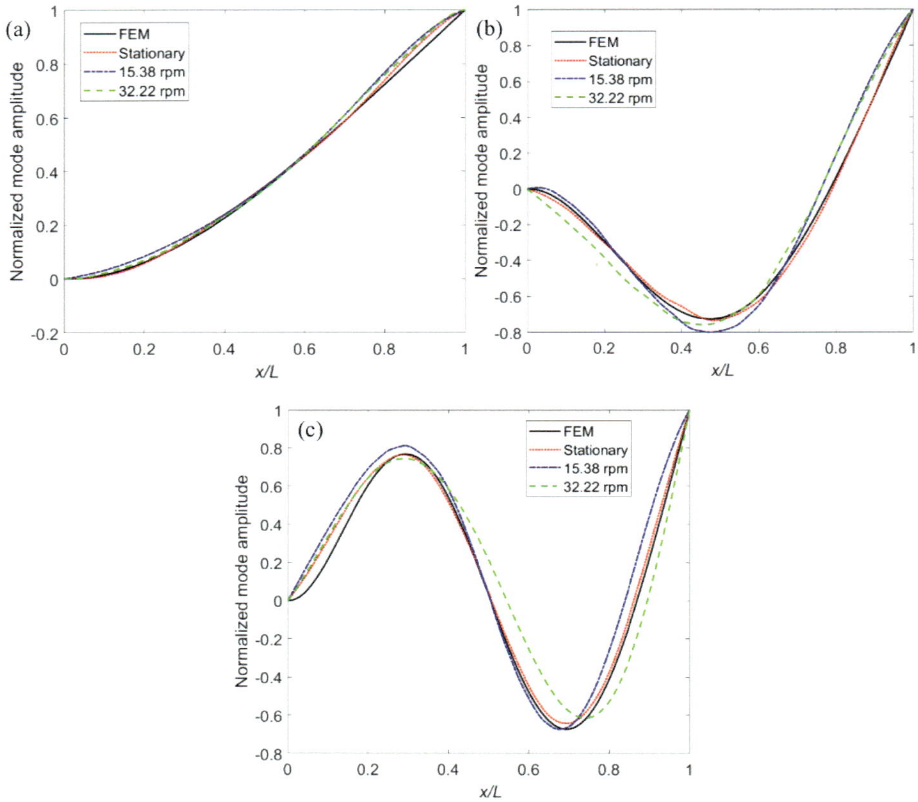

Fig. 10 Normalized estimated (a) first, (b) second, and (c) third undamped mode shapes when scanning the stationary blade, the rotating blade with a speed of 15.38 rpm, the rotating blade with a speed of 32.22 rpm, and the mode shape captured from the FEM results.

Conclusion

A novel image-based long-range tracking CSLDV was developed for tracking and scanning a rotating structure under random excitation with a long distance from the CSLDV. A novel image-processing method for tracking the rotating structure with a complex background was developed. The image-processing method includes a background averaging method and an edge detection method with a distance condition. The background averaging method can obtain the background of captured images of the rotating structure and remove it from captured images. The edge detection method can obtain the location of the rotating structure and track it. A 1D scan scheme was used to generate a straight scan path between its two end points on the structure based on its obtained location. An improved demodulation method was used to estimate damped natural frequencies and undamped mode shapes of the rotating structure. A rotating blade under air flow excitation was tracked and scanned by the image-based long-range tracking CSLDV, and its rotation speed, damped natural frequencies, and 1D undamped mode shapes were successfully estimated and compared with results from the finite element model of the corresponding stationary blade. The image-based long-range tracking CSLDV can be used to measure vibrations of outdoor rotating structures, such as rotating wind turbine blades.

Acknowledgments This research was supported by the National Science Foundation through Grant No. CMMI-1763024. The authors would like to thank Dan Feng for some valuable discussion on the background averaging method.

References

1. S. Rothberg, M. Allen, and P. Castellini, An international review of laser Doppler vibrometry: making light work of vibration measurement, Opt. Lasers Eng. 99(1) 11–22 (2017). https://doi.org/10.1016/j.optlaseng.2016.10.023.
2. B. Stoffregen, and A. Felske, Scanning laser Doppler vibration analysis system. *SAE Transactions*, 94(2) 934–940 (1985).

3. P. Castellini, M. Martarelli, and E. Tomasini, Laser Doppler vibrometry: development of advanced solutions answering to technology's needs. *Mech. Syst.* Signal Process. 20(6) 1265–1285 (2006). https://doi.org/10.1016/j.ymssp.2005.11.015.
4. P. Sriram, S. Hanagud, J. Craig, and N.M. Komerath, Scanning laser Doppler technique for velocity profile sensing on a moving surface, *Appl. Opt.* 29(16) 2409–2417 (1990). https://doi.org/10.1364/AO.29.002409.
5. P. Sriram, S. Hanagud, and J. Craig, Mode shape measurement using a scanning laser Doppler vibrometer, *Int. J. Anal. Exp. Modal Anal.* 7(3) 169–178 (1992).
6. M.S. Allen, and M.W. Sracic, A new method for processing impact excited continuous-scan laser Doppler vibrometer measurements, *Mech. Syst. Signal Process.* 24(3) 721–735 (2010). https://doi.org/10.1016/j.ymssp.2009.11.004.
7. D.M. Chen, Y.F. Xu, and W.D. Zhu, Damage identification of beams using a continuously scanning laser Doppler vibrometer system, *J. Vib. Acoust.* 138(5) 05011 (2016). https://doi.org/10.1115/1.4033639.
8. D.M. Chen, Y.F. Xu, and W.D. Zhu, Identification of damage in plates using full-field measurement with a continuously scanning laser Doppler vibrometer system, *J. Sound Vib.*, 422, 542–567 (2018). https://doi.org/10.1016/j.jsv.2018.01.005.
9. D.M. Chen, Y. F. Xu, and W.D. Zhu, Non-model-based identification of delamination in laminated composite plates using a continuously scanning laser Doppler vibrometer system, *J. Vib. Acoust.* 140(4) 041001 (2018). https://doi.org/10.1115/1.4038734.
10. D.M. Chen, Y.F. Xu, and W.D. Zhu, A comprehensive study on detection of hidden delamination damage in a composite plate using curvatures of operating deflection shapes, *J. Nondestruct. Eval.*, 38 54 (2019). https://doi.org/10.1007/s10921-019-0591-8.
11. L.F., Lyu, and W.D. Zhu, Operational modal analysis of a rotating structure under ambient excitation using a tracking continuously scanning laser Doppler vibrometer system, *Mech. Syst. Signal Process.* 152 107367 (2021). https://doi.org/10.1016/j.ymssp.2020.107367.
12. L.F., Lyu, and W.D. Zhu, Operational modal analysis of a rotating structure subject to random excitation using a tracking continuously scanning laser Doppler vibrometer via an improved demodulation method, *J. Vib. Acoust.* 144(1) 011006 (2021). https://doi.org/10.1115/1.4051178.
13. L.F. Lyu, and W.D. Zhu, Full-field mode shape estimation of a rotating structure subject to random excitation using a tracking continuously scanning laser Doppler vibrometer via a two-dimensional scan scheme, *Mech. Syst. Signal Process.* 169 108532 (2022). https://doi.org/10.1016/j.ymssp.2021.108532
14. L.F. Lyu, G.D. Higgins, and W.D. Zhu, Operational modal analysis of a rotating structure using image-based tracking continuously scanning laser Doppler vibrometry via a novel edge detection method, J. Sound Vib., 525 116797 (2022). https://doi.org/10.1016/j.jsv.2022.116797
15. L.F. Lyu, G.D. Higgins, and W.D. Zhu, A novel robust edge detection method for full-field modal parameter estimation using an image-based tracking continuously scanning laser Doppler vibrometer system. *Mech. Syst. Signal Process.*, 195, 110242 (2023). https://doi.org/10.1016/j.ymssp.2023.110242.
16. A. Stanbridge, and D. Ewins, Modal testing using a scanning laser Doppler vibrometer, *Mech. Syst. Signal Process.* 13(2) 255–270 (1999). https://doi.org/10.1006/mssp.1998.1209.
17. D. Di Maio, and D. Ewins, Continuous scan, a method for performing modal testing using meaningful measurement parameters; Part I, *Mech. Syst. Signal Process.* 25(8) 3027–3042 (2011). https://doi.org/10.1016/j.ymssp.2011.05.018.
18. Y.F. Xu, D.M. Chen, and W.D. Zhu, Damage identification of beam structures using free response shapes obtained by use of a continuously scanning laser Doppler vibrometer system, *Mech. Syst. Signal Process.* 92 226–247 (2017). https://doi.org/10.1016/j.ymssp.2016.12.042.
19. Y.F. Xu, D.M. Chen, and W.D. Zhu, Modal parameter estimation using free response measured by a continuously scanning laser Doppler vibrometer system with application to structural damage identification, *J. Sound Vib.* 485 115536 (2020). https://doi.org/10.1016/j.jsv.2020.115536.
20. S. Yang, and M.S. Allen, Lifting approach to simplify output-only continuous-scan laser vibrometry, *Mech. Syst. Signal Process.* 45(2) 267–282 (2014). https://doi.org/10.1016/j.ymssp.2013.11.010.
21. Y.F. Xu, D.M. Chen, and W.D. Zhu, Operational modal analysis using lifted continuously scanning laser Doppler vibrometer measurements and its application to baseline-free structural damage identification, *J. Vib. Control.* 25(7) 1341–1364 (2019). https://doi.org/10.1177/1077546318821154.
22. R.C. Gonzales, and R.E. Woods, (2018). *Digital image processing* (Pearson, New York, 2018).
23. G.D. Acar, and B.F. Feeny, Bend-bend-twist vibrations of a wind turbine blade. *Wind Energy,* 21(1) 15–28 (2018). https://doi.org/10.1002/we.2141.
24. Y. Chen, A.S.E. Mendoza, D.T. and Griffith, Experimental and numerical study of high-order complex curvature mode shape and mode coupling on a three-bladed wind turbine assembly. *Mech. Syst. Signal Process.* 160 107873 (2021). https://doi.org/10.1016/j.ymssp.2021.107873.
25. D.M. Chen, W.D. Zhu, Investigation of three-dimensional vibration measurement by three scanning laser Doppler vibrometers in a continuously and synchronously scanning mode. *J. Sound Vib.* 498 115950 (2021), https://doi.org/10.1016/j.jsv.2021.115950.
26. K. Yuan, and W.D. Zhu, Estimation of modal parameters of a beam under random excitation using a novel 3D continuously scanning laser Doppler vibrometer system and an extended demodulation method, *Mech. Syst. Signal Process.* 155 107606 (2021). https://doi.org/10.1016/j.ymssp.2021.107606.

Chapter 15
Development and Validation of a Novel Miniaturized High-Precision Laser Vibrometer for Vibration and Ultrasonic Vibration Measurements

Ke Yuan, Zhonghua Zhu, Wei Chen, and Weidong Zhu

Abstract Developing a miniatured laser vibrometer becomes important for many engineering areas, such as experimental and operational modal analyses, model validation, and structural health monitoring. Due to its compact size and light weight, a miniatured laser vibrometer can be attached to various mobilized platforms, such as an unmanned aerial vehicle and a robotic arm whose payloads can usually not be large, to achieve a flexible vibration measurement capability. However, integrating optics into a miniaturized laser vibrometer presents several challenges. These include signal interference from ghost return signals generated by sub-components of integrated photonics, polarization effects caused by waveguide structures, and wavelength drifting due to the semiconductor laser, and poorer noise characteristics of an integrated laser chip compared to a non-integrated circuit. This work proposes a novel chip-based high-precision laser vibrometer by incorporating two or more sets of quadrature demodulation networks into its design. An additional set of quadrature demodulation networks with a distinct reference arm delay line length can be used to conduct real-time compensation to mitigate linear interference caused by temperature and environmental variations. A series of vibration measurements with frequencies ranging from 0.1 Hz to 1 MHz were conducted using the proposed laser vibrometer to show its repeatability and accuracy in vibration and ultrasonic vibration measurements, and its robustness to test surface conditions. The proposed laser vibrometer has the advantage of directly measuring the displacement response of a vibrating structure rather than integrating its velocity response to yield the measured displacement with a laser Doppler vibrometer.

Keywords Miniatured laser vibrometer · Integrated optics · Quadrature demodulation network · Vibration measurement · Ultrasonic vibration measurements

Introduction

A laser Doppler vibrometer (LDV) can measure the velocity response of a vibrating structure based on the Doppler shift principle. As a noncontact vibration measurement device, it can avoid mass loading and is more suitable than a contact-type device, such as an accelerometer and a linear variable differential transducer, for measuring vibration of a light-weight and fragile structure. Scanning LDV and continuously scanning LDV techniques were developed to make LDV measurements automated, rapid, and dense. Experimental and operational modal analyses using a LDV have been widely used in engineering areas, such as model validation, modal identification, structural health monitoring, and damage detection of beams [1], plates [2], pipes [3, 4], bridges [5, 6], buildings [7, 8], turbine blades [9, 10], and sandwich structures [11, 12].

With the development of unmanned system techniques, attaching a LDV on an unmanned system, such as an unmanned aerial vehicle (UAV) [13, 14] and a robotic arm [15–18], for vibration measurement becomes more and more popular. By taking the LDV off a tripod, which provides a common support for the LDV, vibration measurement can be more flexible and practical. However, for such UAV+LDV and Robot+LDV systems, one of major challenges was limitations of payloads

Ke Yuan · Weidong Zhu
Department of Mechanical Engineering, University of Maryland Baltimore County, Baltimore, MD 21250
e-mail: kyuan1@umbc.edu

Zhonghua Zhu · Wei Chen
OmniSensing Photonics, LLC, Columbia, MD 21046
e-mail: zhonghua.zhu@osphotonics.com; wei.chen@osphotonics.com

Janko Slavic et al. (eds.), *Computer Vision & Laser Vibrometry, Vol. 6*, Conference Proceedings of the Society for Experimental Mechanics Series, https://doi.org/10.13052/9788743801511_15

of the UAV and the robot, as higher payloads indicated higher costs and risks during operations. In Ref. [14], the UAV with a sufficiently large payload was selected to hold both the sensor head and the LDV module. In most of the above Robot+LDV systems, industrial robotic arms were used to hold one or multiple laser heads, which needed large spaces for operation and were not practical for most laboratory scenarios. Other possible solutions for reducing robot payloads included separating a laser head and its accessories such as a laser excitation device [17], and using a LDV with a sufficiently small size.

This work aims at developing a new type of miniatured displacement-based laser vibrometers based on integrated optics to resolve the payload limitation issue. Most recent LDVs are designed using discrete optical components. The difference between discrete and integrated optical designs in LDVs can be compared to the difference between vacuum tube circuit and integrated circuit technologies in the electronics world. Main compact LDV techniques include photonic integrated circuits (PIC), laser feedback interferometry, and micro-machined free-space optical interferometers [19–21]. Ref. [20] presented a design on silicon-on-insulator (SOI) for measuring small vibrations at low frequencies. Integrated optics can put an entire optical system into a single chip to reduce its size and weight by thousands of times.

To address above limitations and challenges, this work proposes a novel chip-based high-precision laser vibrometer by incorporating two or more additional sets of quadrature demodulation networks into its design. Major contributions of this work include: (1) an additional set of quadrature demodulation networks with a distinct reference arm delay line length that can mitigate nonlinear interference caused by temperature and environmental variations by conducting real-time compensation; (2) the proposed laser vibrometer can directly measure the vibration displacement of a structure using a phase demodulation method, which is different from a frequency demodulation method used for a conventional LDV, which directly measures the vibration velocity and integrates it to obtain the displacement; and (3) the proposed laser vibrometer has good robustness to test surface conditions.

Working Principles Used in This Work

A laser Doppler vibrometer is developed based on working principles of laser interferometry. A simplified schematic diagram of a typical laser interferometer is shown in Figure 1. A laser diode (LD) with a narrow linewidth is usually preferred as the light source to ensure laser coherence and low phase noise, and a LD stabilizer unit can be used to stabilize its frequency. A beamsplitter is used to divide the laser beam into two: one as the measurement beam and the other as the reference beam. The measurement beam, after reflecting from a device under test (DUT), recombines with the reference beam to generate a beating signal that can be received by a photodetector (PD). Based on the relation between the frequency of the measurement beam and that of the reference beam, laser interferometers can be categorized as homodyne laser interferometers and heterodyne laser interferometers. In a homodyne laser interferometer, the two laser beams have the same frequency, and the displacement of the DUT can be obtained by analyzing the frequency or the phase shift of the beating signal. In a heterodyne laser interferometer, the two laser beams have different frequencies, which can be implemented by using a birefringence-Zeeman dual frequency laser as the laser source or by introducing an acousto-optic modulator (AOM) into its reference arm, as shown in the dashed box in Figure 1. Response of the DUT can then be obtained by analyzing the frequency or the phase shift of the beating signal. The beating signal with a high frequency, which is typically in the range of MHz, can effectively isolate measured response of the DUT from environmental noise to preserve measurement precision even in sub-optimal conditions. Heterodyne interferometers move signal frequencies to the MHz range while homodyne interferometers have signal frequencies to remain at original frequencies, which results in the fact that heterodyne interferometers are less susceptible to low frequency noises than homodyne interferometers. There are several low frequency noises. One is the $1/f$ noise, where f is the driver frequency, from electronics that typically affects the LD driver and causes a high signal noise level in the low frequency range. Also, optical components are susceptible to temperature change, which is usually slow and becomes another major noise source in the low frequency range. Consequently, heterodyne interferometers have a better noise performance. Therefore, strong resistance to environmental noise is one of major advantages of a heterodyne laser interferometer over a homodyne laser interferometer whose reference optical path is susceptible to environmental disruption in a non-integrated setting.

There are two methods to demodulate the vibration signal received by the PD. The first one is the phase demodulation method. It employs demodulation of amplitude changes of the received signal or filtering to directly obtain the envelope of the signal, thereby extracting changes of the signal phase. The second one is the frequency demodulation method. Devices such as lock-in amplifiers can accurately detect changes in the frequency of the signal. Frequency shifts indicate changes in the physical quantity being measured. However, in this configuration, PD loses the complex signal information carried by photons in converting the received light power into the electrical signal. The most straightforward benefit of a complex baseband architecture is the noise-figure improvement achievable by eliminating the imaginary band noise foldback. In the-

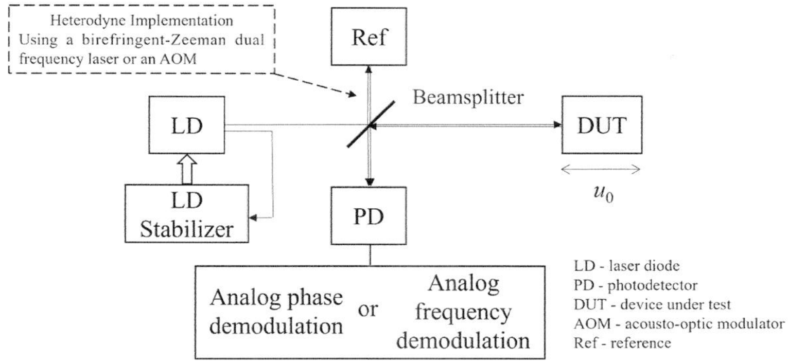

Fig. 1 schematic diagram of a laser interferometer: the homodyne configuration without heterodyne implementation marked by the dashed box, and the heterodyne configuration.

ory, the noise-figure improvement is 3 dB [22]. Environmental changes, mainly the thermal effect, result in design parameter variations of various optical components, especially for an integrated photonic chip. With a complex domain signal, these changes can be compensated in digital decoding [23]. Therefore, the amplitude signal is highly susceptible to system noise and environmental noise. The frequency demodulation method is more precise under conditions with low signal-to-noise ratios (SNRs), and most commercial laser vibrometers are essentially heterodyne laser interferometers using the frequency demodulation method, earning them the name LDVs.

Although the homodyne laser interferometer using the phase demodulation method is susceptible to noise, it is more suitable than the heterodyne laser interferometer for manufacturing a photonic integrated miniatured laser vibrometer due to its simpler configuration. In order to greatly enhance the performance of the homodyne laser interferometer in conditions with low SNRs, quadrature demodulation can be adopted in the design to recover the complete complex-domain photon signal. A schematic of quadrature demodulation is shown in Figure 2. An incoming signal beam can be split into an in-phase component (I) and a quadrature component (Q) through the beamsplitter in a 2-by-2 coupler. Similarly, a reference signal generated by a local oscillator (LO) laser can be split into two beams: one beam maintains its original phase for mixing with the in-phase component, while the other undergoes a (90°) phase shift to mix with the quadrature component. Therefore, quadrature demodulation can also be referred to as a (90°) mixing network. Four outputs from the (90°) mixing network can be expressed as

$$E_1 = \frac{1}{2}(E_{\text{sig}} + E_{\text{LO}}), \tag{1}$$

$$E_2 = \frac{1}{2}(E_{\text{sig}} - E_{\text{LO}}), \tag{2}$$

$$E_3 = \frac{1}{2}(E_{\text{sig}} + iE_{\text{LO}}), \tag{3}$$

$$E_4 = \frac{1}{2}(E_{\text{sig}} - iE_{\text{LO}}), \tag{4}$$

where $i = \sqrt{-1}$, and E_{sig} and E_{LO} are the signals of measurement and reference beams, respectively. Mixed signals can then be received by balanced PDs that can eliminate common-mode noise, yielding enhanced SNRs. Finally, a signal processing unit receives I and Q signals, and demodulates their amplitude and phase information, allowing for the precise detection of the input signal, which can be written as

$$I = R\sqrt{P_{\text{sig}}P_{\text{LO}}}\cos\left(\theta_{\text{sig}}(t) - \theta_{\text{LO}}(t)\right), \tag{5}$$

$$Q = R\sqrt{P_{\text{sig}}P_{\text{LO}}}\sin\left(\theta_{\text{sig}}(t) - \theta_{\text{LO}}(t)\right), \tag{6}$$

where R denotes the responsivity of PDs, P_{sig} and P_{LO} denote the amplitudes of the signals of measurement and reference beams, respectively, and θ_{sig} and θ_{LO} denote the phases of of the signals of measurement and reference beams, respectively.

Incorporating quadrature demodulation can significantly enhance the performance of demodulation by the following aspects. The first one is to enhance phase accuracy. The (90°) mixing network introduced above can produce two signals (I and Q) with a (90°) phase difference, thereby delivering comprehensive phase information. Compared to conventional amplitude-phase demodulation methods, such as a fringe counting method, the I/Q phase demodulation method uses digital

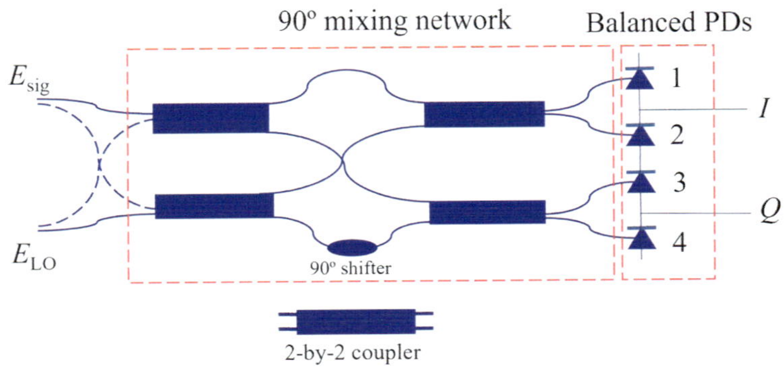

Fig. 2 Schematic of quadrature demodulation.

decoding, which can significantly improve the precision of measurement through subdivision. It ensures that the complete phase information can be accurately reconstructed by a complementary path when a signal path is disrupted by noise or partial distortion. The second one serves to resolve phase ambiguity. In a conventional single-frequency interferometer, phase measurement is typically limited to the range of $[0, 2\pi]$, which can lead to phase "folding" or "wrapping" beyond this range and cause ambiguity. The (90°) mixing network can resolve this issue by accurately computing the angle of two orthogonal signals, enabling accurate and continuous phase determination even beyond the range of $[0, 2\pi]$. Thus, the two orthogonal signals offer a more complete view of phase changes, expanding the dynamic range of measurement. This allows the interferometer to accurately measure the displacement or optical path difference variations over a wider range by digital decoding from a block of data. The last one is to reduce impacts from signal errors and noises. With two independent signal channels processed concurrently, specialized algorithms, such as trending detection and signal averaging, can minimize random noises and enhance the laser vibrometer's responsiveness to genuine signals. This data processing helps cancel the nonlinear errors induced by a variety of noise sources or environmental fluctuations.

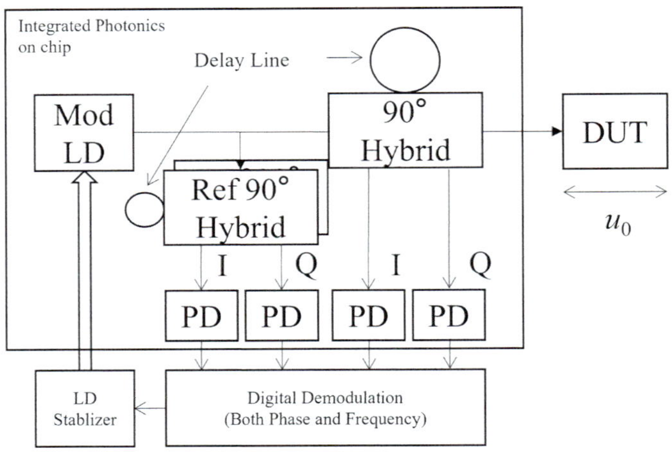

Fig. 3 Schematic of of the proposed chip-based high-precision laser vibrometer.

It is usually difficult to implement quadrature demodulation in a discrete optical device because it requires an accurate (90°) delay on one interference beam in the demodulation network. However, it is relatively easy for photonic integration as the modern lithography processing can accurately control the length of the waveguide. A schematic of the proposed chip-based high-precision laser vibrometer is shown in Figure 3. The laser vibrometer proposed in this work incorporates two or more sets of quadrature demodulation networks in the design. One of the quadrature demodulation networks serves as a main *I*/*Q* demodulator for detecting the interference signal between measurement and reference beams and enabling the extraction of the phase and amplitude information from the signal beam. A second quadrature demodulation network serves as a reference *I*/*Q* demodulator for capturing the phase and amplitude instability of the laser. The demodulated reference signal is used for a closed control loop to lock the laser wavelength, and thus ensure signal stability in the laser vibrometer. The reflection signal from the DUT is only fed into the first quadrature demodulation network, and is mixed with the local

LD source; I/Q signals are then decoded. The input signal of the second quadrature demodulation network is a split LD signal from the LD source after a delay line within the integrated photonic chip. This second quadrature demodulation network is used to catch the interference signal by slow laser center frequency drifting due to temperature change. The signal is considered as the common mode error and deducted by decoded results from the first quadrature demodulation network to retrieve an accurate signal from the DUT. As the delay line of the second quadrature demodulation network is co-printed with the delay line of the first quadrature demodulation network and the path lengths of the second quadrature demodulation network match with those of the first quadrature demodulation network, the common mode error can be removed by the proposed design. The novel laser vibrometer proposed in this work employs a fully digital demodulation approach to capture the comprehensive complex signal information. The reference I/Q demodulator stabilizes the frequency of the semiconductor laser through real-time feedback, reducing phase noise. This strategy allows the regular semiconductor laser with a prone-drifting center wavelength to achieve the performance level comparable to the conventional helium-neon gas laser. In addition, the modern semiconductor laser has a lower laser linewidth and can be easily integrated into a photonic chip.

Experimental Validation

Measurement Repeatability. Experimental validation of the measurement repeatability of the proposed laser vibrometer was conducted through comparisons among vibration responses from three independent measurements of a DUT. The experimental setups of measurement repeatability validation are shown in Figure 4, where the three DUTs were used for measurements with vibration frequencies much lower than, close to, and within the ultrasonic range, respectively. The laser vibrometer was attached to an adjustable frame in measurements to ensure that its laser beam was perpendicular to the surfaces of the DUTs and its laser spot was located at the centers of the DUTs. A signal generator was used to provide sinusoidal excitation to the DUTs. A speaker #1 shown in Figure 4a was used as the DUT that was under sinusoidal excitations with frequencies of 0.1 Hz, 1 Hz, 10 Hz, 100 Hz, and 1000 Hz, which were much lower than the ultrasonic range. A speaker #2 that was smaller than the speaker #1, as shown in Figure 4b, was used as the DUT that was under sinusoidal excitations with frequencies of 5000 Hz and 20,000 Hz, which were close to the ultrasonic range. An ultrasonic vibration source shown in Figure 4c was used as the DUT that was under sinusoidal excitation with a frequency of 1,000,000 Hz, which was within the ultrasonic range. For each excitation frequency, three independent tests were performed with the same data length. The sample results of the three DUTs are shown in Figures 5–7, respectively. One could see that the measured vibration responses from the three datasets for each excitation frequency had good agreement with each other in both the time and frequency domains. To further check the measurement repeatability, the peak values of the vibrations of the DUTs under various excitation frequencies were extracted from their frequency spectra and listed in Table 1, where the percentage differences among them were calculated by using each dataset #1 as a reference. One could see that the maximum difference was lower than 0.5%, meaning that the proposed laser vibrometer could measure vibrations in a wide frequency range with a high level of repeatability.

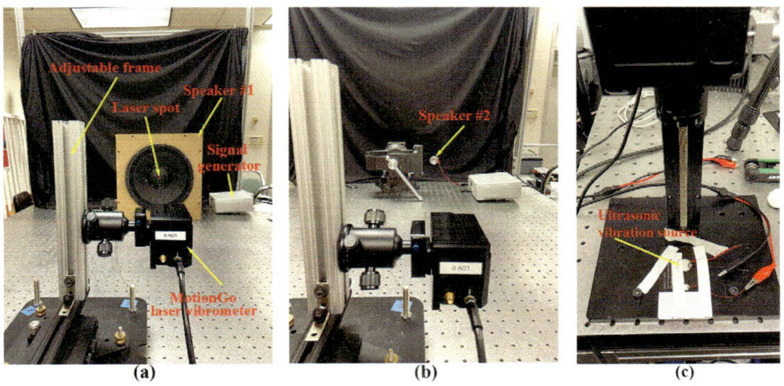

Fig. 4 Experimental setups of the measurement repeatability validation of the proposed laser vibrometer: (**a**) a speaker under excitation frequencies of 0.1 Hz–1000 Hz; (**b**) another speaker under excitation frequencies of 5000 Hz and 20,000 Hz; and (**c**) an ultrasonic vibration source under an excitation frequency of 1,000,000 Hz.

Measurement Accuracy. Experimental validation of the measurement accuracy of the proposed laser vibrometer was conducted through comparisons between vibration responses from the measurements of a DUT using two independent measure-

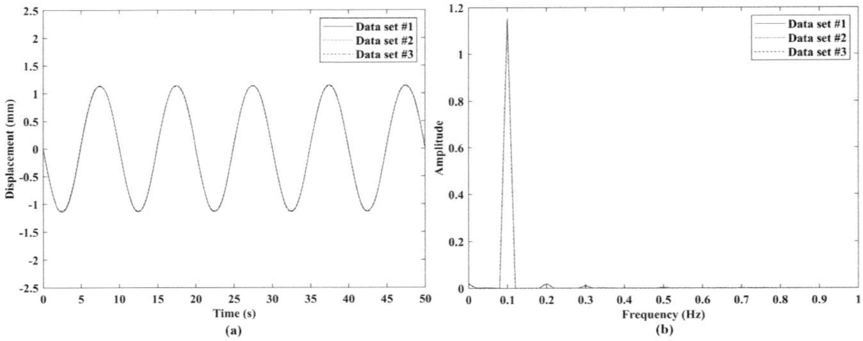

Fig. 5 Vibrations of the speaker #1 under sinusoidal excitation with a frequency of 0.1 Hz: responses from three independent datasets with the same length in the (**a**) time domain and (**b**) frequency domain.

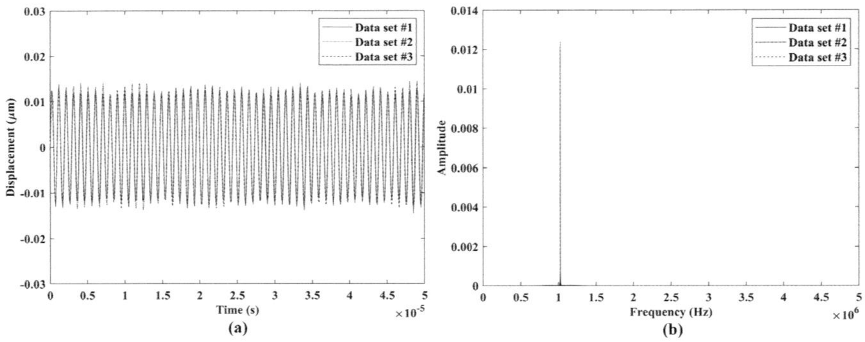

Fig. 6 Vibrations of the speaker #1 under sinusoidal excitation with a frequency of 20,000 Hz: responses from three independent datasets with the same length in the (**a**) time domain and (**b**) frequency domain.

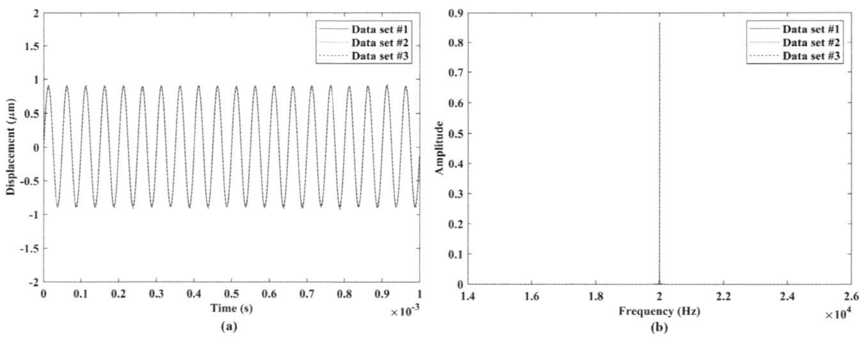

Fig. 7 Vibrations of the speaker #1 under sinusoidal excitation with a frequency of 1,000,000 Hz: responses from three independent datasets with the same length in the (**a**) time domain and (**b**) frequency domain.

ment systems; vibration response measured by a Polytec LDV (PSV-500) (Polytec GmbH, Waldbronn, Baden-Württemberg, Germany), as shown in Figure 8, was used as a reference. Speakers #1 and #2 were used in this work to provide vibrations with frequencies up to 5000 Hz, since the Polytec LDV used in this work had a limited maximum sampling rate and was not an LDV designed for measuring ultrasonic vibration. For each excitation frequency, the tests performed by the two measurement systems had the same data length. The sample results of the two DUTs were shown in Figures 9 and 10. Note that the responses measured by the proposed laser vibrometer were in terms of displacement, while those measured by the Polytec LDV were in terms of velocity. Therefore, velocity responses from the Polytec LDV were integrated to be displacement ones prior to comparison. One can see that the measured vibration responses from the two systems for each excitation frequency had good agreement with each other in both the time and frequency domains. To further check the measurement accuracy, the peak values of the vibrations of the DUTs under various excitation frequencies were ex-

Table 1 Peak values of vibrations of the DUTs under various excitation frequencies, where three independent datasets were measured for each excitation frequency and differences among them were calculated by using each dataset #1 as a reference (unit: μm).

DUT	Excitation frequency (Hz)	Dataset #1	Dataset #2	Dataset #3	Differences
Speaker #1	0.1	1154.4	1149.7	1149.3	0.4% & 0.4%
	1	116.1	116.2	116.3	0.1% & 0.2%
	10	75.3	75.2	75.4	0.1% & 0.1%
	100	30.3	30.3	30.3	0 & 0
	1000	4.9×10^{-1}	4.9×10^{-1}	4.9×10^{-1}	0 & 0
Speaker #2	5000	1.8×10^{-1}	1.8×10^{-1}	1.8×10^{-1}	0 & 0
	20,000	8.7×10^{-1}	8.7×10^{-1}	8.7×10^{-1}	0 & 0
Ultrasonic vibration source	1,000,000	1.2×10^{-2}	1.2×10^{-2}	1.2×10^{-2}	0 & 0

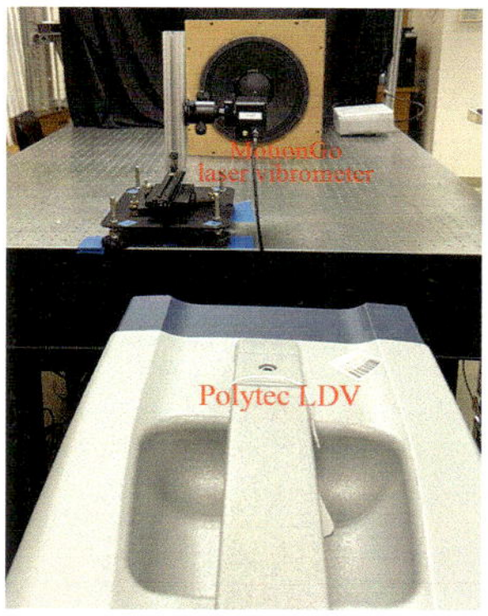

Fig. 8 Experimental setup of the measurement accuracy validation of the proposed laser vibrometer, where a Polytec LDV was used as a reference.

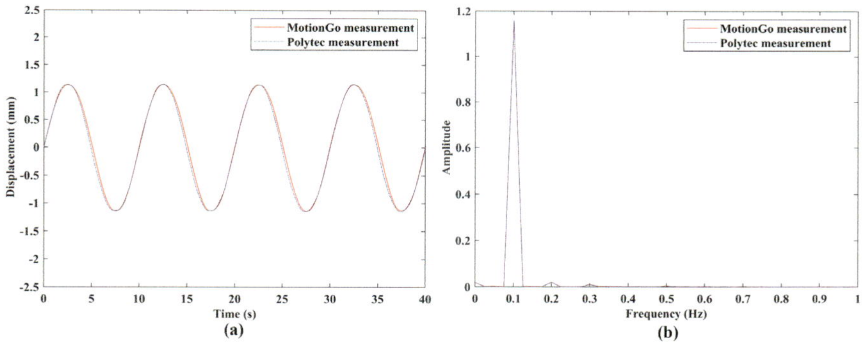

Fig. 9 Vibrations of the speaker #1 under sinusoidal excitation with a frequency of 0.1 Hz: responses from the two independent measurement systems with the same length in the (**a**) time domain and (**b**) frequency domain.

tracted from their frequency spectra and listed in Table 2. One could see that the maximum difference was lower than 1%, meaning that the proposed laser vibrometer could measure vibrations in0 a wide frequency range with a high level of accuracy.

Robustness to Test Surface Conditions. This section aimed to assess the performance of the proposed laser vibrometer for different test surface conditions, where the SNR was used as an indicator. The experimental setup is shown in Figure 11, where the left subfigure was the speaker #1 with a piece of reflective tape on its surface and the right subfigure was the speaker #1 with its natural surface. It could be seen that the natural surface of the speaker #1 was dark and rough, meaning that it had weak reflection. For a conventional LDV without a signal enhancement technique, attaching a reflective tape on the test structure surface was an important and necessary step in the test preparation stage. A signal enhancement technique,

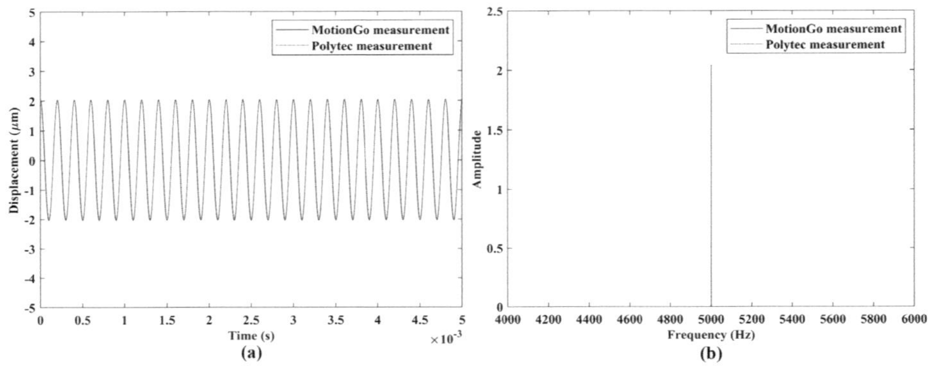

Fig. 10 Vibrations of the speaker #2 under sinusoidal excitation with a frequency of 5000 Hz: responses from the two independent measurement systems with the same length in the (**a**) time domain and (**b**) frequency domain.

Table 2 Peak values of vibrations of the DUTs under various excitation frequencies, where vibration responses were measured for each excitation frequency using the two independent measurement systems and differences among them were calculated by using each data from the Polytec LDV as a reference (unit: μm).

DUT	Excitation Frequency (Hz)	Polytec Measurement	MotionGo Measurement	Differences
Speaker #1	0.1	1157.99	1157.97	¡¡0.01%
	1	600.95	595.17	0.96%
	10	75.50	74.97	0.70%
	100	44.67	44.61	0.13%
	1000	4.97×10^{-1}	4.95×10^{-1}	0.40%
Speaker #2	5000	2.03	2.04	0.49%

such as the QTec technology developed by Polytec, usually used multiple channels, while the laser vibrometer proposed in this work was based on a single-channel detection. For fair comparison purposes, a conventional Polytec LDV without the QTec technology signal enhancement was used in this comparison. Vibration responses of the speaker #1 under sinusoidal excitations with frequencies of 10 Hz and 1000 Hz, measured by the Polytec LDV, which was a conventional model without the QTec technology, were shown in Figures 12a and 13a, respectively. One could see that the SNRs of the responses slightly decreased from 53.69 to 51.59 for the case of 10 Hz excitation, and significantly decreased from 11.42 to 7.46 for the case of 1000 Hz excitation, when removing the reflective tape. Vibration responses of the speaker #1 under sinusoidal excitations with frequencies of 10 Hz and 1000 Hz, measured by the proposed laser vibrometer, are shown in Figures 12b and 13b, respectively. The SNRs of the responses remained approximately constant for enhanced and natural surfaces, meaning that the proposed laser vibrometer had a good robustness to the test surface conditions. Note that the SNR results from the two measurement systems were not compared to each other, as they had different measurement conditions, such as measurement distances and optical diameters.

Conclusion

This work proposes a novel chip-based high-precision laser vibrometer that can directly measure the vibration displacement of a structure through the phase demodulation method. The feasibility for the productization and the features for the vibration measurement of the proposed miniatured laser vibrometer are discussed in this work. Some conclusions are listed as follows:

1. The laser vibrometer proposed in this work incorporates two or more sets of quadrature demodulation networks in the design. One quadrature demodulation network serves as a main *I*/*Q* demodulator for signal measurement. A reference quadrature demodulation network can capture the phase and amplitude instability of the laser. Correlation between the signals from the first and second quadrature networks can be computed to mitigate signal interference caused by ghost reflectance or scattering signals. A third quadrature demodulation network set with a distinct reference arm delay line length can be included to further compensate the temperature effect by capturing temperature or environmental variations that impact its delay line.

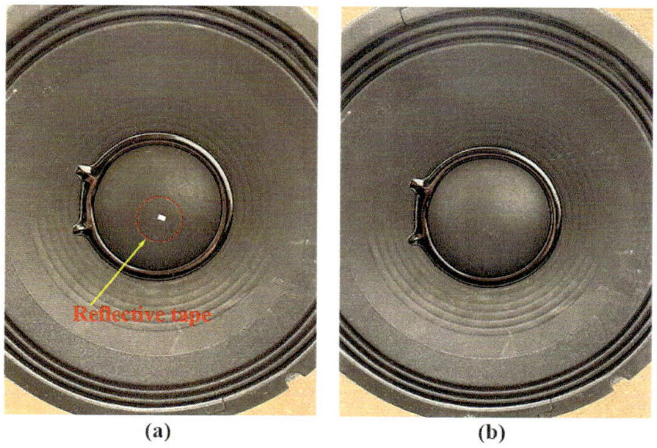

Fig. 11 Experimental setup for assessing the robustness of the proposed laser vibrometer to test surface conditions using the speaker #1 with (**a**) a surface enhanced by a reflective tape, and (**b**) a natural surface.

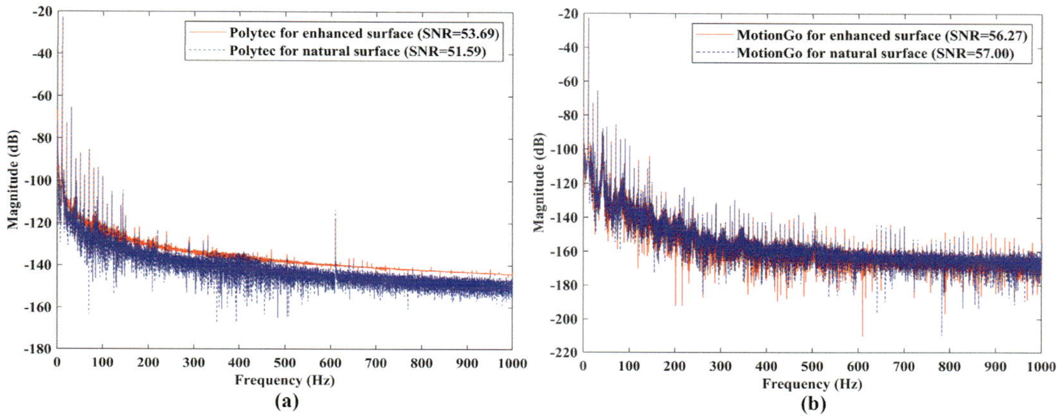

Fig. 12 SNRs of measured responses of the speaker #1 with enhanced and natural surfaces under sinusoidal excitation with a frequency of 10 Hz using (**a**) the Polytec LDV, and (**b**) the proposed laser vibrometer.

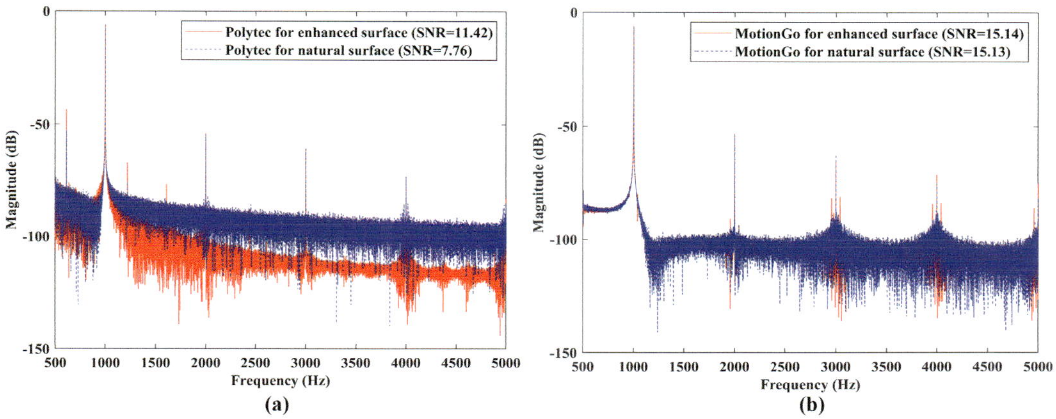

Fig. 13 SNRs of measured responses of the speaker #1 with enhanced and natural surfaces under sinusoidal excitation with a frequency of 1000 Hz using (**a**) the Polytec LDV, and (**b**) the proposed laser vibrometer.

2. Experimental validation of the measurement repeatability of the proposed laser vibrometer was conducted through comparisons among the vibration responses from the three independent measurements of the DUTs under sinusoidal excitations with frequencies from 0.1 Hz to 1,000,000 Hz. The maximum difference among measurements is lower than 0.5%, meaning that the proposed laser vibrometer can measure vibrations in a wide frequency range with a high level of repeatability.

3. Experimental validation of the measurement accuracy of the proposed laser vibrometer was conducted through comparisons between vibration responses from measurements of the DUTs using the proposed laser vibrometer and a Polytec LDV. The maximum difference between the results from the two measurement systems was lower than 1%, meaning that the proposed laser vibrometer could measure vibrations in a wide frequency range with a high level of accuracy.
4. SNRs of responses measured by the proposed laser vibrometer could remain approximately constant for surfaces with and without reflective tapes, meaning that it had better robustness to test surface conditions than a conventional LDV.

Acknowledgments This research was funded by the Maryland Industrial Partnerships Program through the grant number 7207 and the National Science Foundation through the grant number 1763024.

References

1. Lyu, L., Yuan, K., and Zhu, W. "A novel demodulation method with a reference signal for operational modal analysis and baseline-free damage detection of a beam under random excitation". *Journal of Sound and Vibration*, 571:118068 (2024)
2. Yuan, K. and Zhu, W. "In-plane operating deflection shape measurement of an aluminum plate using a three-dimensional continuously scanning laser doppler vibrometer system". *Experimental Mechanics*, pages 667–676 (2022)
3. Rezaei, D. and Taheri, F. "A novel application of a laser doppler vibrometer in a health monitoring system". *Journal of Mechanics of Materials and Structures*, 5(2):289–304 (2010)
4. Yuan, K. and Zhu, W. "A novel mirror-assisted method for full-field vibration measurement of a hollow cylinder using a three-dimensional continuously scanning laser doppler vibrometer system". *Mechanical Systems and Signal Processing*, 216:111428 (2024)
5. Nassif, H.H., Gindy, M., and Davis, J. "Comparison of laser doppler vibrometer with contact sensors for monitoring bridge deflection and vibration". *Ndt & E International*, 38(3):213–218 (2005)
6. Yu, T., Tang, Q., and Vinayaka, S. "Identifying structural properties of a steel railway bridge for structural health monitoring using laser doppler vibrometry". *Automation in Construction*, 160:105320 (2024)
7. Roozen, N.B., Labelle, L., Rychtáriková, M., and Glorieux, C. "Determining radiated sound power of building structures by means of laser doppler vibrometry". *Journal of Sound and Vibration*, 346:81–99 (2015)
8. Huang, Z., Xie, M., Chen, C., Du, Y., and Zhao, J. "Engineering application of a safety-state evaluation model for hidden frame-supported glass curtain walls based on remote vibration". *Journal of Building Engineering*, 26:100915 (2019)
9. Yuan, K. and Zhu, W. "Identification of modal parameters of a model turbine blade with a curved surface under random excitation with a three-dimensional continuously scanning laser doppler vibrometer system". *Measurement*, 214:112759 (2023)
10. Zabihi, A., Aghdasi, F., Ellouzi, C., Singh, N.K., Jha, R., and Shen, C. "Non-contact wind turbine blade crack detection using laser doppler vibrometers". *Energies*, 17(9):2165 (2024)
11. Yuan, K. and Zhu, W. "Modeling of welded joints in a pyramidal truss sandwich panel using beam and shell finite elements". *Journal of Vibration and Acoustics*, 143(4):041002 (2021)
12. Hameury, C., Ferrari, G., Franchini, G., and Amabili, M. "An experimental approach to multi-input multi-output nonlinear active vibration control of a clamped sandwich beam". *Mechanical Systems and Signal Processing*, 216:111496 (2024)
13. Garg, P., Moreu, F., Ozdagli, A., Taha, M.R., and Mascareñas, D. "Noncontact dynamic displacement measurement of structures using a moving laser doppler vibrometer". *Journal of Bridge Engineering*, 24(9):04019089 (2019)
14. Garg, P., Nasimi, R., Ozdagli, A., Zhang, S., Mascarenas, D.D.L., Reda Taha, M., and Moreu, F. "Measuring transverse displacements using unmanned aerial systems laser doppler vibrometer (uas-ldv): Development and field validation". *Sensors*, 20(21):6051 (2020)
15. Bertini, L., Neri, P., Santus, C., and Guglielmo, A. "Automated experimental modal analysis of bladed wheels with an anthropomorphic robotic station". *Experimental Mechanics*, 57:273–285 (2017)
16. Margerit, P., Gobin, T., Lebée, A., and Caron, J.F. "The robotized laser doppler vibrometer: On the use of an industrial robot arm to perform 3d full-field velocity measurements". *Optics and Lasers in Engineering*, 137:106363 (2021)
17. Ahmed, H., Mohsin, A., Hong, S.C., Lee, J.R., and Ihn, J.B. "Robotic laser sensing and laser mirror excitation for pulse-echo scanning inspection of fixed composite structures with non-planar geometries". *Measurement*, 176:109109 (2021)
18. Brethee, K.F., Uwayed, A.N., and Alden Qwam, A.Y. "A novel index for vibration-based damage detection technique in laminated composite plates under forced vibrations: Experimental study". *Structural Health Monitoring*, 22(5):3109–3125 (2023)
19. Li, Y., Verstuyft, S., Yurtsever, G., Keyvaninia, S., Roelkens, G., Van Thourhout, D., and Baets, R. "Heterodyne laser doppler vibrometers integrated on silicon-on-insulator based on serrodyne thermo-optic frequency shifters". *Applied optics*, 52(10):2145–2152 (2013)
20. Li, Y. and Baets, R. "Homodyne laser doppler vibrometer on silicon-on-insulator with integrated 90 degree optical hybrids". *Optics Express*, 21(11):13342–13350 (2013)
21. Li, Y., Dieussaert, E., and Baets, R. "Miniaturization of laser doppler vibrometers—a review". *Sensors*, 22(13):4735 (2022)
22. Ramasubramanian, K. and Instruments, T. "Using a complex-baseband architecture in fmcw radar systems". *Texas Instruments*, 19 (2017)
23. Nguyen, T.H., Scalart, P., Gay, M., Bramerie, L., Peucheret, C., Gomez-Agis, F., Sentieys, O., Simon, J.C., and Joindot, M. "New metric for iq imbalance compensation in optical qpsk coherent systems". *Photonic Network Communications*, 36:326–337 (2018)

Chapter 16
Capture of Thin Film Vibromorphology using Laser Vibrometry and Frequency-Domain Viscoelastic Vibroacoustics

Kent X. Eng, Steven Kao, Natalie J. White, Nakul Deshpande, Jeffery D. Tippmann, and Eric C. Bryant

Abstract Applications of thin film vibromorphology are increasingly relevant because of the emergent use of engineered polymers for device design (*e.g.*, of Triboelectric Nanogenerators (TENGs)). Modern research applications of thin-film technology like TENGs focus on harvesting mechanical energy for power generation or sensing technologies. However, there is an existing research gap in the numerical and experimental investigation of thin film vibromorphology that compliments design efforts for modern technological applications.

To address this gap, the present work presents a combined experimental and numerical investigation of the vibroacoustics of thin polymer films. The vibroacoustic morphology (*i.e.*, so-called "vibromorphology") of one-or-more contacting thin polymer films during deformation under acoustic excitation is characterized. Specifically, variations in experimental frequency response curves (FRCs) under the films' structural-dynamics parameterization (*viz.*, a considered subset of influence from size, initial tensioning, thickness, multiplicity, etc.) are described via a combination of experimental and numerical methods.

In doing so, experimental optical inference of the films' viscoelastic deformations under acoustic excitation via 3D laser vibrometry is complimented with finite element analysis (FEA) of the films' structural dynamics. Influence of the films' geometry and fixture (*e.g.*, film size and initial tensioning) are described by modification of numerical mesh and material data. Novel FEA-based simulation methods for film contact are exercised, particularly with regard to treatment of FEA as a forward problem for material data inversion.

Keywords Film vibromorphology · Finite element analysis (FEA) · Laser vibrometry · Model validation · Optimization

Kent X. Eng · Steven Kao · Natalie J. White
National Security Education Center (NSEC), Los Alamos National Laboratory, Los Alamos, NM 87545
e-mail: kente@lanl.gov; tundra@lanl.gov; njwhite@lanl.gov

Kent X. Eng
Department of Civil and Environmental Engineering, Princeton University, Princeton, NJ 08544
e-mail: kente@lanl.gov

Steven Kao
Department of Electrical and Computer Engineering, The University of New Mexico, Albuquerque, NM 87131
e-mail: tundra@lanl.gov

Natalie J. White
Department of Mechanical Engineering, Brigham Young University, Provo, UT 84602
e-mail: njwhite@lanl.gov

Nakul Deshpande · Eric C. Bryant
National Security Earth Science (EES-17), Los Alamos National Laboratory, Los Alamos, NM 87545
e-mail: njwhite@lanl.gov ; ecbryant@lanl.gov

Jeffery D. Tippmann
Space Remote Sensing and Data Science (ISR-6), Los Alamos National Laboratory, Los Alamos, NM 87545
e-mail: jtippmann@lanl.gov

Janko Slavic et al. (eds.), *Computer Vision & Laser Vibrometry, Vol. 6*, Conference Proceedings of the Society for Experimental Mechanics Series, https://doi.org/10.13052/9788743801511_16

Introduction

Triboeletic Nanogenerators (TENGs) comprise a new mechanical energy harvesting technology based on contact electrification and electrostatic induction. TENGs are a multi-thin-flim system: a parallel-plate capacitor formed between a tribo-charged dielectric material surface and two electrode materials [1]. When a TENGs vibrates, opposed surfaces are understood to experience effective charge polarization inducing electric discharge at the electrodes; charge strength/polarity yielded depends on material properties, straining, temperature, *etc.* [2]. Resurgent interest in applications of thin film vibromorphology is motivated by the emergence of engineered polymers in TENG design [3, 4] for: harvesting mechanical energy for power generation [5, 6] and use in sensor technologies [7, 8], among other applications. Currently, a research gap exists in experimental and numerical investigation of multi-layer polymer, thin-film vibromorphology for engineering design of TENG devices.

In this work, measurements are designed to enable *separately tensioned* thin films to be tested (1) *individually*, (2) *in proximity*, and (3) *in contact* to describe vibromorphological sensitivity. In addition, our numerical work establishes a principled approach to characterize multi-thin-film deformation during so-called "*en echelon*" contact under frequency-domain Finite Element Analysis (FEA). Complicating this task, manufacturer-provided polymer properties may be unreliable due to inconsistent manufacturing processes (*i.e.*, batch variability [9]) and/or infeasibility of standard testing regimes due to specimen thinness. As such, the additional challenge of inverse methods for film flexural parameters is also addressed.

Research Objectives

To address the research gap identified above, our current procedure combines experimental and numerical investigation of thin polymer films' vibroacoustic morphology. Our work provides:

- Vibroacoustic experimentation to quantify impact of thin films' initial tensioning, multiplicity, and/or contact distance
- Numerical simulation of thin films' vibroacoustic excitement via frequency-domain FEA
- Optimization-based method for inference from thin films' structural dynamics
- Integrated framework of experimental, FEA-based, and optimization methods to infer thin films' flexural properties

Substantive prior research has been undertaken to address experimental/numerical challenges of TENG design, as indicated above, and is now distinguished from the present work. In addition, our unique research emphasis is identified: development of an analysis framework to integrate experimental and numerical results of multi-layer-film vibromorphology.

Research Background

Thin polymers, as used in TENG design, are nonstructural materials and manufactured absent a codified standard for flexural material performance. As a consequence, flexural material properties – Young's modulus, E, Poisson's Ratio, ν, *etc.* – can differ significantly between samples due to batch variability [9]. Moreover, standard evaluation methods may fail because of the films' compliance to out-of-plane deformation. To address this issue, later nondestructive methods in membrane dynamics [10, 11, 12] and frequency spectra [13] have been introduced. Our novel contribution is to describe multi-layer polymers' vibromorphology with respect to (abbreviated "w.r.t.") unstudied characteristics like film tensioning – ultimately, with the aim to improve TENG operational efficiency by tuning to specific natural frequencies [13].

Laser Doppler Vibrometry (LDV), a nondestructive testing method, is used in our thin-film dynamics analysis due to its advantage of detecting resonant out-of-plane motion [14]. Extant LDV methods for resonant frequency analysis focus on a single layer [12, 15] or clamped multi-layers [16]. Experimentation on clamped multi-layers with engineering initial tensioning is outside the present literature, with current state-of-art for initial tensioning remaining single-layer [17, 18]. In this work, testing is conducted on multi-layer polymers with separate clamping on each layer: the layers may be tested separately, together, and in presence or absence of tensioning. In addition, our experimentation incorporates improved multi-layer-film fixtures – *i.e.*, new composite film "holders" – enabling vibrometry under parameterization of contact distance.

Mathematical Description

Researchers have developed diverse FEA strategies to model the behavior of single thin films, such as elastic modulus, residual stress [17], and channel cracking [19], *etc.* However, frequency-domain FEA for multi-layer polymer assemblages is less broad, particularly if layer interfaces are understood as potential sites of viscous dissipation (as here). Researchers have developed other FEA methods for TENGs, focusing mainly on sliding-mode theory [20], topology modeling [21], and electrostatic behavior [22]. There is a research gap on FEA for dynamic interactions between polymer films – particularly combined with experiential determinations on the applicability of frequency-domain analysis.

In-line with [23], a variational principle for frequency-domain structural analysis incorporating effects of *en echelon* structure-structure interaction is developed. The material under investigation is viscoelastic. Thus, the time-domain stress and strain tensors and displacement vector will be complex-valued fields. Under cyclic loading with frequency ω at time t, these fields are described by the mapping:

$$\tilde{\boldsymbol{\sigma}} = \boldsymbol{\sigma}\, e^{i\omega t} \quad \text{(stress)} \tag{1}$$

$$\tilde{\boldsymbol{\epsilon}} = \boldsymbol{\epsilon}\, e^{i\omega t} \quad \text{(strain)} \tag{2}$$

$$\tilde{\boldsymbol{u}} = \boldsymbol{u}\, e^{i\omega t} \quad \text{(displacement)} \tag{3}$$

where the real parts of fields $\tilde{\boldsymbol{\sigma}}, \tilde{\boldsymbol{\epsilon}}, \tilde{\boldsymbol{u}}$ approximate the actual material state. Time-independent fields $\boldsymbol{\sigma}, \boldsymbol{\epsilon}, \boldsymbol{u}$ are likewise complex-valued, with displacement $\boldsymbol{u}$ ultimately discretized as the solution vector for frequency-domain FEA.

Multi-Body Potential

The above notation is now generalized to a two-body assemblage. A complex-valued displacement vectorfield, $\boldsymbol{u}_\alpha \in \mathbb{C}^3$, is restricted to the primary solid domain, $\mathscr{B}_\alpha \subseteq \mathbb{R}^3$. The displacement vectorfield, $\boldsymbol{u}_\beta \in \mathbb{C}^3$, is likewise restricted to a secondary solid domain, $\mathscr{B}_\beta \subseteq \mathbb{R}^3$. Under isothermal conditions, this variational principle is

$$\Phi(\boldsymbol{u}_\alpha, \boldsymbol{u}_\beta) = \Phi_\alpha(\boldsymbol{u}_\alpha) + \Phi_\beta(\boldsymbol{u}_\beta) + \Phi_{\alpha\beta}(\boldsymbol{u}_\alpha, \boldsymbol{u}_\beta) \tag{4}$$

where the solid continuum terms are

$$\Phi_\iota(\boldsymbol{u}_\iota) = \int_{\mathscr{B}_\iota} (U_\iota - K_\iota)\mathrm{d}V - \int_{\partial\mathring{\mathscr{B}}_\iota} u_{\iota\, i} \mathring{S}_{\iota\, ij} n_j \mathrm{d}S, \quad \iota = \alpha, \beta \tag{5}$$

with $\mathring{S}_{\iota\, ij} n_j$ prescribed on boundaries $\partial\mathring{\mathscr{B}}_\iota$ for $\iota = \alpha, \beta$. On each of $\mathscr{B}_\alpha$ and $\mathscr{B}_\beta$, nonsymmetric second-order stress tensor $\boldsymbol{S}_\iota$ is

$$\boldsymbol{S}_\iota = \frac{\partial U_\iota}{\partial \nabla \boldsymbol{u}_\iota} = \underbrace{\frac{\partial U_\iota}{\partial \boldsymbol{\epsilon}_\iota}}_{=\boldsymbol{\sigma}_\iota} : \frac{\partial \boldsymbol{\epsilon}_\iota}{\partial \nabla \boldsymbol{u}_\iota}, \quad \iota = \alpha, \beta \tag{6}$$

where $\boldsymbol{\sigma}_\iota$ and $\boldsymbol{\epsilon}_\iota$ are symmetric static and kinematic tensors, respectively.

Inter-Body Potential

The interface coupling term specifies *en echelon* motion via the particularization:

$$\Phi_{\alpha\beta}(\boldsymbol{u}_\alpha, \boldsymbol{u}_\beta) = \int_{\partial\mathscr{B}_{\alpha\beta}} \left(\frac{k_N^*}{2} [\![u]\!]_{N\,i} [\![u]\!]_{N\,i} + \frac{k_T^*}{2} [\![u]\!]_{T\,i} [\![u]\!]_{T\,i} \right) \mathrm{d}S = \Phi_{\beta\alpha}(\boldsymbol{u}_\alpha, \boldsymbol{u}_\beta) \tag{7}$$

for total, normal, and tangential jump quantities given, respectively, by

$$[\![\boldsymbol{u}]\!]_{\alpha\beta} = \boldsymbol{u}_\beta - \boldsymbol{u}_\alpha, \quad [\![\boldsymbol{u}]\!]_N|_{\partial\mathscr{B}_{\alpha\beta}} = \boldsymbol{N}|_{\partial\mathscr{B}_{\alpha\beta}} \cdot [\![\boldsymbol{u}]\!]_{\alpha\beta}, \quad [\![\boldsymbol{u}]\!]_T|_{\partial\mathscr{B}_{\alpha\beta}} = \boldsymbol{T}|_{\partial\mathscr{B}_{\alpha\beta}} \cdot [\![\boldsymbol{u}]\!]_{\alpha\beta} = [\![\boldsymbol{u}]\!]_{\alpha\beta} - [\![\boldsymbol{u}]\!]_N|_{\partial\mathscr{B}_{\alpha\beta}}. \tag{8}$$

Second-order projections tensors $\boldsymbol{N}$ and $\boldsymbol{T}$ are such that

$$N_{ij} = n_i n_j, \qquad T_{ij} = \delta_{ij} - n_i n_j.$$

As per usual for FEA models, $\partial\mathscr{B}_{\alpha\beta}$ is described by unit outward normal $\boldsymbol{n}|_{\partial\mathscr{B}_{\alpha\beta}}$ vector from primary solid to secondary solid domain; $\partial\mathscr{B}_{\beta\alpha}$ is described by unit outward normal $\boldsymbol{n}|_{\partial\mathscr{B}_{\beta\alpha}}$ vector from secondary to primary solid domain.

Intra-Body Potential

Variations in the uncoupled potentials, $\Phi_\iota(\boldsymbol{u}_\iota)$ for $\iota = \alpha, \beta$, are now developed w.r.t. displacement vectorfield $\boldsymbol{u}_\iota$, as well as variations in the coupled potential $\Phi(\boldsymbol{u}_\alpha, \boldsymbol{u}_\beta)$. To do so, a specification of U_ι's function dependence on $\boldsymbol{\epsilon}_\iota$ is required. As for frequency-domain structural analysis, U_ι is taken as quadratic in $\boldsymbol{\epsilon}_\iota$. Consistent with [23], U_ι may be equivalently rewritten as

$$U_\iota = \frac{\lambda_\iota^*}{2}(\mathbf{1} : \boldsymbol{\epsilon}_\iota)^2 + \mu_\iota^* \boldsymbol{\epsilon}_\iota : \boldsymbol{\epsilon}_\iota = \frac{\lambda_\iota^*}{2}(\varepsilon_{\iota\,kk})^2 + \mu_\iota^* \varepsilon_{\iota\,ij}\varepsilon_{\iota\,ij}, \quad \iota = \alpha, \beta \tag{9}$$

for $\lambda_\iota^*, \mu_\iota^* \in \mathbb{C}$ denoting the complex-valued Lame's and shear moduli. Likewise after [23], λ_ι^* and μ_ι^* may be related to common engineering measurements of complex-valued bulk and shear compliance. λ_ι^* and μ_ι^*'s real parts are the materials' real Lame's and real shear moduli; conversely, their imaginary parts likely express dependence on frequency ω. Departing from [23], K_ι is written as

$$K_\iota = \frac{\rho_\iota^* \omega^2}{2} \boldsymbol{u}_\iota \cdot \boldsymbol{u}_\iota = \frac{\rho_\iota^* \omega^2}{2} u_{\iota\,i} u_{\iota\,i}, \quad \iota = \alpha, \beta \tag{10}$$

for $\rho_\iota^* \in \mathbb{C}$ denoting the complex-valued solid density. ρ_ι^*'s real part is the solid density; conversely, its imaginary part may express dependence on frequency ω. Our departure – *i.e.*, introducing a complex-valued density – indicates: *Raleigh damping is incorporated as a specialization of complex-valued material data.*

Stationarity

Strong-form governing equations are recovered from the stationary point,

$$\underset{\boldsymbol{u}_\iota}{\delta} \{\Phi(\boldsymbol{u}_\alpha, \boldsymbol{u}_\beta)\} = 0 \Leftrightarrow S_{\iota\,ij,j} + \rho_\iota^* \omega^2 u_{\iota\,i} = 0, \tag{11}$$

as are boundary conditions given as

$$\underset{\boldsymbol{u}_\iota}{\delta} \{\Phi(\boldsymbol{u}_\alpha, \boldsymbol{u}_\beta)\} = 0 \Leftrightarrow S_{\iota\,ij} n_j|_{\partial \mathring{\mathscr{B}}_\iota} = \mathring{S}_{\iota\,ij} n_j|_{\partial \mathring{\mathscr{B}}_\iota} \text{ and } S_{ij} n_j|_{\partial \mathscr{B}_{\alpha\beta}} = (k_N^* [\![u]\!]_{N\,i} + k_T^* [\![u]\!]_{T\,i})|_{\partial \mathscr{B}_{\alpha\beta}}, \tag{12}$$

which comprise, as desired, standard prescribed traction and desired interface relations. Based on the stated variational principle, multi-thin-film FEA of interface interactions can integrate with experimental results exhibiting *en echelon* deformation.

Methods

Our overall framework forms a bridge between experimental and numerical multi-layer-film vibromorphology – by comparing measured resonant frequencies from experimental results to predicted resonant frequencies from FEA, optimized to minimize the difference thereof (Fig. 1). In each iteration of the optimization procedure, the optimizer receives predicted resonant frequencies from FEA and determines a normalized square difference from experimental measurements, a.k.a., a so-called "loss function." The optimizer then outputs an updated estimation of modal parameters, using the ADAM algorithm [24] as implemented in PyTorch. Updated parameters are used to update elastic flexural moduli used as FEA inputs, with FEA then rerun to predict resonant frequencies. This cycle repeats until the optimizer minimizes the loss function.

Experimental Procedure

LDV was used to measure the resonant frequencies of the thin films, both individually and as contacting multi-film assemblages. In particular, care is required in fixturing films for LDV due to compliance under out-of-plane deformation. To control the measurement, a circular sample of polymer was held in place with a designed fixture described in Sec. . The ring-shaped fixture clamped only the sample edges leaving the center free to vibrate, defining the boundary conditions. Fixture was clamped by fastening six screws and nuts along the perimeter. To make boundary conditions as uniform as possible, nuts were tightened using a torque wrench to the same torque value.

Once clamped in the fixture, tension could be optionally applied by attaching the "base piece." The base attached via screws and nuts in four locations. Distance between the base and clamping pieces was kept uniform by using stops or measuring with calipers. This distance governed tension applied to the sample: increased proximity of base to clamping pieces

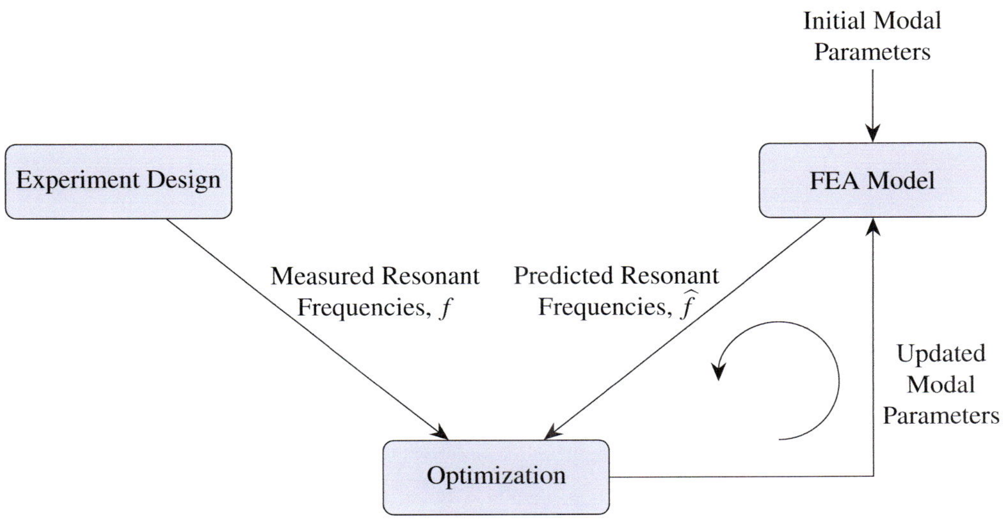

Fig. 1 Thin film model framework.

induced greater tensioning, because the tube on the base piece pressed against the sample. To avoid plastic deformation, strains of less than 2% were applied (the estimated elastic limit). Images of the sample were captured before and after tensioning for use in digital image correlation (DIC) for strain quantification, Sec. .

For testing two samples in a multi-thin-film assemblage, each sample would first be clamped and tensioned in its own fixture. Then, the two base pieces would be held together such that the samples face each other. If the samples were to be in contact, then the base pieces are clamped together with four screws at each of four tabs. If multi-thin-film samples were in close proximity, then a thin "stop" was between the tabs to create the space. Once sample(s) were secured in the fixture, fixture was held by lab clamps and oriented towards the vibrometer's laser and camera, approximately 90° to the camera view.

A nearby speaker was used as sound source for film acoustic excitation during LDV. Speaker was placed parallel to the assemblage, so that sample edges are approximately equidistant from the speaker. Speaker, along with the polymer, fixture, and LDV scanner, is shown in Fig. 2. A chirp ranging from 0-2 kHz was emitted for each scan point; this range should be reevaluated if new materials or geometries are used. LDV measured sample surface velocity at each scan point for computer data acquisition. Scan duration, usually 5 minutes or less, was determined by the number of scan points on the sample face. Data was viewed via Polytec™ software and exported for post-processing. Average sample surface velocity as a function of excitation frequency was examined for frequencies at which surface velocity peaked, which constituted the natural, or "resonant," frequencies. Coincident surface velocity distributions revealed modal shapes to confirm resonant frequencies.

Fixture Design

Two primary objectives motivated new experimental fixture design: testing films under tension and testing two films layered together. Applying tension is of interest to "tune" resonance to specific natural frequencies, which may maximize the efficiency of multi-thin-film assemblages in devices like TENGs. Additionally, measurements on multi-thin-film assemblages under contact parameterization promise similar capability for performance enhancement. A constraint was that film specimens should be clamped and tensioned independently, in order to test films separately (with or without tension) and then in contact – *viz.*, without being removed from the fixture and re-clamped. Fixture design was actuated with Polylactic Acid (PLA) 3D printing, an inexpensive and widely available manufacturing process.

Prior work [12] described a method of applying tension using a circular tube. A membrane was held in place by two rings, or flanges (similar to the aforementioned existing fixture). These two rings were moved towards a supporting back panel via threaded rods. A circular tube protruding from the back panel pressed against the membrane as the two rings moved closer to the panel, inducing tension. Fixture design implemented a similar concept from [12] by introducing a third base piece in addition to the two clamping rings: this base piece featured a circular tube with rounded edges to mitigate sample deformation. Instead of threaded rods, the base piece was penetrated by screws attaching to tabs on the lower clamping ring. Distance between base piece and clamping rings determined applied tension, *i.e.*, from tightening of screws controlled by stops.

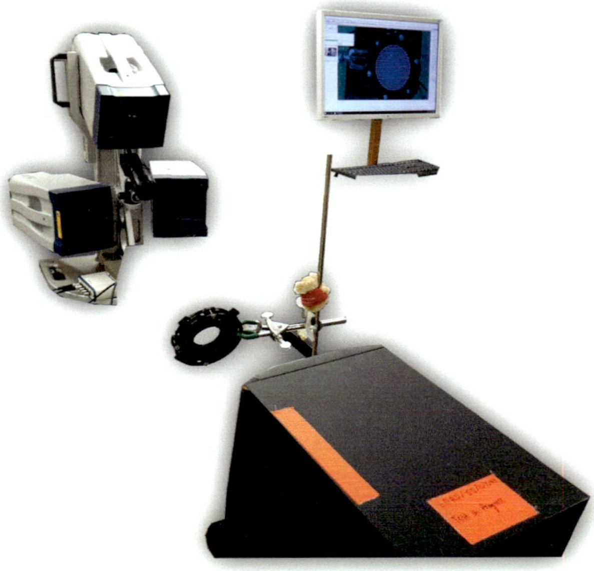

Fig. 2 Experimental set-up for LDV scans is shown. A speaker, in the foreground, provided acoustic excitation for the polymer, which is held in front of the LDV scanner. A laptop for data acquisition is in the top right corner.

(a) Top view (b) Isometric view (c) Cross-sectional isometric view

Fig. 3 The existing fixture, consisting of two clamping rings that secure the sample, as shown from various angles.

Centrally, our new design allowed for two independently clamped films to be fixtured in contact. Previously, this was infeasible because the existing fixture's clamping rings obstructed contact. In the new design, the rings clamped the edge of the sample at an angle of 7°, mitigating warping resultant from a more extreme angle. Independent symmetric fixtures were then conjoined with screws penetrating through tabs along the edges of the base piece, resulting in contact between the thin-film samples. With these alterations, both identified design objectives were met: *LDV scans could be performed on (1) samples under tension and (2) separately clamped samples (either tensioned or untensioned) in contact.*

Digital Image Correlation (DIC)

Strain quantification in the films was infeasible using physical instrumentation like strain gauges. Digital image correlation (DIC) was explored as an alternative, using the Ncorr MATLAB method [25]. Each film sample was speckled with spray paint, with image capture before and after tensioning to perform 2D DIC analysis. Image capture was as consistent as possible: any difference between current and reference images would be interpreted as deformation by the DIC algorithm,

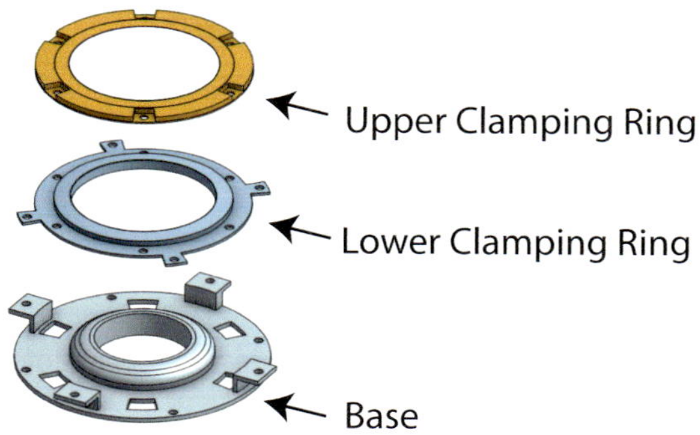

Fig. 4 Exploded view of the experimental fixture.

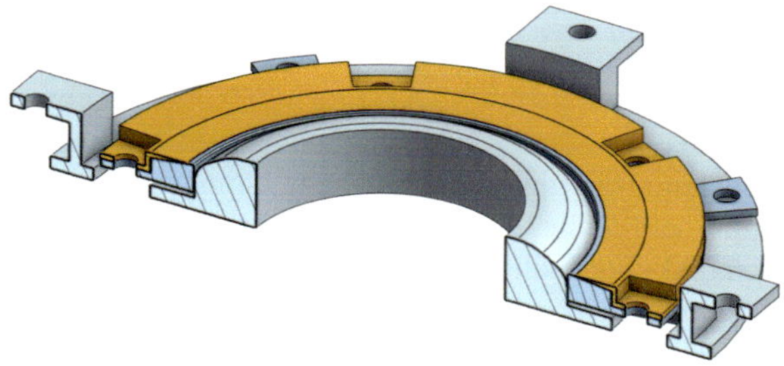

Fig. 5 Cross-sectional view of the experimental fixture.

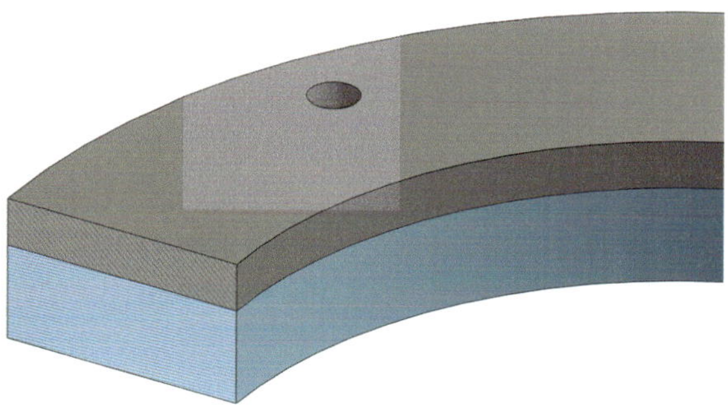

Fig. 6 Cross-sectional view of Gen 0 (initial fixture).

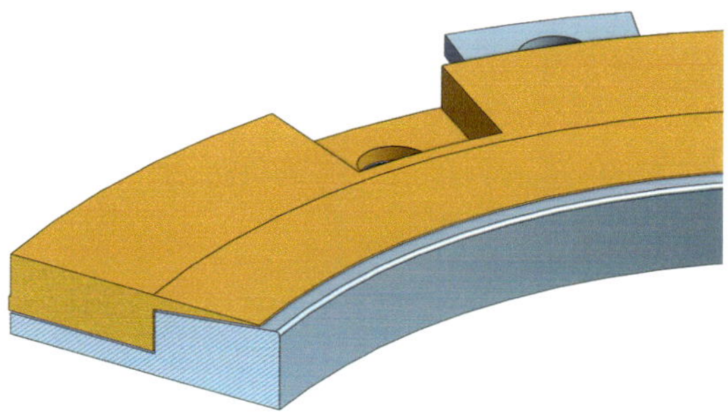

Fig. 7 Cross-sectional view of the clamping rings for Gen 4, showing angled clamping.

which calculated transformation for subsets within the Region Of Interest (ROI). DIC camera was mounted on a tripod, and fixture with the sample was mounted on a small 3D-printed stand (Fig. 8), placed in the same orientation before and after tensioning. Camera height was adjusted so the ROI filled a majority of the frame, maximizing resolution.

Fig. 8 Set-up for image capture for DIC. A camera on a tripod was used to photograph the polymer with fine speckled pattern.

DIC reference image was the sample image before tensioning, and ROI was the sample within the inner diameter of the base piece. ROI was chosen because the deformation in this region can be reasonably assumed to be 2D; if the ROI included the edge of the sample, which lay at an angle when tensioned, the change in angle would be misinterpreted by the algorithm as a 2D deformation. Furthermore, the only strain of interest was in the stretched region under vibroacoustic observation.

Strain was calculated from DIC displacements, by a least squares plane fit over displacement data for a window of sampled points in the ROI [25]. Because the strain calculation involves differentiation, a small strain window will result in a noisy strain field. Strain window size was increased until minimum and maximum strains converged, yielding a smoothing effect.

FEA Model

FEA was formulated in MOOSE using PETSc for eigensystem solution [26], with visualization in Paraview. FEA was based on the mathematical model that was discussion in Sec. . The mathematical model was implemented as cantilever beam, pretensioned circular membrane, and circular plate, and verified with the theoretical solution. For the model of pretensioned circular membrane, circular plate, and the thin film model, the boundary conditions were clamped which were assumed to be fixed. As particularly illustrative for the related optimization task, only circular plate verification is detailed.

For the circular plate verification [27], classically the transverse (*i.e.*, out-of-plane) displacement of a plate is described by a biharmonic equation written in the "flexural modulus,"

$$D = \frac{Eh^3}{12(1-\nu^2)}, \tag{13}$$

where E is the Young's modulus, h is the thickness, and ν is the Poisson ratio. Material is understood to have density ρ. The boundary of the circular plate is permanently clamped at its outer radius a. The jth natural frequency is:

$$\lambda_j^2 = \omega_j a^2 \sqrt{\frac{\rho h}{D}} \quad \Rightarrow \quad \lambda_j^4 = \underbrace{\omega_j^2}_{=f_j} a^4 \left(\frac{\rho h}{D}\right) \tag{14}$$

such that, by Eq. (13),

$$f_1 = \frac{\lambda_1^4 D}{a^4 \rho h} = \frac{\lambda_1^4 h^2}{12 a^4 \rho}\left(\frac{E}{1-\nu^2}\right) \tag{15}$$

where a is the radius of circular plate. λ_1 is the coefficient constant for the 1st natural frequency with $\lambda_1^2 = 10.21$.

Optimization

Our optimization algorithm is investigated in two case studies: a nonconvex problem to directly optimize material properties vs. a convex problem to optimize projected parameters as a mapping of material properties. To validate the optimization algorithm with simulated noise, six target frequencies $f_j \in \boldsymbol{f}$ were obtained via FEA provided target parameters, $\boldsymbol{\theta}$, adding Gaussian noise (with variance one-tenth of the resonant frequency). Fig. 9 shows distribution of $f_j \in \boldsymbol{f}$.

Predicted frequencies, each $\widehat{f}_i$, were obtained from FEA-based determination of predicted parameters, $\widehat{\boldsymbol{\theta}}$, obtained from PyTorch Adam optimization. If the optimization problem is strictly convex, target $\boldsymbol{\theta}$ will match the unique global minimum solution for prediction $\widehat{\boldsymbol{\theta}}$.

Case Study 1: Nonconvex Problem

In the first case study, the goal is to optimize the material parameters directly. Let m and n be number of samples and number of considered resonant frequencies. Define the loss function as

$$L_1 = \frac{1}{m}\sum_{i=1}^{m}\sum_{j=1}^{n}\left(A_{jj}^{(i)}\left(f_j^{(i)} - \widehat{f}_j\right)\right)^2 \tag{16}$$

where $f_j^{(i)}$ is the jth component (resonant frequency) of the ith sample, and normalization matrix $A^{(i)}$ is

$$A_{jk}^{(i)} = \begin{cases} \frac{1}{f_j^{(i)}} & \text{when } j = k \\ 0 & \text{when } j \neq k. \end{cases} \tag{17}$$

$A^{(i)}$ is applied to weight each resonant frequency equally in the loss function: mentioned in Sec. , the 1st resonant frequency is nearly four times smaller than the 6th resonant frequency.

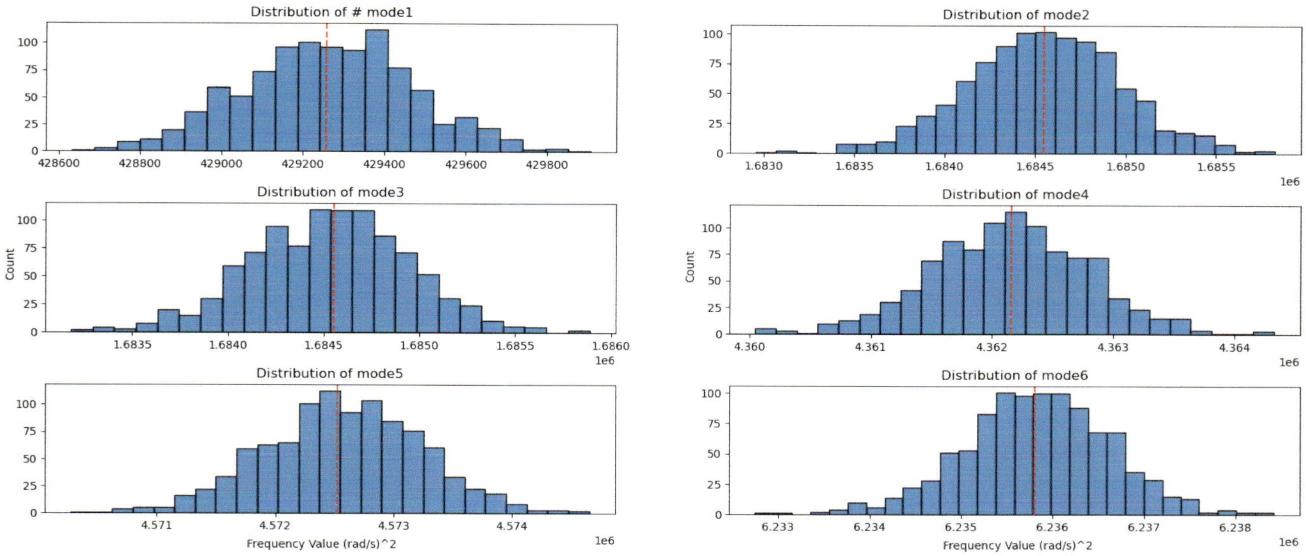

Fig. 9 Distribution of the 1st six resonant frequencies. The red line presents the outputted resonant frequency from the target model. The histogram shows the distribution of the frequency after adding Gaussian noise to form 1000 samples.

A FEA forward pass obtains $\widehat{\boldsymbol{f}}$ given $\widehat{\boldsymbol{\theta}}$. Let $\widehat{\boldsymbol{X}}$ represent PyTorch optimization variables, and let $\mathbf{C}$ represent scaling from $\widehat{\boldsymbol{X}}$ to FEA parameters $\widehat{\boldsymbol{\theta}}$. Their relation is written as

$$\widehat{\theta} = \mathbf{C}\widehat{\boldsymbol{X}} = \begin{bmatrix} c_1 & 0 \\ 0 & c_2 \end{bmatrix} \begin{bmatrix} x_1 \\ x_2 \end{bmatrix} = \begin{bmatrix} c_1 x_1 \\ c_2 x_2 \end{bmatrix}. \tag{18}$$

Since optimized via gradient descent (*i.e.*, in order to use FEA result $\widehat{\boldsymbol{f}}$ to update $\widehat{\boldsymbol{\theta}}$), the loss function's gradient is needed:

$$\frac{\partial L}{\partial \widehat{\boldsymbol{X}}} = \frac{\partial L}{\partial \widehat{\boldsymbol{f}}} \frac{\partial \widehat{\boldsymbol{f}}}{\partial \widehat{\boldsymbol{X}}}. \tag{19}$$

$\partial L / \partial \widehat{\boldsymbol{f}}$, viable to express analytically, is *computed by PyTorch Autograd*. Conversely, as $\widehat{\boldsymbol{f}}$ is output from the FEA forward pass, the latter matrix-valued partial derivative is computed by finite differencing of FEA with step size h:

$$\frac{\partial \widehat{\boldsymbol{f}}}{\partial \widehat{X}_j} = \frac{\boldsymbol{f}(c_j(\widehat{X}_j + h)) - \boldsymbol{f}(c_j(\widehat{X}_j - h))}{2h} \tag{20}$$

where $\boldsymbol{f}(\cdot)$ represents the frequency vector resultant from FEA. The gradient is used to update $\widehat{\boldsymbol{X}}$ until optimality.

In this case study, the target model is parameterized by shear modulus, G, bulk modulus, K, and density, ρ. Since ρ is assumed to be known, $G, K \in \boldsymbol{\theta}$. The goal is to find optimized $\widehat{G}, \widehat{K} \in \widehat{\boldsymbol{\theta}}$ yielding the minimum loss function. However, this optimization problem is not strictly convex: optimization may give a possibly non-global minimizer $\widehat{\boldsymbol{\theta}}$, as shown in Sec. , different from true $\boldsymbol{\theta}$. To address this issue, $\boldsymbol{\theta}$ is mapped to describe a strictly convex optimization problem with a unique, global minimum.

Case Study 2: Convex Problem

In Sec. , the optimization problem was not strictly convex lacking a unique global solution. In order to reform the loss function as convex, an appropriate mapping to $\boldsymbol{\theta}$ should be selected. From Eq. (15), the 1st resonant frequency was linear w.r.t $E/(1-\nu^2)$. Elastic modulus E and Poisson's ratio ν could be related to Lamé's constant λ and P-wave modulus M, based on the relationship between material properties,

$$E = \frac{(M-\ell)(M+2\ell)}{M+\ell}, \quad \nu = \frac{\ell}{M+\ell}, \quad f_1 = \frac{\lambda^4 h^2}{12a^4\rho}\left(-\frac{\ell^2}{M} + M\right). \tag{21}$$

From Eq. (21), the 1st resonant frequency was linear w.r.t ℓ^2/M, and M which can describe target parameters $\boldsymbol{\theta}$. In addition, a penalty term was added to Eq. (16) to ensure strict convexity of the loss function. M_c represented the experimental P-wave modulus, to be obtained from non-destructive experimentation. L_2 optimization's loss function was defined by:

$$L_2 = \frac{1}{m}\sum_{i=1}^{m}\sum_{j=1}^{n}\left[\left(A_{jj}^{(i)}\left(f_j^{(i)} - \widehat{f}_j\right)\right)^2\right] + \frac{k}{2}\left(\widehat{M} - M_c\right)^2. \tag{22}$$

$\widehat{\boldsymbol{X}}$ is the optimization variable such that:

$$\begin{bmatrix} \widehat{\ell}^2/\widehat{M} \\ \widehat{M} \end{bmatrix} = \mathbf{C}\widehat{\boldsymbol{X}} = \begin{bmatrix} c_1 x_1 \\ c_2 x_2 \end{bmatrix}. \tag{23}$$

Assuming $c_1 = c_2 = c > 0$:

$$\widehat{\ell} = \sqrt{\widehat{M} c_1 x_1} = c\sqrt{x_1 x_2}, \quad \widehat{M} = c_2 x_2 = c x_2. \tag{24}$$

Iterative parameters $\widehat{G}, \widehat{\ell} \in \widehat{\boldsymbol{\theta}}$ are defined by:

$$\widehat{\boldsymbol{\theta}} = \begin{bmatrix} \widehat{G} \\ \widehat{\ell} \end{bmatrix} = \begin{bmatrix} (\widehat{M} - \widehat{\ell})/2 \\ \widehat{\ell} \end{bmatrix} = c\begin{bmatrix} (x_2 - \sqrt{x_1 x_2})/2 \\ \sqrt{x_1 x_2} \end{bmatrix}. \tag{25}$$

The optimization algorithm is shown in Algorithm 1.

Algorithm 1 Parameter optimization with FEA

Input: target frequency $\boldsymbol{f}$, scaling constant $\mathbf{C}$, experimental P-wave modulus M_c, penalty constant k, step-size h, tolerance ε_t, learning rate γ
Initialize variables: $\widehat{\boldsymbol{X}}_t$
while $\Delta L_2 > \varepsilon_t$ **do**
    Compute variable transformation: $\widehat{\boldsymbol{X}}_t \Rightarrow \widehat{\boldsymbol{\theta}}_t$
    Execute FEA with iterative parameters, $\widehat{\boldsymbol{\theta}}_t$, to compute predicted resonant frequencies, $\widehat{\boldsymbol{f}}$
    Compute $L_2(\widehat{\boldsymbol{X}}_t)$ as Eq. (22)
    Execute FEA to compute $\partial\widehat{\boldsymbol{f}}/\partial\widehat{\boldsymbol{X}}$ as Eq. (20)
    Compute $\partial L_2/\partial\widehat{\boldsymbol{X}}$
    Update $\widehat{\boldsymbol{X}}_{t+1}$ by Adam optimizer
    Compute $\Delta L_2 = L_2(\widehat{\boldsymbol{X}}_t)$ - $L_2(\widehat{\boldsymbol{X}}_{t+1})$
end while
return Optimal parameters $\widehat{\boldsymbol{X}}$

After the change in iterative parameters and optimization variables, Eq. (22) is strictly convex so the loss function has a unique and global minimum.

Proof: By assuming $n = 1$ in Eq. (22), combining Eq. (15) and Eq. (21) yields (as verified in Sec.) that:

$$\widehat{f}_1 = \frac{\lambda_1^4 h^2}{12a^4\rho}\left(-\frac{\ell^2}{M} + M\right). \tag{26}$$

The Hessian matrix of L_2 with assumption of $n = 1$:

$$\mathbf{H} = \begin{bmatrix} \frac{\partial^2 L_2}{\partial x_1^2} & \frac{\partial^2 L_2}{\partial x_1 x_2} \\ \frac{\partial^2 L_2}{\partial x_2 x_1} & \frac{\partial^2 L_2}{\partial x_2^2} \end{bmatrix} = \begin{bmatrix} 2\left(A_{11}^{(i)}\right)^2\left(\frac{\lambda_1^4 h^2 c}{12a^4\rho}\right)^2 & -2\left(A_{11}^{(i)}\right)^2\left(\frac{\lambda_1^4 h^2 c}{12a^4\rho}\right)^2 \\ -2\left(A_{11}^{(i)}\right)^2\left(\frac{\lambda_1^4 h^2 c}{12a^4\rho}\right)^2 & 2\left(A_{11}^{(i)}\right)^2\left(\frac{\lambda_1^4 h^2 c}{12a^4\rho}\right)^2 + kc^2 \end{bmatrix}. \tag{27}$$

If $\mathbf{H}$ is positive definite, the loss function will be strictly convex. $\mathbf{H}$ is positive definite if and only if $\mathbf{H}$ is symmetric and all its eigenvalues are real and positive. As $\mathbf{H}$ is symmetric and real-valued, its eigenvalues are also real-valued. Assuming the product of $\mathbf{H}$'s eigenvalues, $\det\mathbf{H} > 0$, its eigenvalues are either both positives or both negative. To eliminate the possibility of both negative eigenvalues, $\mathbf{z}\mathbf{H}\mathbf{z}^{\mathrm{T}} \nless 0$ must be satisfied for arbitrary $\mathbf{z}$. But $\mathbf{z}\mathbf{H}\mathbf{z}^{\mathrm{T}} \nless 0$ satisfies when $\mathbf{z} = \begin{bmatrix} 1 & 0 \end{bmatrix}$ because

$$H_{11} = 2(A_{11}^{(i)})^2(\lambda_1^4 h^2 c)^2/(12a^4\rho)^2 > 0,$$

and both eigenvalues must be positive. Thus, **H** is positive definite as

$$\det \mathbf{H} = \left(2\left(A_{11}^{(i)}\right)^2\left(\frac{\lambda_1^4 h^2 c}{12a^4\rho}\right)^2\right)\left(2\left(A_{11}^{(i)}\right)^2\left(\frac{\lambda_1^4 h^2 c}{12a^4\rho}\right)^2 + kc^2\right) - \left(2\left(A_{11}^{(i)}\right)^2\left(\frac{\lambda_1^4 h^2 c}{12a^4\rho}\right)^2\right)^2 = 2k\left(A_{11}^{(i)}\right)^2\left(\frac{\lambda_1^4 h^2 c^2}{12a^4\rho}\right)^2,$$

which is positive if and only if $k > 0$ (as required). Summarizing:

$$L_2 \text{ has a unique global minimum} \Leftrightarrow L_2 \text{ is strictly convex} \Leftrightarrow \mathbf{H} \text{ is positive definite} \Leftrightarrow \det\mathbf{H} > 0 \Leftrightarrow k > 0. \quad (28)$$

Results

Using an initial fixture design shown in Fig. 3, LDV was conducted on two samples of the same material (PVDF 254 μm). The correlation coefficient for the two datasets was 0.71. The 1st resonant frequency occurs near 500 Hz for both samples, as visualization of the velocity across the sample confirmed peaks at lower frequencies to result from noise. Discrepancy exists between the location of this 1st peak (and later peaks as well), even between two samples of the same material.

Using the redesigned fixture, LDV scans were conducted on two samples of the same material (PVDF 254 μm). Fig. 11 shows the results when no tension is applied to the sample. Results are plotted on a log scale. Correlation coefficient for this

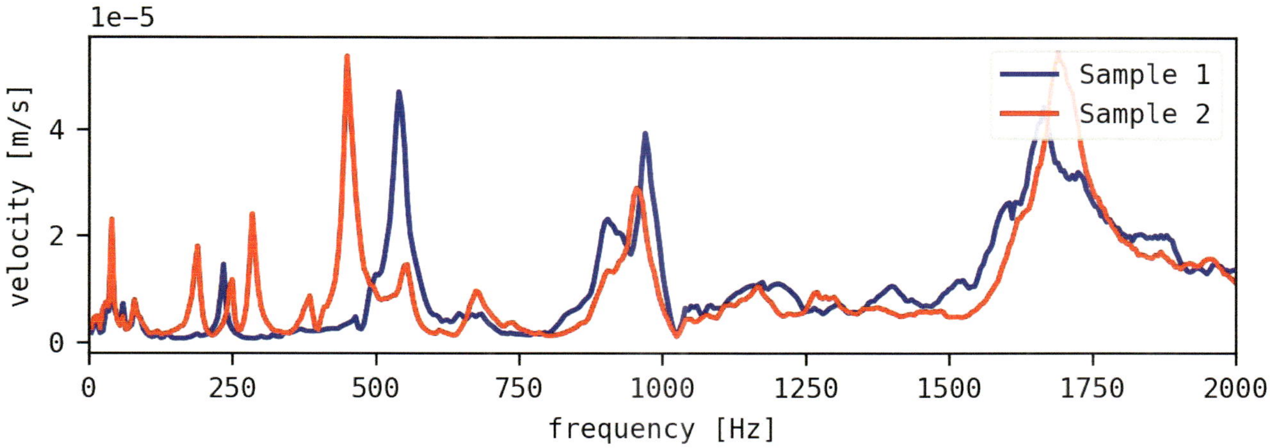

Fig. 10 Correlation coefficient was 0.71 for two samples of the same material when tested with the initial fixture.

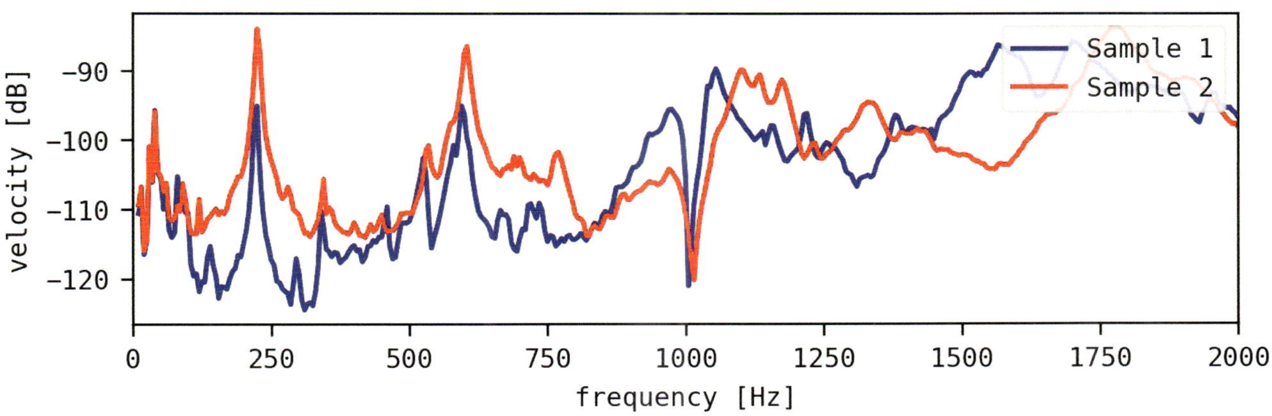

Fig. 11 Correlation coefficient was 0.71 for two samples of the same material when tested with the redesigned fixture and no tension was applied.

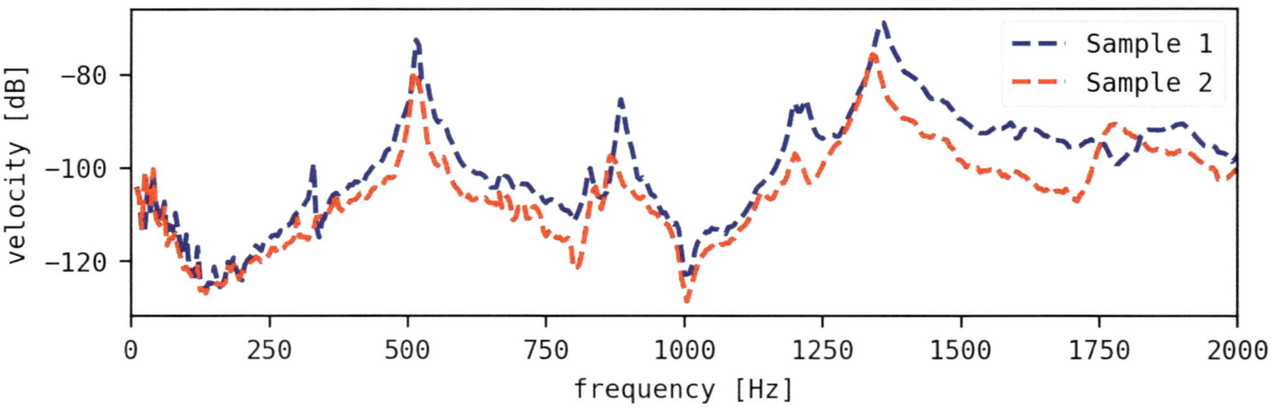

Fig. 12 Correlation coefficient was 0.93 for two samples of the same material when tested with the redesigned fixture and tension was applied.

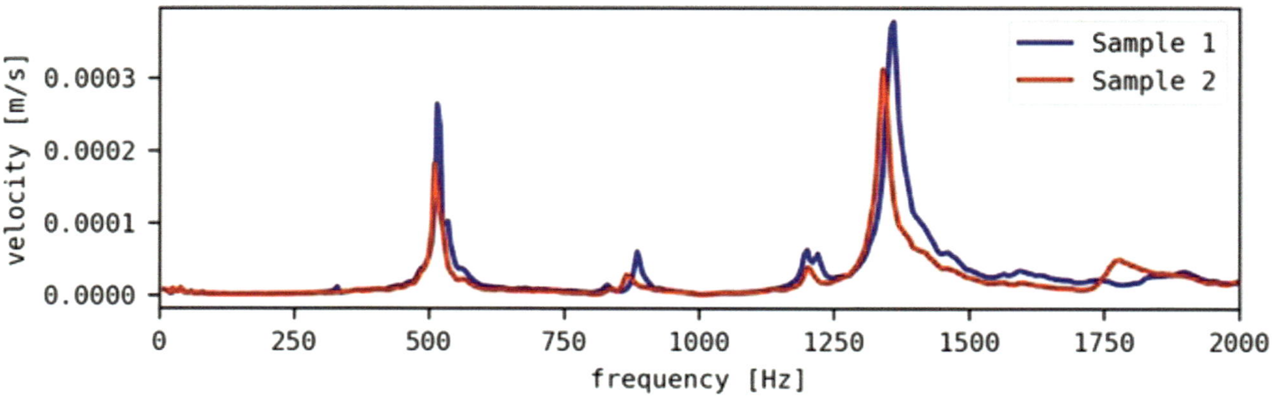

Fig. 13 Correlation coefficient was 0.81 for two samples of the same material when tested with the redesigned fixture and tension was applied.

data was found to be 0.71. Then, tension was applied to both samples. Distance between the clamping rings and the base piece, which controls the level of tension, was a uniform value of 0.1 inches for both samples. LDV scans were reconducted, and results are shown (with a log scale) in Fig. 12. Correlation coefficient in the tensioned case was found to be 0.93. As a baseline, tension was applied to two samples of PVDF 254 μm, with seperate data collection for each. Correlation was 0.81, and results are shown in Figure 13. This correlation coefficient differs to the results in Fig. 12, though the experiment was of the same nature. When these two films were held in contact by attaching the two fixtures to each other, and data was collected again, correlation increased. Correlation coefficient was 0.99, as seen in Fig. 14.

DIC

Sample 1 and sample 2 were both analyzed with DIC. Sample 1 showed a maximum strain of 0.10% in the x-direction and 0.10% in the y-direction, as shown in Fig. 15. Sample 2 showed a maximum strain of 0.23% in the x-direction and 0.18% in the y-direction, as shown in Fig. 16. Maximum strain in the x-direction occurred at the leftmost or rightmost areas within the ROI for both samples, along the outer edge of the ROI. Likewise, maximum strain in the y-direction occurred along the top or bottom edges of the ROI for both samples. In both samples, the minimum value for strain (ε_{xx} and ε_{yy}) is negative, indicative of an actual or perceived contraction.

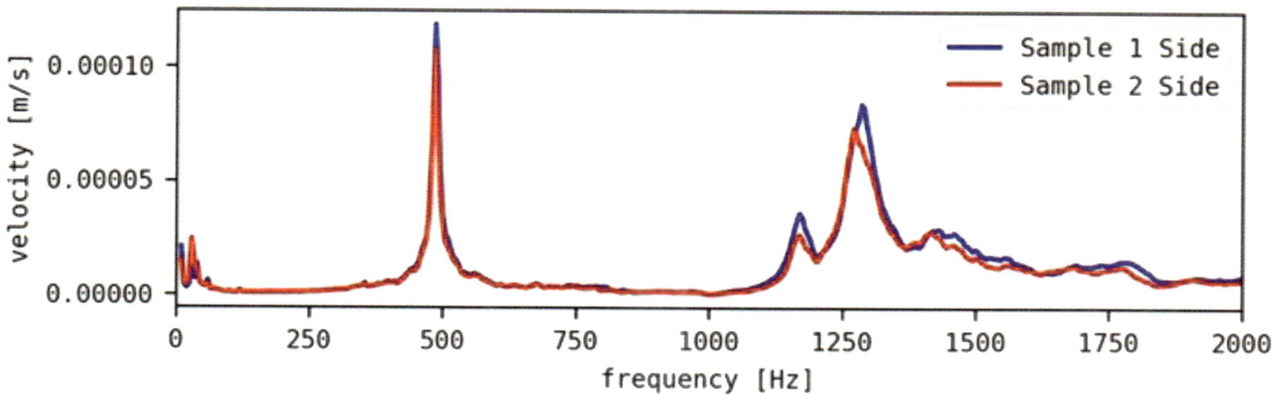

Fig. 14 Correlation coefficient was 0.99 for two samples of the same material held in contact. The films were tested with the redesigned fixture and tension was applied.

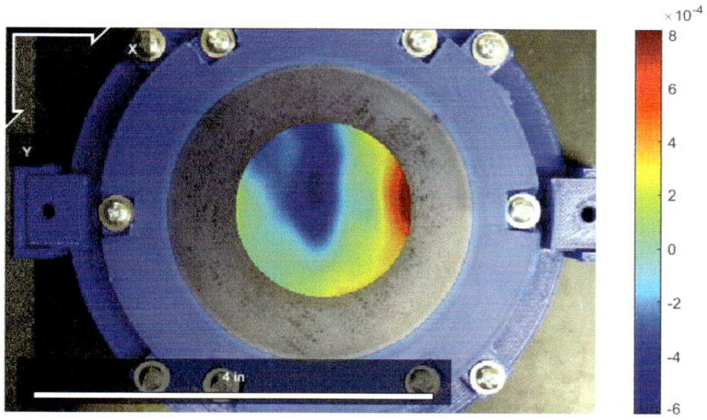

(a) ε_{xx}

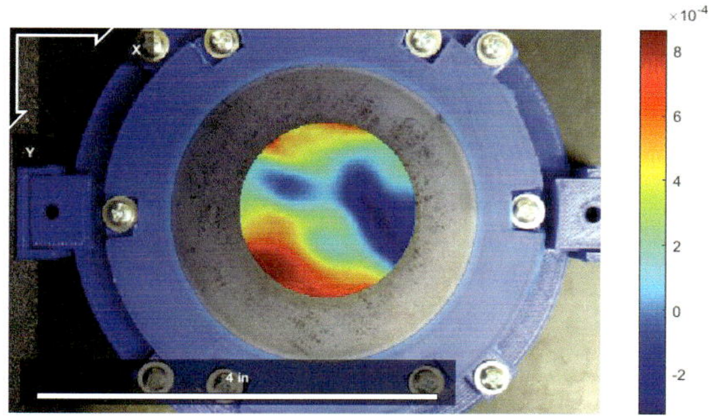

(b) ε_{yy}

Fig. 15 Strain fields in Sample 1 after tensioning.

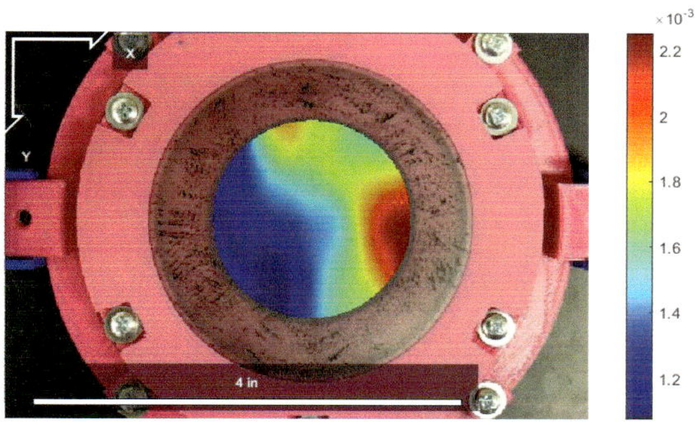

(a) ε_{xx}

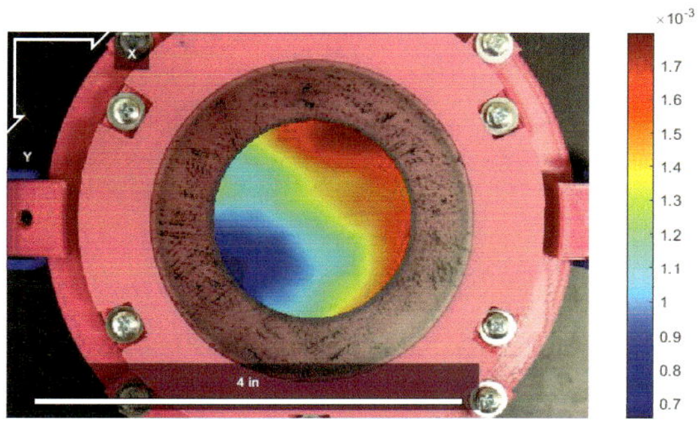

(b) ε_{yy}

Fig. 16 Strain fields in Sample 2 after tensioning.

FEA Model

For the clamped circular plate validation model, the modal parameters are shown in Table 1. Comparison of 1st modes between the FE clamped circular plate validation model and Eq. (15) is shown in Table 2. The 1st mode's error between theoretical equation and FEA is 0.05%.

Table 1 Clamped circular plate validation model's modal parameters.

a (m)	h (m)	ρ (kg/m^3)	E (MPa)	ν
200	10	7.8e-9	2.1	0.28

Table 2 1st mode accuracy of circular plate and circular membrane validation model.

Type of Validation	Theoretical Equation (Hz)	Validation FEA (Hz)	Error (%)
Clamped Circular Plate	17.86	17.95	0.05
Clamped Pretensioned Circular Membrane	881	883	0.19

Optimization

The optimization algorithm was examined with 10 random initializations of $\widehat{\boldsymbol{X}}$ from a uniform distribution, $U(0, 10)$. The same initialization was used on both case study to demonstrate and validate the difference between the two case studies. For the target model, model parameters and the specific parameters, $\boldsymbol{\theta}$, are shown in Table 3.

Table 3 Target model's model parameters and specific parameters, $\boldsymbol{\theta}$.

a (m)	h (mm)	ρ (kg/m^3)	G (MPa)	ℓ (MPa)	K (MPa)
0.036	0.076	1.45e3	828	7448	8000

Fig. 17 shows convergence iteration results under 10 random initializations. The top plot shows the convergence of the loss function under either L_1 or L_2. The second and the third plots show the convergence of the iterative parameters, $\widehat{\boldsymbol{\theta}}$. Solid line represents the specific parameters, $\boldsymbol{\theta}$, which serve as ground truth. Dashed lines represent iterative parameter values, $\widehat{\boldsymbol{\theta}}$, across iterations. Average of the 10 converged iterative parameters, $\widehat{\boldsymbol{\theta}}$, is shown in Table 4, comparing error w.r.t. $\boldsymbol{\theta}$.

For case study 1 under the L_1 loss function, the average of the 10 converged value of $\widehat{G}$ and $\widehat{K}$ is 970 MPa and 6090 MPa, respectively. Comparing to G and K, the percentage error of $\widehat{G}$ and $\widehat{K}$ is 17.1% and -23.9%, respectively. For case study 2 under the L_2 loss function, the converged value of $\widehat{G}$ and $\widehat{\ell}$ is 829 MPa and 7450 MPa, respectively. Comparing to G and ℓ, the percentage error of $\widehat{G}$ and $\widehat{\ell}$ is 0.12% and 0.03%, respectively.

Table 4 Percentage error of the iterative parameters, $\widehat{\boldsymbol{\theta}}$.

Loss function	K (MPa)	$\widehat{K}$ (MPa, %)	G (MPa)	$\widehat{G}$ (MPa, %)	ℓ (MPa)	$\widehat{\ell}$ (MPa, %)
L_1	8000	6090 (-23.9%)	828	970 (+17.1%)	–	–
L_2	–	–	828	829 (+0.12%)	7448	7450 (+0.03%)

Discussion

LDV scans with the initial fixture show discrepancies in the results for two samples of the same material (Fig. 10). When LDV scans were conducted with the redesigned fixture, the results for two samples of the same material still showed discrepancies (Fig. 11). In both cases, correlation coefficient was equal to 0.71. However, for the case with the redesigned fixture, such discrepancy appeared to come from the amplitude of the peaks rather than the location.

When tension was applied to both samples, LDV scans showed an increase in correlation, with correlation coefficient increasing to 0.93 (Fig. 12). These results indicate that applying a tension to sample of the same material makes the results between the different samples more consistent with one another and likely more repeatable.

In a separate experiment, documented in Fig. 13 and Fig. 14, two films in tension were scanned first separately and then while held in mutual contact. Correlation coefficient increased from 0.81 in the case of separation to 0.99 in the case of contact. This indicates an interaction between films in contact under fixturing: the two films vibrate *en echelon* under acoustic excitation.

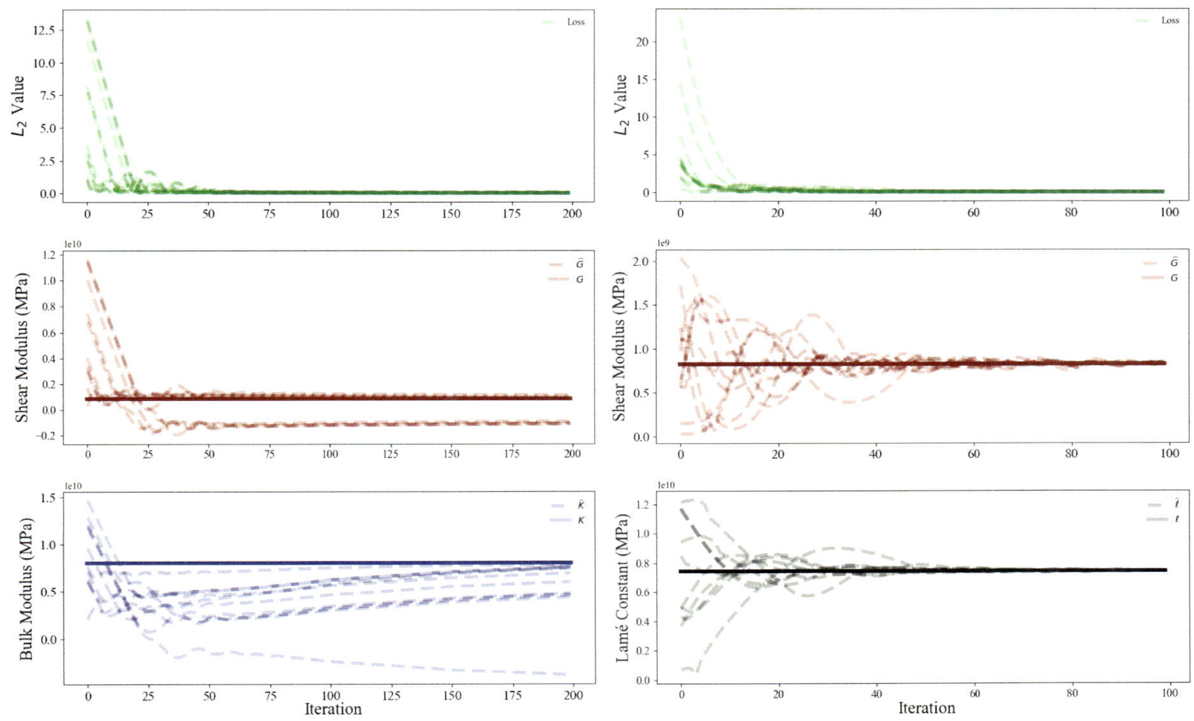

(a) Iteration result under L_1

(b) Iteration result under L_2

Fig. 17 L_1 and L_2 convergence result

DIC

DIC results demonstrated reasonable magnitudes of peak strain. Uncertainties associated with the tensioning include uncertainties in the material of the fixture, uncertainties in the material properties of the two films, and uncertainties associated with the tightening procedure between base piece and clamping pieces. Strains in both samples (Sec.) indicate: the strain in Sample 2 is approximately twice the magnitude of that in Sample 1 in both horizontal and vertical directions. Also, the largest value of maximum strain was found to be 0.23%; since this is less than 2%, the sample is assumed to not undergo plastic deformation.

Additionally, the shape of the strain fields as visualized in Fig. 15 and Fig. 16 are similar. In both samples, the maximum strain in the x-direction occurs along the sides of the sample, and maximum strain in the y-direction occurs along the top or bottom of the sample. This is consistent with what should be expected from a uniform radial strain applied to a circular region. However, a symmetric strain field was expected because applied strain should be uniform throughout the sample. For future work, symmetry would be a promising indication of improved DIC technique.

Uncertainties prevent the use of 2D DIC to fully characterize strain in the sample. For example, the assumption that the sample can be approximated as 2D introduces uncertainty. This assumption is an over-simplification and may account for the negative minimum strain values, which indicate a contraction. Though the contraction is small, it is not negligible as on the same order of magnitude as the maximum strain values. The results indicate that DIC is a reasonable technique for quantifying strain in a sample, but more advanced instrumentation is desirable.

FEA Model

Although the variational principle in Sec. includes interface interaction, the mathematical formulation remains valid for uncoupled solid domains. Table 2 shows error between the theoretical equations and the FEA verification, for benchmarking on single solid domains. Notably (though not detailed herein), error analysis shows that FEA is valid regardless of solid-domain geometry – although cantilever beam flexure significantly differs from circular membrane/plate deformation. Therefore, methods in this study could extend to other applications.

Optimization

Fig. 17 compares convergence of the L_1 and L_2 optimization algorithms. As shown in Fig. 17a, $\widehat{G}$ and $\widehat{K}$ do not converge to the optimal values, G and K, under L_1 optimization with random initialization. Moreover, different initialization leads to different converged values: as L_1 is demonstrably nonconvex, $\widehat{\boldsymbol{\theta}}$ exhibits multiple solutions due to local minima. Indeed, although the L_1 loss function apparently converges, percentage error of average $\widehat{G}$ and $\widehat{K}$ remains more than 15% as shown in Table 4. Given the engineering objective is to determine film flexural properties from experimentation, our results indicate that L_1 optimization is algorithmically inefficient for this purpose.

As shown in Fig. 17b, $\widehat{G}$ and $\widehat{\ell}$ converge to the optimal values, G and ℓ, under the L_2 optimization algorithm with random initialization. Percentage error of the two parameters is less than 0.2% as shown in Table 4. Since L_2 is reformulated for convexity, the loss function numerically exhibits an accurate, unique, and global minimum. Therefore, the L_2 optimization algorithm satisfies the research objective, *i.e.*, to recover the material properties of the tested thin film. The convergence result also indirectly shows alignment between FEA and theoretical formulation. Convexity of L_2 was proved based on the theoretical formulation as shown in Eq. (28). Therefore, FEA preserved the convexity of L_2 indicated by the theoretical formulation.

Conclusion

This study develops novel experimental and FEA-based optimization framework for estimating material parameters of the tested thin films. By the advancement of the fixture design, the thin films can be tested under tension separately and in contact. Meanwhile, the vibration response of thin films becomes more reproducible after tensioning. DIC is used to preliminarily extract initial tensile strains of thin films. FEA precision is verified with the theoretical formulation. The optimization algorithm is also proven with modeled scenarios under random initializations. Overall, this study begins to fill the research gap in the numerical and experimental investigation of thin film vibromorphology, with the following contributions:

- A novel modal parameter optimization framework based on modal response from experiments and numerical models
- Advanced fixture design for repeatable experiment on thin films under tension
- Investigation of tensioning effect on vibroacoustic behavior (vibromorphology)
- Established and verified 3D frequency-domain FEA coupled to PyTorch optimizer
- Reformulated convex optimization algorithm by parameter mapping

Acknowledgments The authors would like to thank Thomas Zane Box and Professor Ricardo Mejia for advising and contributing to the experimental setup, and all the managers and mentors from the 25th Los Alamos Dynamics Summer School for providing resources and hosting the program. This research was funded by Los Alamos National Laboratory (LANL) through the Engineering Institute's Los Alamos Dynamics Summer School. The Engineering Institute is a research and education collaboration between LANL and the University of California San Diego's Jacobs School of Engineering. This collaboration seeks to promote multidisciplinary engineering research that develops and integrates advanced predictive modeling, novel sensing systems, and new developments in information technology to address LANL mission relevant problems.

References

1. Simiao Niu and Zhong Lin Wang. Theoretical systems of triboelectric nanogenerators. *Nano Energy*, 14:161–192, May 2015.
2. Syed Nasimul Alam, Arka Ghosh, Pankaj Shrivastava, Uddeshya Shukla, Kamal Garg, Abhay Chowdary Edara, and Nityananada Sahoo. An introduction to triboelectric nanogenerators. *Nano-Structures & Nano-Objects*, 34:100980, April 2023.
3. Aihua Chen, Chen Zhang, Guang Zhu, and Zhong Lin Wang. Polymer Materials for High-Performance Triboelectric Nanogenerators. *Advanced Science*, 7(14):2000186, 2020.
4. Mina Shanbedi, Haleh Ardebili, and Alamgir Karim. Polymer-based triboelectric nanogenerators: Materials, characterization, and applications. *Progress in Polymer Science*, 144:101723, September 2023.
5. Jianxiong Zhu, Minglu Zhu, Qiongfeng Shi, Feng Wen, Long Liu, Bowei Dong, Ahmed Haroun, Yanqin Yang, Philippe Vachon, Xinge Guo, Tianyiyi He, and Chengkuo Lee. Progress in TENG technology—A journey from energy harvesting to nanoenergy and nanosystem. *EcoMat*, 2(4):e12058, 2020.
6. Yaokun Pang, Yunteng Cao, Masoud Derakhshani, Yuhui Fang, Zhong Lin Wang, and Changyong Cao. Hybrid Energy-Harvesting Systems Based on Triboelectric Nanogenerators. *Matter*, 4(1):116–143, January 2021.
7. Ping Sun, Nixin Cai, Xiaodi Zhong, Xiaojiong Zhao, Lei Zhang, and Saihua Jiang. Facile monitoring for human motion on fireground by using MiEs-TENG sensor. *Nano Energy*, 89:106492, November 2021.

8. Qitao Zhou, Jing Pan, Shujun Deng, Fan Xia, and Taesung Kim. Triboelectric Nanogenerator-Based Sensor Systems for Chemical or Biological Detection. *Advanced Materials*, 33(35):2008276, 2021.
9. Luke B. Beardslee, Marcel C. Remillieux, and T. J. Ulrich. Determining material properties of components with complex shapes using Resonant Ultrasound Spectroscopy. *Applied Acoustics*, 178:108014, July 2021.
10. Yunlong Li, Inkyu Oh, Jiehao Chen, and Yuhang Hu. A new design of dielectric elastomer membrane resonator with tunable resonant frequencies and mode shapes. *Smart Materials and Structures*, 27(6):065029, May 2018.
11. Merve K Ozturk, Klara Kalinova, Banu Nergis, and Cevza Candan. Comparison of resonance frequency of a nanofibrous membrane and a homogeneous membrane structure. *Textile Research Journal*, 83(20):2204–2210, December 2013.
12. M. Kamper and A. Bekker. Non-contact experimental methods to characterise the response of a hyper-elastic membrane. *International Journal of Mechanical and Materials Engineering*, 12(1):15, July 2017.
13. Md Salauddin, R. M. Toyabur, P. Maharjan, M. Salauddin Rasel, Hyunok Cho, and Jae Yeong Park. Design and experimental analysis of a low-frequency resonant hybridized nanogenerator with a wide bandwidth and high output power density. *Nano Energy*, 66:104122, December 2019.
14. Paolo Chiariotti, Christian Rembe, Paolo Castellini, and Matt Allen. Laser Doppler Vibrometry Measurements in Structural Dynamics. In R. Allemang and Peter Avitabile, editors, *Handbook of Experimental Structural Dynamics*, pages 1–45. Springer, New York, NY, 2020.
15. Daria A. Derusova, Vladimir P. Vavilov, Nikolay V. Druzhinin, Victor Y. Shpil'noi, and Alexey N. Pestryakov. Detecting Defects in Composite Polymers by Using 3D Scanning Laser Doppler Vibrometry. *Materials*, 15(20):7176, January 2022.
16. N. B. Roozen, L. Labelle, Q. Leclère, K. Ege, and S. Alvarado. Non-contact experimental assessment of apparent dynamic stiffness of constrained-layer damping sandwich plates in a broad frequency range using a Nd:YAG pump laser and a laser Doppler vibrometer. *Journal of Sound and Vibration*, 395:90–101, May 2017.
17. Shujun Ma, Shiliang Wang, Francesca Iacopi, and Han Huang. A resonant method for determining the residual stress and elastic modulus of a thin film. *Applied Physics Letters*, 103(3):031603, July 2013.
18. Nathaniel M. Roberts and James S. Sharp. Piezoelectric excitation and acoustic detection of thin film polymer membrane vibrations. *Physical Review E*, 109(1):014802, January 2024.
19. R. Huang, J. H. Prévost, Z. Y. Huang, and Z. Suo. Channel-cracking of thin films with the extended finite element method. *Engineering Fracture Mechanics*, 70(18):2513–2526, December 2003.
20. Simiao Niu, Ying Liu, Sihong Wang, Long Lin, Yu Sheng Zhou, Youfan Hu, and Zhong Lin Wang. Theory of Sliding-Mode Triboelectric Nanogenerators. *Advanced Materials*, 25(43):6184–6193, 2013.
21. Jinkai Chen, Junchao Wang, Weipeng Xuan, Shurong Dong, and Jikui Luo. Universal Triboelectric Nanogenerator Simulation Based on Dynamic Finite Element Method Model. *Sensors*, 20(17):4838, January 2020.
22. Zirui Zhao, Xiaoli Wang, Yanqiang Hu, Zhihao Li, and Lizhou Li. Electromechanical modeling for triboelectric nanogenerators considering the distribution of polymer transfer film. *Nano Energy*, 93:106895, March 2022.
23. Subrata Mukherjee. Variational principles in dynamic thermoviscoelasticity. *International Journal of Solids and Structures*, 9(10):1301–1316, October 1973.
24. Diederik P. Kingma and Jimmy Ba. Adam: A Method for Stochastic Optimization, January 2017.
25. J. Blaber, B. Adair, and A. Antoniou. Ncorr: Open-Source 2D Digital Image Correlation Matlab Software. *Experimental Mechanics*, 55(6):1105–1122, July 2015.
26. Guillaume Giudicelli, Alexander Lindsay, Logan Harbour, Casey Icenhour, Mengnan Li, Joshua E. Hansel, Peter German, Patrick Behne, Oana Marin, Roy H. Stogner, Jason M. Miller, Daniel Schwen, Yaqi Wang, Lynn Munday, Sebastian Schunert, Benjamin W. Spencer, Dewen Yushu, Antonio Recuero, Zachary M. Prince, Max Nezdyur, Tianchen Hu, Yinbin Miao, Yeon Sang Jung, Christopher Matthews, April Novak, Brandon Langley, Timothy Truster, Nuno Nobre, Brian Alger, David Andrš, Fande Kong, Robert Carlsen, Andrew E. Slaughter, John W. Peterson, Derek Gaston, and Cody Permann. 3.0 - MOOSE: Enabling massively parallel multiphysics simulations. *SoftwareX*, 26:101690, 2024.
27. Arthur W. Leissa. Circular plates. Scientific and Technical Information Division, National Aeronautics and Space Administration, 1969.

Chapter 17
Evaluation of Camera-Denoising Techniques towards Force Reconstruction

Sean Collier, Jonathan Young, Hruday Shah, Nicholas Vlajic, and Tyler Dare

Abstract Due to an abundance of spatial points, camera-based surface vibrations promise reduced testing times, higher resolution in response measurements, and the data necessary to over-constrain historically poor inverse problems in structural dynamics. For example, initial studies and formulations of a camera-based, modal force reconstruction (MFR) technique show great success and utility, but is highly sensitive to measurement noise—something abundant in camera data. The improvement of camera data has become a devoted effort in recent years, theoretically providing the path necessary to facilitate the use of MFR on realistic data. Such methods proposed range from simple singular value decomposition to those enforcing physical constraints on the measured response using additional accelerometers. To that end, this paper explores a variety of these camera-denoising techniques, their efficacy, and their utility towards improving measured displacements of a simple, pinned plate and cantilevered beam for use in modal force reconstruction.

Keywords Vibration · Denoising · Camera-Based · Decomposition · Substructuring

Introduction

Here we discuss camera noise and its role as a major drawback of a purely vision-based analysis. Indeed, while the efficacy of camera-based surface vibration measurement has seen increasing interest and characterization, its utility is far more limited compared to a laser or accelerometer in terms of noise floor and frequency detection. This limitation is clear from much of the authors' work alone [1–3], but is evident across the literature simply by the amount of work dedicated to the refinement of a purely camera-based vibration-measurement technique [4–10]. The aim of this work is to compare the utility of such camera denoising techniques towards the eventual use within camera-based modal force reconstruction, which has been observed to be plausible but prone to over-fitting of higher order modes in the presence of typical measurement noise. While regularization with a reduced number of modes has shown some success, further improvement to the technique requires improvement to the measurement quality. Thankfully, much work has been done for the improvement of noisy high speed camera data.

The first approach mentioned here is one that follows naturally from one-dimensional signal processing. If one follows the standard neighborhood assumptions of optical flow—especially if the object is captured with high resolution—then a local averaging around the pixel of interest is easy to implement and rather effective [10–14]. For example, it was shown in [1] that a local average around a pixel of interest was able to reduce the noise floor well enough to resolve motions originally obscured. While this approach is highly effective at interrogating a single point, such averaging may not be ideal when high spatial resolution is needed (e.g., spatial derivatives), or in the case where the measurement is not spatially dense across the structure. Thus, maintaining the spatial benefit of the measurement while also improving signal-to-noise requires more capable and general techniques.

One such technique comes from the use of data reduction methods, and the tendency of noise to fall within higher-order singular values of a decomposition [15]. Two methods, in particular, have been recently introduced for the direct

Sean Collier · Jonathan Young · Nicholas Vlajic · Tyler Dare
The Pennsylvania State University, University Park, PA 16804
e-mail: smc604@psu.edu; juy218@psu.edu; nav5000@psu.edu; tpd10@psu.edu

Hruday Shah
Georgia Institute of Technology, Atlanta, GA 30332
e-mail: hshah316@gatech.edu

Janko Slavic et al. (eds.), *Computer Vision & Laser Vibrometry, Vol. 6*, Conference Proceedings of the Society for Experimental Mechanics Series, https://doi.org/10.13052/9788743801511_17

de-noising of camera data for a modal decomposition [16, 17]. The first, presented at IMAC 2022 by Trainotti et al., uses the spatial density of the information to remove incoherent noise by use of low-rank reconstructions from the singular value decomposition (SVD) [16, 18]. In this work, the data is arranged in three different orientations leading to PRANK: a standard "FRF" orientation with drive points by accelerometers by frequency; a "flattened response" or "principle response function" orientation (PR-), where the data is rearranged to be frequency by all motion; and a "Hankel" form, where each drive-accelerometer response is rearranged into a Hankel matrix (-ANK). The SVD on each of these matrices produces a different effective filtering, helping to force outliers into higher singular values while maintaining prominent features. The second comes from the fluids community, where Scherl et al. leveraged Robust Principle Component Analysis (RPCA) [19] towards the modal decomposition in corrupted and noisy measurements of fluid flows. Here, RPCA aims to separate the raw image frames into a low-rank representation and a sparse matrix of outliers, much in the same aim as PRANK. Perhaps the most attractive aspect of RPCA is the rigorously defined minimization problem that underlies the regularization—the"robust" portion of RPCA—through which the rank to maintain is found with no user input. Given the proof of concept for RPCA was done on particle image velocimetry, the results should prove analogous for structural vibration measurements with digital image correlation and optical flow.

The final approach mentioned here shows high promise by use of regularization and known, trusted sparse measurements [20–22]. In these works, noisy high-speed camera data is combined with a sparse set of high-fidelity sensor data to resolve motion originally below the noise floor. In the initial approach by Javh et al. [20], an accelerometer is used to identify eigenvalues while the full-field camera data is used to identify mode shapes/modal constants. When used together in a Least-Squares Complex Frequency or Least-Squares Frequency-Domain method approach, the combined information was used to reconstruct the frequency response functions, showing modes originally below the noise floor of the camera. However, while resonances were captured, the neighboring regions in the real and imaginary parts experienced some issue. This work was improved soon after in [21, 22], which followed a similar motivation. In addition to merging data from parallel modal processes, a substructuring approach known as system equivalent model mixing (SEMM) is used. Typically, SEMM involves a model and a sparse data set from the structure; these are combined to update the dynamics of the model. Here, the full-field camera data is used as the parent model, and the sparse measurements come from more traditional, high fidelity sensors like laser doppler vibrometers and accelerometers. This technique showed improvements to [20], correcting the apparent phase issues.

These works provide tractable avenues for camera denoising, and merit use and observation on typical displacement measurements. The paper proceeds by motivating the theory briefly for each of these techniques. Next, we move to comparing their performance—from a user's perspective—on both temporal and spatial denoising efficacy for synthetic and experimental data. Further work and conclusions follow towards camera-based modal force reconstruction.

Background

The three methods investigated are outlined herein: PRANK [16], RPCA [17], and SEMM [22].

PRANK [16] can be thought of as an extension to the well practiced idea that noise is represented by the higher singular values of a decomposition, meaning a low-rank representation can be reconstructed which removes noise and improves quality. As discussed in the introduction, PRANK leverages continuity and smoothness across multiple data orientations. Within the PRF orientation, the data are represented as input-output FRFs across frequency, where the low rank representation can be related to the surface averaged receptance. If there exists an apparent resonance in a single input-output FRF, this would be removed in a low-rank representation. Likewise, spuriosities caused by typical camera-based displacement errors (e.g., super-pixel motion for optical flow), would also be evident in this representation, as the clipping of the displacement conveys broadband features not present across all FRFs. The Hankel orientation overlaps sections of an individual input-output FRF, where the low-rank representation acts as a moving average. This behaviour is due to the high-ranking of random measurement noise over increasing frequency; in addition, the low-rank representation may not provide a hankel matrix, and so enforcing this form by a diagonal average of assumed-equal quantities provides additional denoising. These two orientations can be done in series or parallel to varying effect, as described by the authors. PRANK can be understood pictorially in Figure 1.

RPCA [17] again can be understood to be a regularized implementation of the SVD, where the impacts of outliers in the regression is minimized. Specifically, RPCA aims to separate principal components, $\mathbf{L}$, from a sparse matrix of outliers, $\mathbf{S}$ that together form the input $\mathbf{X}$. This specific implementation described is a probabilistic proxy for the intractable optimization problem, resulting in the form

$$\min_{\mathbf{L},\mathbf{S}} \|\mathbf{L}\|_* + \lambda_0 \|\mathbf{S}\| \text{ subject to } \mathbf{L} + \mathbf{S} = \mathbf{X} \quad (1)$$

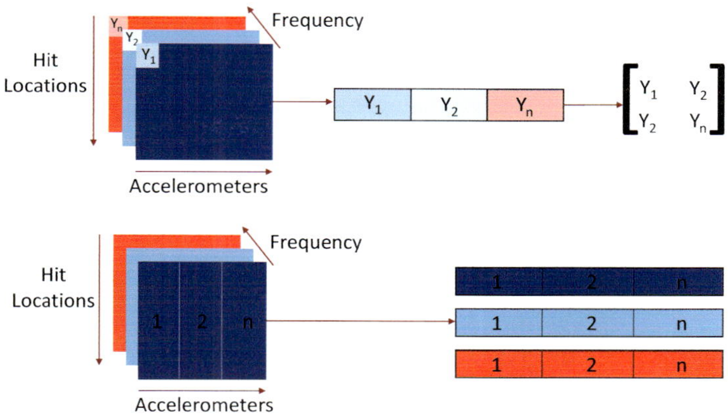

Fig. 1 PRANK Orientations: Hankel (top) and PRF (bottom).

where $\|\cdot\|_*$ represents the nuclear norm, and λ_0 is the hyper parameter given by $\lambda_0 = \lambda/\sqrt{max(n,m)}$ for an input matrix $\mathbf{X}$ of size $m \times n$. The proxy converges to the true optimization problem if $\lambda = 1$, given that $\mathbf{L}$ is not sparse and $\mathbf{S}$ is not low rank. Scherl et al. found these assumptions may only be partially valid for PIV data, such that λ may need to be varied slightly, in practice.

SEMM [22] aims to regularize the noisy camera data by the introduction of a number of trusted, high signal-to-noise measurements (e.g., accelerometers) at coincident degrees of freedom (DoF). The technique, pulled from dynamic substructuring, attempts to enforce compatibility at such shared DoF between a parent (camera data) and overlay model (accelerometer). Here, this is done using the reconstructed FRFs, found using the combined camera/accelerometer method from Javh et al. [20] and the polyreference least-squares complex frequency-domain method [23].These reconstructions serve as the parent model and the same accelerometer FRFs as the overlay. The aim, then, is the accelerometer will provide final corrections to the reconstructed camera FRFs as it would if the parent model were a finite element solution. The final calculation for a hybrid SEMM model is performed using the Lagrange Multiplier Frequency-Based Substructuring (LM FBS) single line formulation

$$\mathbf{Y}^{SEMM} = \mathbf{Y}_{gg} - \mathbf{Y}_{gg}^{par}(\mathbf{Y}_{bg}^{rem})^{+}(\mathbf{Y}_{bb}^{rem} - \mathbf{Y}_{bb}^{ov})(\mathbf{Y}_{gb}^{rem})^{+}\mathbf{Y}_{gg}^{par} \tag{2}$$

where $\mathbf{Y}^{par}$ is the parent model (camera mobilities), $\mathbf{Y}^{ov}$ is the overlay model (sensor mobilities), $\mathbf{Y}^{rem}$ the removed DoF from the parent model, and $\mathbf{Y}^{SEMM}$ the final hybrid model.

Analytical Plate

It was of interest to replicate a typical measurement scenario on a structure more complex than a one-dimensional beam or mulitiple degree-of-freedom oscillator. The main model used was a forced thin rectangular plate, whose equation of motion goes as

$$D\nabla^4 w(x,y,t) + \rho h \frac{\partial^2 w(x,y,t)}{\partial t^2} = F(x,y,t) \tag{3}$$

for D, w, ρ, h, and F being the flexural rigidity, displacement, mass density, thickness, and forcing fuction, respectively. Simple supports were assumed so to model the plate analytically. Time harmonic, steady state solutions in the homogeneous case follow as

$$w(x,y,t) = \Phi(x,y)T(t) = \sum_{m,n} A_{mn} \sin\frac{m\pi x}{L_x} \sin\frac{n\pi y}{L_y} e^{j\omega_{mn}t} \tag{4}$$

for some mode number m, n, with modal participation $A_{m,n}$ and natural frequency ω_{mn} given by

$$\omega_{mn} = \sqrt{\frac{D}{\rho h}}\Big[\Big(\frac{m\pi}{L_x}\Big)^2 + \Big(\frac{n\pi}{L_y}\Big)^2\Big].$$

This plate was driven over twenty unique points, whose response was composed of the first 175 modes. Panel area discretization was 33×35, allowing for a sufficient number of points per wavelength up to the modes of interest. Artificial damping and noise were applied, and the plate was allowed to decay over the record. These displacements were then converted to mobility for use in the methods. For reference, this matrix is 1155 outputs $\times$ 20 inputs $\times$ 2000 frequency bins.

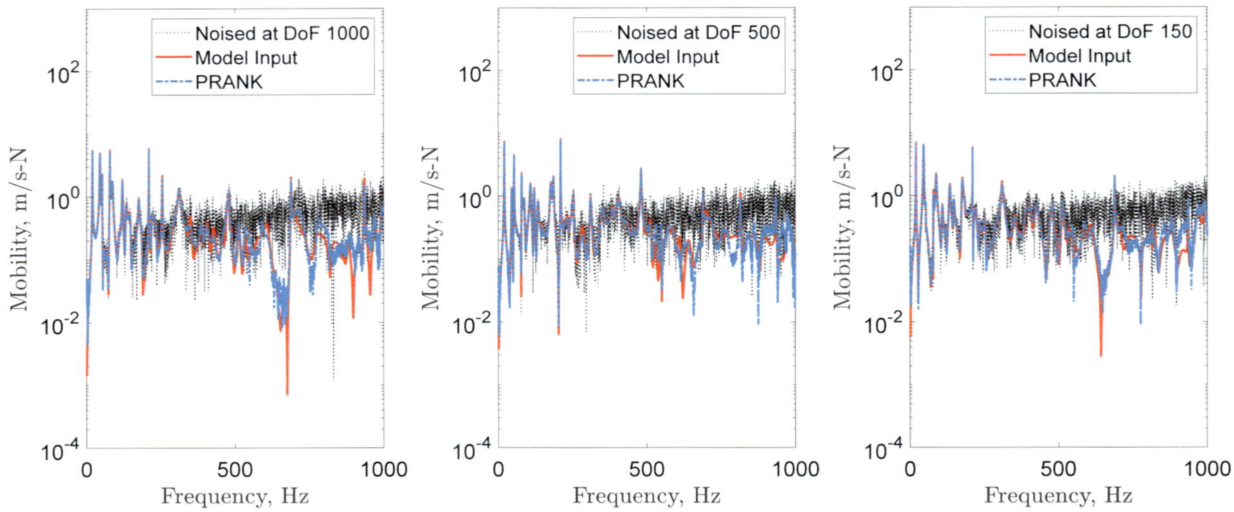

Fig. 2 Random mobilities compared between noised (model output), model input, and PRANK output.

Looking first at the results from PRANK, Figure 2 illustrates extremely good visual agreement between the analytical model input and PRANK output over multiple DoF. While only a subset is shown, the effect over the surface may be summarized using the surface averaged mobility given in Figure 3. Again, PRANK is seen to reveal all of the underlying dynamics of the structure, including those which are highly damped and an order of magnitude below the modeled camera noise floor. The frequency response assurance criterion ($FRAC$) is used to assess the degree of similarity between the surface averages. For PRANK, the $FRAC$ is found to be $FRAC = 0.9979$, indicating high similarity between the two as in the visual comparison.

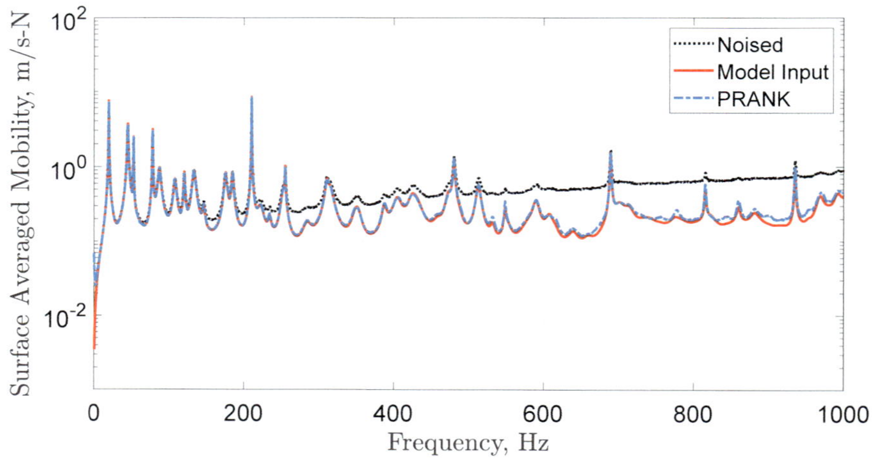

Fig. 3 Surface averaged mobilities compared between noised (model output), model input, and PRANK output.

Moving now to the RPCA results, Figure 4 shows the denoised mobilities using a $\lambda = 0.33$—a value found to work well on other camera datasets, and in agreement with Scherl et al.'s need to reduce λ for realistic data. It is clear that this value made for too aggressive of a filter, reducing magnitudes by more than half. That said, dynamics of similar contour are revealed below the noise floor. Figure 5 shows this effect more clearly. Given the reduction of the noise floor, the surface average is now much closer in contour to the model input albeit lower than the true magnitude. Here, the $FRAC$ calculation gives $FRAC = 0.6021$, likely due to the similar curvature but varied magnitudes. Clearly the results suggest an increase in the hyper parameter to reduce the aggressive filtering and improve agreement.

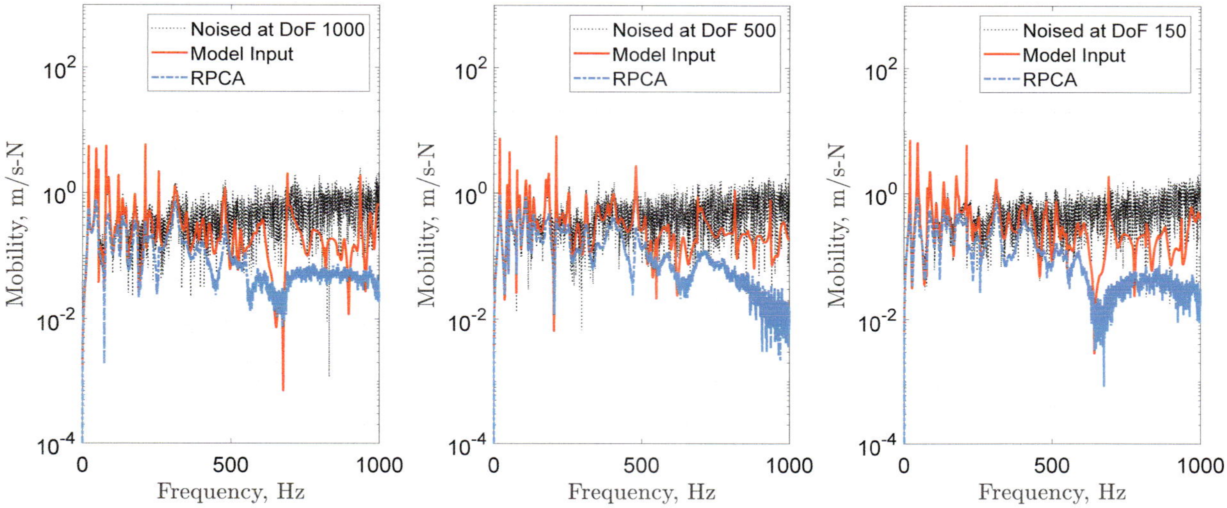

Fig. 4 Random mobilities compared between noised (model output), model input, and RPCA output ($\lambda = 0.33$).

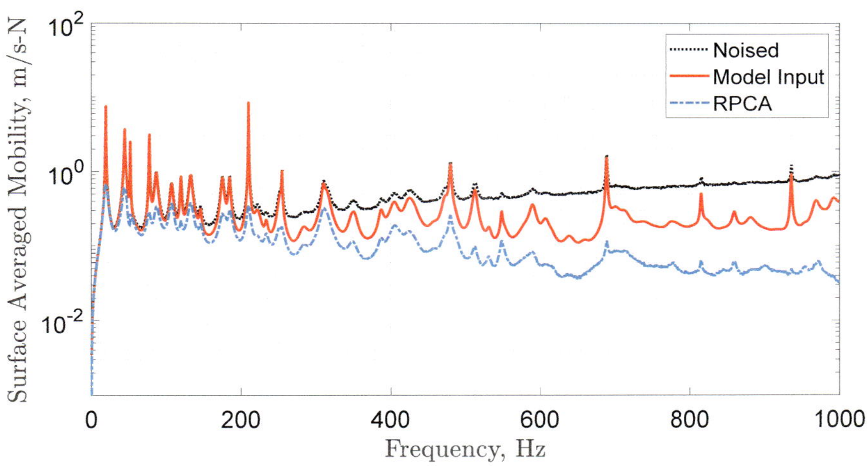

Fig. 5 Surface averaged mobilities compared between noised (model output), model input, and RPCA output ($\lambda = 0.33$).

Hence, we now consider RPCA using a $\lambda = 1$, to both reduce the impact of the filter as well as investigate the ideal case. Figure 6 shows the improvement of the data reduction using the higher λ value. Now, the performance of RPCA is comparable to PRANK, though it is observed that the relative magnitudes are still affected and there remains a considerable amount of noise at higher frequencies. The surface averaged mobility of Figure 7 again shows improvement using the higher value, now falling well on the model. Here, the $FRAC$ value gives $FRAC = 0.9780$—still below that of PRANK, but much improved over the more aggressive filtering. However, RPCA on datasets of this size is computationally quite expensive, with this data reduction being on the order of 6-12 hours (1000 to 2000 iterations), compared to the few minutes needed for PRANK. Given the regression is dependent on the number of observations, it is not tractable to run a subset to estimate the ideal λ. Indeed, the authors were able to rerun the estimate, but such a repeat is non-ideal. These results imply utility, but also impracticality of RPCA for large camera-based datasets when compared to PRANK.

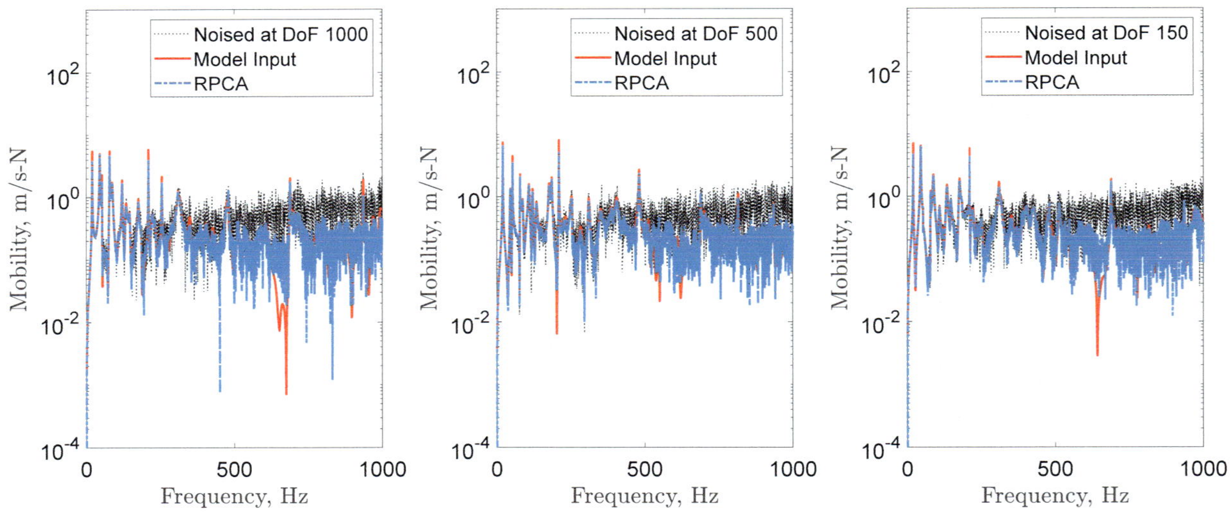

Fig. 6 Random mobilities compared between noised (model output), model input, and RPCA output ($\lambda = 1$).

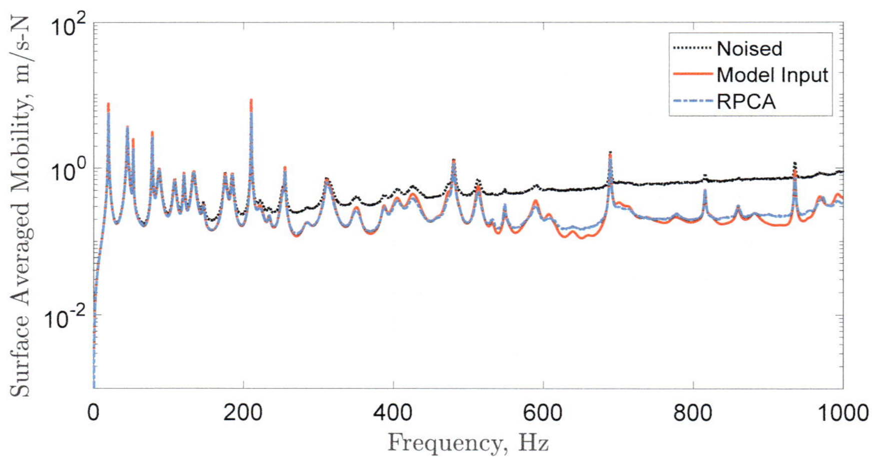

Fig. 7 Surface averaged mobilities compared between noised (model output), model input, and RPCA output ($\lambda = 1$).

Finally, we consider the camera-accelerometer hybrid reconstructions and subsequent corrections done using SEMM. Certainly, the technique is the most technically involved: a modal parameter estimation must be done using the overlay, reference signals; mode shapes must then be calculated from the estimated modal parameters and noisy camera data; reconstructions of the FRFs are then created; finally allowing for their use in SEMM to refine the reconstructions. With this all being so, these techniques are seen to visually perform the best compared to PRANK and RPCA. As before, the random mobilities are presented in Figures 8, 10 and surface averaged mobility in Figures 9, 11. While the surface average between PRANK, the hyrbid reconstruction, and SEMM are visually quite close, the subtle out-perfomance in estimating the zeros, in parallel to the tighter fit of damping and magnitude show these last techniques to be the most robust, if being the most user involved.

Considering first the hyrbid method alone, the $FRAC$ between the hybrid results and model input gives $FRAC = 0.9995$, indicating almost perfect agreement, and a quantitatively better reconstruction than PRANK.

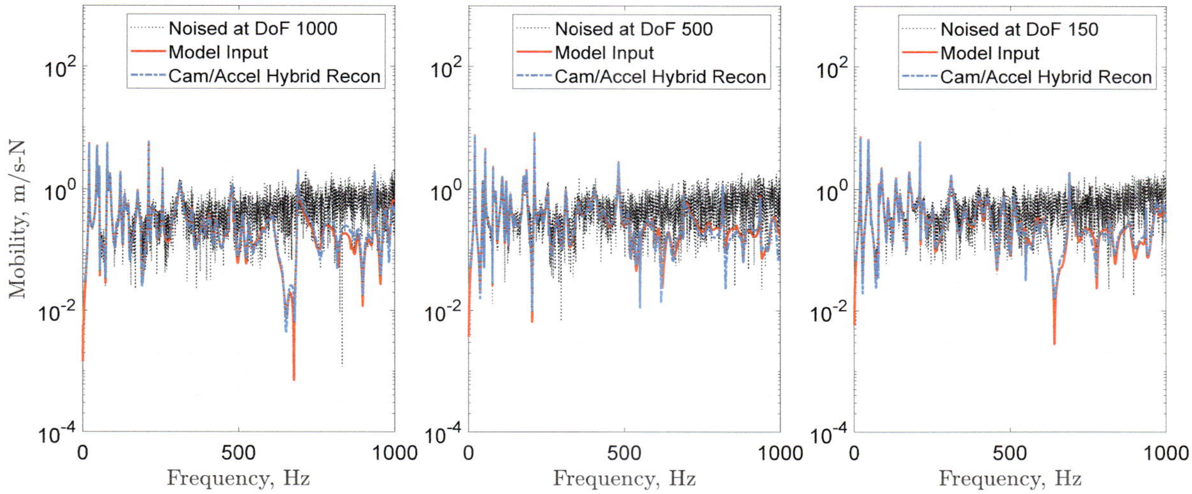

Fig. 8 Random mobilities compared between noised (model output), model input, and hybrid output.

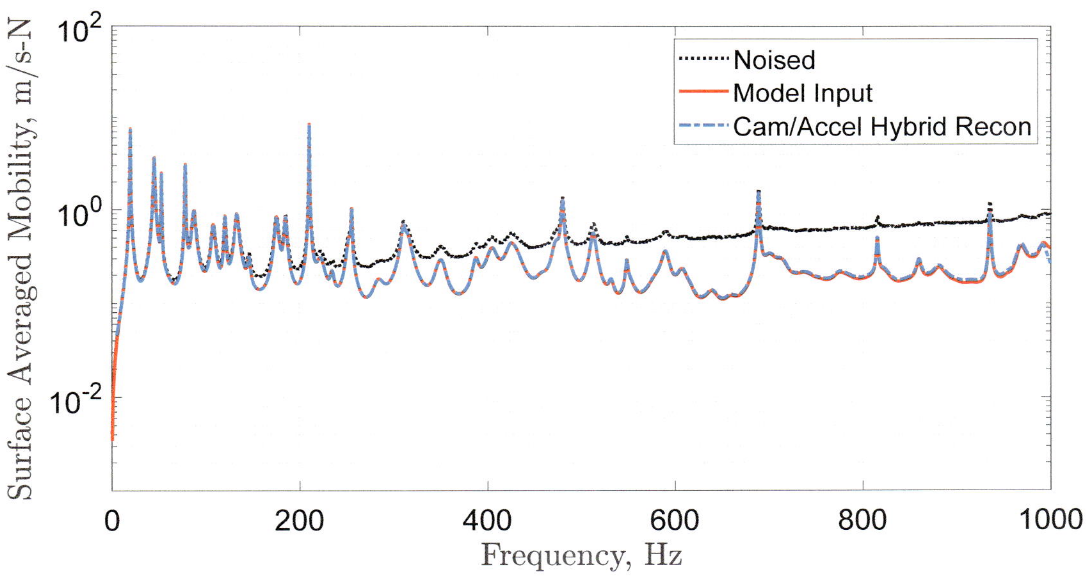

Fig. 9 Surface averaged mobilities compared between noised (model output), model input, and hybrid output.

Interestingly, the $FRAC$ between SEMM and the model input is slightly lower, with $FRAC = 0.9968$. This slight reduction in performance can be seen visually in both the random mobilities and the surface average. Further, it is clear that much of refinement of SEMM has been in amplitude, and not adjustment of the dynamics approximated by the FRF reconstructions. It is possible that the performance was biased by the accelerometer locations chosen, and this thought is supported by the varied performance in the random mobilitie. If the overlay accelerometers chosen happen to have all missed a portion of dynamics, the over-compensation in the sub-structuring stage would follow logically. In future work, such possible sensitivity to the performance of SEMM will be explored.

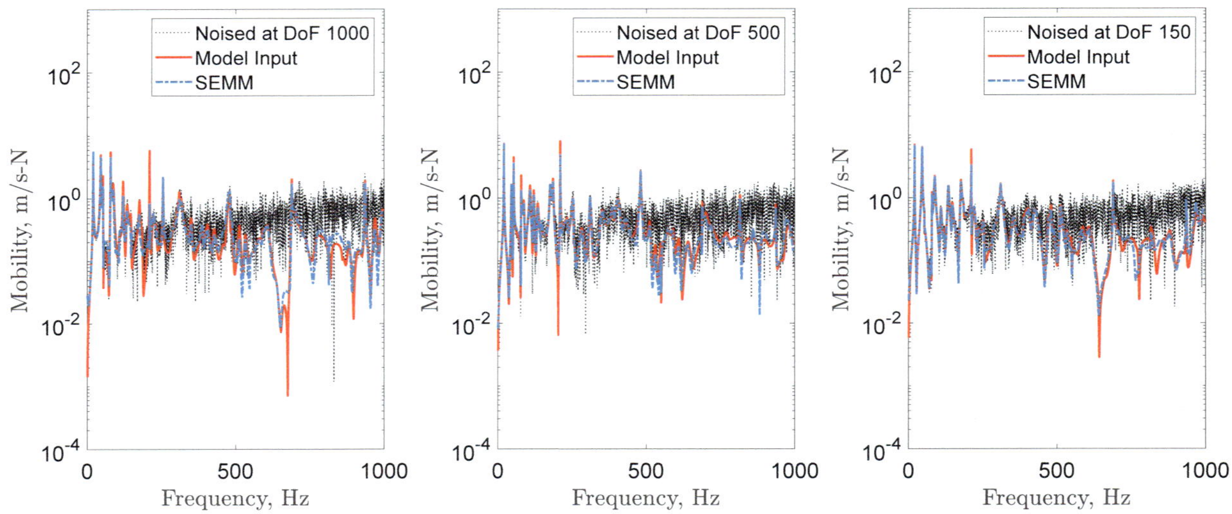

Fig. 10 Random mobilities compared between noised (model output), model input, and SEMM output.

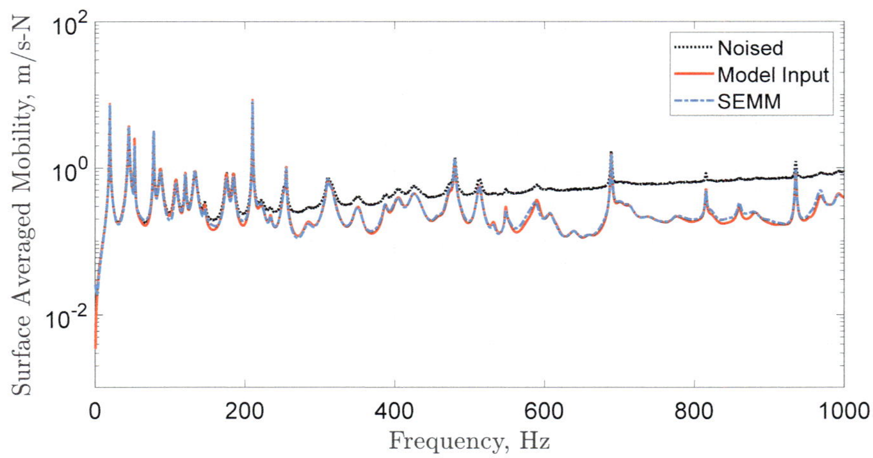

Fig. 11 Surface averaged mobilities compared between noised (model output), model input, and SEMM output.

Experimental Beam

For the experimental dataset, a 0.43 m aluminum cantilever beam ($E = 68$ GPa) was struck with an instrumented modal hammer, as seen in Figure 12. A Phantom v1212 high-speed camera was used to record video data at a resolution of 800×1280 and rate of 5000 frames per second. Constant lighting was used to preserve brightness constancy and a black background was used to promote good contrast at the edge for motion extraction. Camera and setup parameters led to a calibration factor of 0.4658 $\frac{mm}{px}$. Further details of the setup can be found in [1].

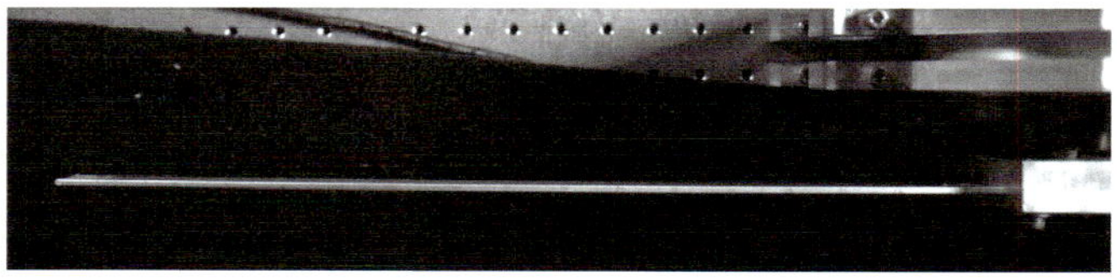

Fig. 12 Video frame of cantilevered beam, showing sharp contrast

A laser doppler vibrometer (LDV) aimed near the tip of the beam serves here as the reference and overlay model. As before, each method was applied to the matrix of mobilities from a SIMO test of the beam. For reference, this matrix is

905 outputs × 1 input × 2000. It is worth noting that this implies substantially fewer observations for the data reduction techniques than in the analytical case. Further, we do not consider SEMM here, as the single reference is not expected to improve results of the camera-accelerometer hybrid.

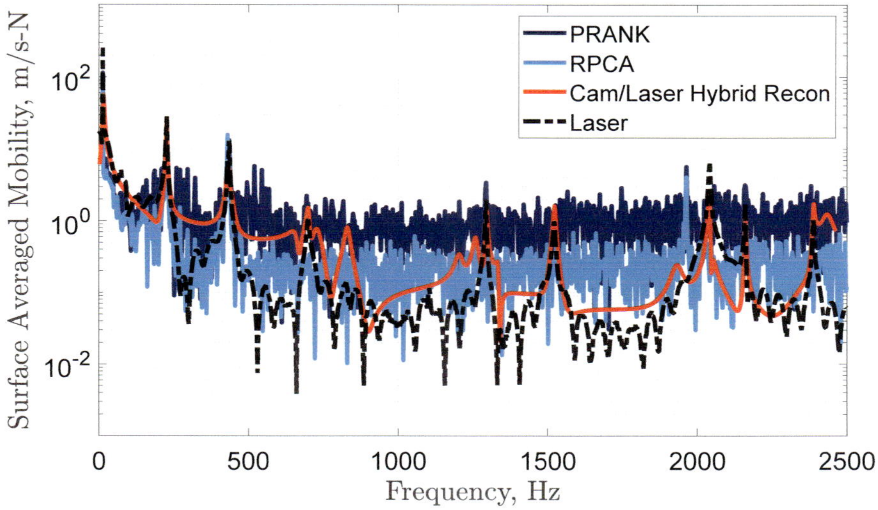

Fig. 13 Mobility comparison between the reference laser and denoising techniques, showing superior performance from the camera-accelerometer hybrid.

Figure 13 presents the results of the three techniques on the beam SIMO data. It is evident that the effect is lost here, with both PRANK and RPCA failing to reveal much if any of the underlying dynamics beflow the camera noise floor. Indeed, where PRANK was almost identical in result to SEMM in the analytical case, the overall lack of observations is shown to be a detriment to the method. This follows intuitively, and should be of no surprise. Further while RPCA is seen to fail at identifying any new dynamics, it does reduce the camera noise floor by an order of magnitude, which may be beneficial for time-domain analysis of lower-frequency behavior, such as below 500 Hz. When considering SEMM, the initial, visual results appear quite good: the noise has been removed and pole-zero behaviour is now apparent. Both of these factors are important for accurate and stable camera-based modal force reconstruction.

Turning to Figure 14, similar trends may be observed, where RPCA and SEMM appear to agree decently well to each other and the raw data below 500 Hz, with PRANK providing the least denoising. Overall, RPCA and PRANK have lowered

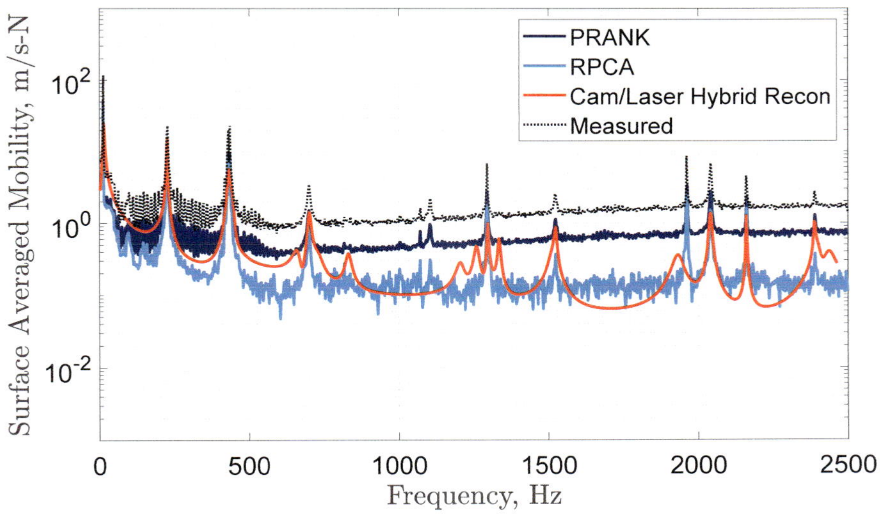

Fig. 14 Surface averaged mobility comparison between the reference laser and denoising techniques, showing superior performance from the camera-accelerometer hybrid.

the noise floor, but reveal no dynamics that are not already apparent in the raw surface average. What is also clear by comparing these techniques is the lack of magnitude retention between the raw data and any of the methods—particularly at the first resonance. While these curves may be visually appealing, and possibly sufficient to render higher-quality mode shapes, the modification of true magnitude will directly hinder the force reconstruction process. Indeed, such error in modal participation caused by noisy data is the exact problem the authors are attempting to solve.

Conclusion

Herein, we have aimed to evaluate and compare three recently introduced camera denoising techniques, and their efficacy for noise reduction towards a modal force reconstruction approach. Application to MIMO data showed great utility across all three methods, revealing underlying dynamics (e.g., modes, pole-zero behaviour) which were otherwise obscured by the camera noise floor. What was most encouraging was the ability of PRANK to reveal almost identical dynamics to the more-involved SEMM, both being in excellent agreement to the pure, analytical model input. When moving to SIMO data, the data reduction techniques were shown to struggle due to the overall lack of observations with good signal-to-noise. SEMM appeared again the most successful, though not without fault in terms of absolute magnitude. General conclusions from the users' perspective may be summarized as follows: PRANK and SEMM are comparable in cases with an abundance of inputs, outputs, and frequency bins, while RPCA appears to perform slightly worse and is computationally more expensive; Data reduction techniques such as PRANK and RPCA will inherently struggle when there are few datums of quality; and SEMM appears the most robust and trustworthy given the incorporation of a physics-informed regularization. As work progresses on modal force reconstruction—an inherently SIMO related process—the authors hope and are excited to see improvements to these techniques and their eventual application to worthwhile engineering problems.

Acknowledgments This work was partially supported by the PIPELINE: Penn State Intern PipelinE LInks to Navy Engineering program, ONR grant #N000142312656. The Penn State PIPELINE Program motivates and connects students and faculty to careers and research opportunities with the Navy technical workforce. The authors would also like to thank the ARL Walker Assistantship program, which provided support for acquiring the data used here.

References

1. Collier, S. and Dare, T. "Accuracy of phase-based optical flow for vibration extraction". *J. Sound Vib.* (2022)
2. Collier, S. and Dare, T. "Informed pixel pushing: A new method of large-motion handling for phase-based optical flow". *Measurement* (2023)
3. Collier, S. and Dare, T. "Regime sorting for multiscale vibration and phase-based motion extraction". In Di Maio, D. and Baqersad, J., editors, *Rotat. Mach., Opt. Methods & Scanning LDV Methods, Volume 6*, Cham (2023) Springer Int. Publ.
4. Baqersad, J., Poozesh, P., Niezrecki, C., and Avitabile, P. "Photogrammetry and optical methods in structural dynamics – a review". *Mech. Syst. Signal Process.*, 86:17–34 (2017)
5. Poozesh, P., Sarrafi, A., Mao, Z., Avitabile, P., and Niezrecki, C. "Feasibility of extracting operating shapes using phase-based motion magnification technique and stereo-photogrammetry". *J. Sound Vib.*, 407:350–366 (2017)
6. Rohe, D. and Reu, P. "Experimental modal analysis using phase quantities from phase-based motion processing and motion magnification". *Exp. Tech.*, 45:297–312 (2021)
7. Chen, J.G., Wadhwa, N., Cha, Y.J., Durand, F., Freeman, W.T., and Buyukozturk, O. "Modal identification of simple structures with high-speed video using motion magnification". *J. Sound Vib.*, 345:58–71 (2015)
8. Diamond, D., Heyns, P., and Oberholster, A. "Accuracy evaluation of sub-pixel structural vibration measurements through optical flow analysis of a video sequence". *Measurement*, 95:166–172 (2017)
9. Javh, J., Slavič, J., and Boltežara, M. "The subpixel resolution of optical-flow-based modal analysis". *Mech. Syst. Signal Process.*, 88:89–99 (2017)
10. Barron, J.L., Fleet, D.J., and Beauchemin, S.S. "Performance of optical flow techniques". *Int. J. Comput. Vis.*, 12:43–77 (1994)
11. Hay, J. "High dynamic range imaging for the detection of motion". *PhD Thesis,*, University of Louisville (Unpublished results)
12. Fleet, D.J. and Jepson, A.D. "Computation of component image velocity from local phase information". *Int. J. Comput. Vis.*, 5:77–104 (1990)
13. Wadhwa, N., Rubinstein, M., Durand, F., and Freeman, W.T. "Phase-based video motion processing". *ACM Trans. Graph.*, 32(4) (2013)
14. Lucas, B.D. and Kanade, T. "An iterative image registration technique with an application to stereo vision". In *Proc. 7th Int. Jt. Conf. Artif. Intell. - Volume 2*, IJCAI'81, page 674–679, San Francisco, CA, USA (1981) Morgan Kaufmann Publ. Inc.
15. Epps, B. and Krivitzky, E. "Singular value decomposition of noisy data: noise filtering". *Exp. Fluids*, 60 (2019)
16. Trainotti, F., Klaassen, S.B., Bregar, T., and Rixen, D.J. "A singular value based filtering strategy for noise reduction on measured response data". In *41th Int. Mod. Anal. Conf.*, pages 1–4. Unpublished (2022)

17. Scherl, I., Strom, B., Shang, J.K., Williams, O., Polagye, B.L., and Brunton, S.L. "Robust principal component analysis for modal decomposition of corrupt fluid flows". *Phys. Rev. Fluids*, 5:054401 (2020)
18. Trainotti, F., Slimak, T., and Rixen, D. "Towards full-field frequency based substructuring: an application case" (2024)
19. Candès, E.J., Li, X., Ma, Y., and Wright, J. "Robust principal component analysis?". *J. ACM*, 58(3) (2011)
20. Javh, J., Slavič, J., and Boltežar, M. "High frequency modal identification on noisy high-speed camera data". *Mechanical Systems and Signal Processing*, 98:344–351 (2018)
21. Bregar, T., Zaletelj, K., Čepon, G., Slavič, J., and Boltežar, M. "Improved frf estimation from noisy high-speed camera data using semm". In Di Maio, D. and Baqersad, J., editors, *Rotat. Mach., Opt. Methods & Scanning LDV Methods, Volume 6*, Cham (2020) Springer Int. Publ.
22. Bregar, T., Zaletelj, K., Čepon, G., Slavič, J., and Boltežar, M. "Full-field frf estimation from noisy high-speed-camera data using a dynamic substructuring approach". *Mechanical Systems and Signal Processing*, 150:107263 (2021)
23. Peeters, B., Van der Auweraer, H., Guillaume, P., and Leuridan, J. "The polymax frequency-domain method: A new standard for modal parameter estimation?". *Shock and Vibration*, 11(3-4):523692 (2004)

For Product Safety Concerns and Information please contact our
EU representative GPSR@taylorandfrancis.com Taylor & Francis
Verlag GmbH, Kaufingerstraße 24, 80331 München, Germany